ADVANCES IN CHEMICAL ENGINEERING
Volume 22

Intelligent Systems in Process Engineering

Part II: Paradigms from Process Operations

ADVANCES IN
CHEMICAL ENGINEERING

Series Editors

JAMES WEI
School of Engineering and Applied Science
Princeton University
Princeton, New Jersey

JOHN L. ANDERSON
Department of Chemical Engineering
Carnegie Mellon University
Pittsburgh, Pennsylvania

KENNETH B. BISCHOFF
Department of Chemical Engineering
University of Delaware
Newark, Delaware

MORTON M. DENN
College of Chemistry
University of California at Berkeley
Berkeley, California

JOHN H. SEINFELD
Department of Chemical Engineering
California Institute of Technology
Pasadena, California

GEORGE STEPHANOPOULOS
Department of Chemical Engineering
Massachusetts Institute of Technology
Cambridge, Massachusetts

ADVANCES IN CHEMICAL ENGINEERING
Volume 22

Intelligent Systems in Process Engineering
Part II: Paradigms from Process Operations

Edited by

GEORGE STEPHANOPOULOS
CHONGHUN HAN

*Laboratory for Intelligent Systems in Process Engineering
Department of Chemical Engineering
Massachusetts Institute of Technology
Cambridge, Massachusetts*

ACADEMIC PRESS
San Diego   New York   Boston   London   Sydney   Tokyo   Toronto

TP
145
.D7
v.22
1995

This book is printed on acid-free paper. ∞

Copyright © 1995 by ACADEMIC PRESS, INC.

All Rights Reserved.
No part of this publication may be reproduced or transmitted in any form or by any means, electronic or mechanical, including photocopy, recording, or any information storage and retrieval system, without permission in writing from the publisher.

Academic Press, Inc.
A Division of Harcourt Brace & Company
525 B Street, Suite 1900, San Diego, California 92101-4495

United Kingdom Edition published by
Academic Press Limited
24-28 Oval Road, London NW1 7DX

International Standard Serial Number: 0065-2377

International Standard Book Number: 0-12-008522-4

PRINTED IN THE UNITED STATES OF AMERICA
95　96　97　98　99　00　QW　9　8　7　6　5　4　3　2　1

"All virtue is one thing: Knowledge"
                        PLATO

However,

"The chance of the quantum theoretician is not the ethical freedom of the Augustinian"
                                                NORBERT WIENER

George Stephanopoulos dedicates
this editorial work to
Eleni-Nikos-Elvie
with love and gratitude

Chonghun Han dedicates
this editorial work to
Jisook and *Albert*

CONTENTS VOLUME 22

Contributors to Volume 22 xi
Prologue xix

Nonmonotonic Reasoning:
The Synthesis of Operating Procedures in Chemical Plants

Chonghun Han, Ramachandran Lakshmanan, Bhavik Bakshi,
and George Stephanopoulos

I. Introduction 314
   A. Previous Approaches to the Synthesis of Operating Procedures 316
   B. The Components of a Planning Methodology 318
   C. Overview of the Chapter's Structure 324
II. Hierarchical Modeling of Processes and Operations 324
   A. Modeling of Operations 325
   B. Modeling of Process Behavior 329
III. Nonmonotonic Planning 334
   A. Operator Models and Complexity of Nonmonotonic Planning 336
   B. Handling Constraints on the Temporal Ordering of Operational Goals 337
   C. Handling Constraints on the Mixing of Chemicals 339
   D. Handling Quantitative Constraints 343
   E. Summary of Approach for Synthesis of Operating Procedures 348
IV. Illustrations of Modeling and Nonmonotonic Operations Planning 351
   A. Construction of Hierarchical Models and Definition of Operating States . . . 351
   B. Nonmonotonic Synthesis of a Switchover Procedure 359
V. Revamping Process Designs to Ensure Feasibility of Operating Procedures . . . 368
   A. Algorithms for Generating Design Modifications 369
VI. Summary and Conclusions 374
   References 375

Inductive and Analogic Learning:
Data-Driven Improvement of Process Operations

Pedro M. Saraiva

I. Introduction 378
II. General Problem Statement and Scope of the Learning Task 381
III. A Generic Framework to Describe Learning Procedures 384
   A. A Generic Formalism 385
   B. Major Departures from Previous Approaches 385
IV. Learning with Categorical Performance Metrics 389
   A. Problem Statement 389
   B. Search Procedure, S 391

C. Case Study: Operating Strategies for Desired Octane Number 394
V. Continuous Performance Metrics 396
   A. Problem Statement 396
   B. Alternative Problem Statements and Solutions 398
   C. Taguchi Loss Functions as Continuous Quality Cost Models 401
   D. Learning Methodology and Search Procedure, S 403
   E. Case Study: Pulp Digester 405
VI. Systems with Multiple Operational Objectives 408
   A. Continuous Performance Variables 409
   B. Categorical Performance Variables 409
   C. Case Study: Operational Analysis of a Plasma Etching Unit 413
VII. Complex Systems with Internal Structure 417
   A. Problem Statement and Key Features 417
   B. Search Procedures 424
   C. Case Study: Operational Analysis of a Pulp Plant 426
VIII. Summary and Conclusions 431
   References 432

Empirical Learning through Neural Networks: The Wave-Net Solution

ALEXANDROS KOULOURIS, BHAVIK R. BAKSHI, AND GEORGE STEPHANOPOULOS

I. Introduction 438
II. Formulation of the Functional Estimation Problem 441
   A. Mathematical Description 444
   B. Neural Network Solution to the Functional Estimation Problem 449
III. Solution to the Functional Estimation Problem 451
   A. Formulation of the Learning Problem 451
   B. Learning Algorithm 465
IV. Applications of the Learning Algorithm 471
   A. Example 1 471
   B. Example 2 474
   C. Example 3 477
V. Conclusions 479
VI. Appendices 480
   A. Appendix 1 480
   B. Appendix 2 481
   C. Appendix 3 482
   References 483

Reasoning in Time: Modeling, Analysis, and Pattern Recognition of Temporal Process Trends

BHAVIK R. BAKSHI AND GEORGE STEPHANOPOULOS

I. Introduction 487
   A. The Content of Process Trends: Local in Time and Multiscale 488
   B. The Ad Hoc Treatment of Process Trends 490

	C. Recognition of Temporal Patterns in Process Trends		492
	D. Compression of Process Data		493
	E. Overview of the Chapter's Structure		494
II.	Formal Representation of Process Trends		495
	A. The Definition of a Trend		496
	B. Trends and Scale-Space Filtering		500
III.	Wavelet Decomposition: Extraction of Trends at Multiple Scales		507
	A. The Theory of Wavelet Decomposition		508
	B. Extraction of Multiscale Temporal Trends		516
IV.	Compression of Process Data through Feature Extraction and Functional Approximation		527
	A. Data Compression through Orthonormal Wavelets		527
	B. Compression through Feature Extraction		530
	C. Practical Issues in Data Compression		530
	D. An Illustrative Example		532
V.	Recognition of Temporal Patterns for Diagnosis and Control		535
	A. Generating Generalized Descriptions of Process Trends		538
	B. Inductive Learning through Decision Trees		541
	C. Pattern Recognition with Single Input Variable		543
	D. Pattern Recognition with Multiple Input Variables		544
VI.	Summary and Conclusions		545
	References		546

Intelligence in Numerical Computing: Improving Batch Scheduling Algorithms through Explanation-Based Learning

Matthew J. Realff

I.	Introduction		550
	A. Flowshop Problem		552
	B. Characteristics of Solution Methodology		553
II.	Formal Description of Branch-and-Bound Framework		555
	A. Solution Space Representation—Discrete Decision Process		555
	B. The Branch-and-Bound Strategy		557
	C. Specification of Branch-and-Bound Algorithm		563
	D. Relative Efficiency of Branch-and-Bound Algorithms		564
	E. Branching as State Updating		566
	F. Flowshop Lower-Bounding Scheme		568
III.	The Use of Problem-Solving Experience in Synthesizing New Control Knowledge		570
	A. An Instance of a Flowshop Scheduling Problem		570
	B. Definition and Analysis of Problem-Solving Experience		573
	C. Logical Analysis of Problem-Solving Experience		578
	D. Sufficient Theories for State-Space Formulation		579
IV.	Representation		581
	A. Representation for Problem Solving		583
	B. Representation for Problem Analysis		588
V.	Learning		593
	A. Explanation-Based Learning		594

      B. Explanation 598
      C. Generalization of Explanations 601
  VI. Conclusions 607
      References 608

INDEX 611
CONTENTS OF VOLUMES IN THIS SERIAL 621

CONTRIBUTORS VOLUME 22

Numbers in parentheses indicate the pages on which the authors' contributions begin.

BHAVIK R. BAKSHI, *Department of Chemical Engineering, Ohio State University, Columbus, Ohio 43210* (313, 347, 485)

CHONGHUN HAN, *Laboratory for Intelligent Systems in Process Engineering, Department of Chemical Engineering, Massachusetts Institute of Technology, Cambridge, Massachusetts 02139* (313)

ALEXANDROS KOULOURIS, *Laboratory for Intelligent Systems in Process Engineering, Department of Chemical Engineering, Massachusetts Institute of Technology, Cambridge, Massachusetts 02139* (437)

RAMACHANDRAN LAKSHAMANAN, *Department of Chemical Engineering, University of Edinburgh, Edinburgh, Scotland, United Kingdom* (313)

MATTHEW J. REALFF, *School of Chemical Engineering, Georgia Institute of Technology, Atlanta, Georgia 30332* (549)

PEDRO M. SARAIVA, *Department of Chemical Engineering, University of Coimbra, 3000 Coimbra, Portugal* (377)

GEORGE STEPHANOPOULOS, *Laboratory for Intelligent Systems in Process Engineering, Department of Chemical Engineering, Massachusetts Institute of Technology, Cambridge, Massachusetts 02139* (437, 485)

CONTENTS VOLUME 21

CONTRIBUTORS TO VOLUME 21 xvii
PROLOGUE xix

Modeling Languages: Declarative and Imperative Descriptions of Chemical Reactions and Processing Systems

CHRISTOPHER J. NAGEL, CHONGHUN HAN, AND GEORGE STEPHANOPOULOS

I. Introduction 2
   A. The Five Premises of a Modeling System 3
   B. Review of Modeling Systems for Process Simulation 7
   C. Modeling Systems in Chemistry 10
II. LCR: A Language for Chemical Reactivity 13
   A. Modeling Elements of LCR 13
   B. Semantic Relations among Modeling Elements in LCR 26
   C. Syntax of LCR 33
III. Formal Construction of Representations for Chemicals and Reactions . . . 36
   A. Extension of LCR's Modeling Objects 36
   B. The "Model-Class Decomposition Digraph" (MCDD) 50
   C. Generation and Representation of Reaction Pathways 53
   D. Creation of Contextual Reaction Models 58
   E. Case Study: Ethane Pyrolysis 64
IV. MODEL. LA.: A Modeling Language for Process Engineering 73
   A. Basic Modeling Elements 73
   B. Semantic Relationships 75
   C. Hierarchies of Modeling Subclasses 76
   D. Syntax 78
V. Phenomena-Based Modeling of Processing Systems 78
   A. The "Chemical Engineering Science" Hierarchies of Modeling Elements . . . 79
   B. Formal Construction of Models 82
   C. Multifaceted Modeling of Processing Systems 82
   D. Computer-Aided Implementation of MODEL.LA. 87
   References 90

Automation in Design: The Conceptual Synthesis of Chemical Processing Schemes

CHONGHUN HAN, GEORGE STEPHANOPOULOS, AND JAMES M. DOUGLAS

I. Introduction 94
   A. Conceptual Design of Chemical Processing Schemes 96
   B. Issues in the Automation of Conceptual Process Design 98

II. Hierarchical Approach to the Synthesis of Chemical Processing Schemes:
   A Computational Model of the Engineering Methodology 103
   A. Hierarchical Planning of the Process Design Evolution 104
   B. Goal Structures: Bridging the Gap between Design Milestones 107
   C. Design Principles of the Computational Model 117
III. HDL: The Hierarchical Design Language 122
   A. Multifaceted Modeling of the Process Design State 123
   B. Modeling the Design Tasks 129
   C. Elements for Human—Machine Interaction 134
   D. Object-Oriented Failure Handling 138
   E. Management of Design Alternatives 138
IV. Concept Designer: The Software Implementation 139
   A. Overall Architecture 139
   B. Implementation Details 143
V. Summary 144
   References 145

Symbolic and Quantitative Reasoning: Design of Reaction Pathways through Recursive Satisfaction of Constraints

Michael L. Mavrovouniotis

I. Reaction Systems and Pathways 148
II. Catalytic Reaction Systems 151
   A. Basic Concepts, Terminology, and Notation 151
   B. Previous Work on the Construction of Mechanisms 154
   C. Structure of the Algorithm 155
   D. Features of the Algorithm 159
   E. Examples 160
III. Biochemical Pathways 169
   A. Features of the Pathway Synthesis Problem 173
   B. Formulation of Constraints 175
   C. Algorithm 176
   D. Examples 179
IV. Properties and Extensions of the Synthesis Algorithm 183
V. Summary 185
   References 185

Inductive and Deductive Reasoning: The Case of Identifying Potential Hazards in Chemical Processes

Christopher Nagel and George Stephanopoulos

I. Introduction 188
   A. Predictive Hazard Analysis 190
   B. Incompleteness of Conventional Hazard Analysis Methodologies 192
   C. Premises of Traditional Approaches 193
   D. Overview of Proposed Methodology 194
II. Reaction-Based Hazards Identification 195
   A. System Foundations 196
   B. Modeling Languages and Their Role in Hazards Identification 198

C. Generations of Reactions and Evaluation of Thermodynamic States		205
III. Inductive Identification of Reaction-Based Hazards		209
A. Hazards Identification Algorithm		211
B. Properties of Reaction-Based Hazards Identification		214
C. An Example in Reaction-Based Hazard Identification: Aniline Production		217
IV. Deductive Determination of the Causes of Hazards		221
A. Methodological Framework		222
B. Variables as "Causes" or "Effects"		225
C. Construction of Variable-Influence Diagrams		227
D. Characterization of Variable-Influence Pathways		232
E. Assessment of Hazards-Preventive Mechanisms		235
F. Fault-Tree Construction		238
G. An Example of Reaction-Based Hazard Identification: Reaction Quench		241
V. Conclusion		253
References		254

Searching Spaces of Discrete Solutions: The Design of Molecules Possessing Desired Physical Properties

KEVIN G. JOBACK AND GEORGE STEPHANOPOULOS

I. Introduction		258
A. Brief Review of Previous Work		260
B. General Framework for the Design of Molecules		264
II. Automatic Synthesis of New Molecules		267
A. The Generate-and-Test Paradigm		267
B. The Search Algorithm		271
C. Case Study: Automatic Design of Refrigerants		283
D. Case Study: Automatic Design of Polymers as Packaging Materials		284
III. Interactive Synthesis of New Molecules		290
A. Illustration of Interactive Design		291
B. Case Study: Interactive Design of Refrigerants		296
C. Case Study: Interactive Design of an Extraction Solvent		299
D. Case Study: Interactive Design of a Pharmaceutical		301
IV. The *Molecule-Designer* Software System		304
A. General Description		304
B. Interactive-Design-Relevant Sections		305
V. Concluding Remarks		307
References		309

COMBINED INDEX APPEARS AT THE END OF VOLUME 22	611
CONTENTS OF VOLUMES IN THIS SERIAL	621

CONTRIBUTORS VOLUME 21

Numbers in parentheses indicate the pages on which the authors' contributions begin.

JAMES M. DOUGLAS, *Department of Chemical Engineering, University of Massachusetts, Amherst, Massachusetts 01003* (43)

CHONGHUN HAN, *Laboratory for Intelligent Systems in Process Engineering, Department of Chemical Engineering, Massachusetts Institute of Technology, Cambridge, Massachusetts 02139* (1, 43)

KEVIN G. JOBACK, , *Molecular Knowledge Systems, Inc., Nashua, New Hampshire 03063* (257)

MICHAEL L. MAVROVOUNIOTIS, *Department of Chemical Engineering, Northwestern University, Evanston, Illinois 60208* (147)

CHRISTOPHER J. NAGEL, *Molten Metal Technology, Inc., Waltham, Massachusetts 02154* (1, 187)

GEORGE STEPHANOPOULOS, *Laboratory for Intelligent Systems in Process Engineering, Department of Chemical Engineering, Massachusetts Institute of Technology, Cambridge, Massachusetts 02139* (1, 43, 187, 257)

PROLOGUE

The adjective "intelligent" in the term "intelligent systems" is a misnomer. No one has ever claimed that an intelligent system in an engineering application possesses the kind of intelligence that allows it to *induce* new knowledge, (1) or "to contemplate its creator, or how it evolved to be the system that it is". (2) Åström and McAvoy (3) have suggested terms such as "knowledgeable" and "informed" to accentuate the fact that these software systems depend on large amounts of (possibly) fragmented and unstructured knowledge. For the purposes of this book, the term "intelligent system" always implies a computer program, and although the quotation marks around the adjective intelligent may be dropped occasionally, no one should perceive it as a computer program with attributes of human-like intelligence. Instead, the reader should interpret the adjective as characterizing a software artifact that possesses a computational procedure, an algorithm, which attempts to "model and emulate," and thus automate an engineering task that used to be carried out *informally* by a human. Whether or not this models the actual cognitive process in a human is beyond the scope of this book.

In the wide spectrum of engineering activities, collectively known as *process engineering* and encompassing tasks from product and process development through process design and optimization to process operations and control, so-called intelligent systems have played an important role. Ten years ago the broad introduction of knowledge-based expert systems created a pop culture that started affecting many facets of process engineering work. Expert systems were followed by their cousins, fuzzy systems, and the explosion in the use of neural networks. During the same period, the object-oriented programming (OOP) paradigm, one of the most successful "products" of artificial intelligence, has led to a revolutionary rethinking of programming practices, so that today OOP is the paradigm of choice in software engineering. After 10 years of work, 15 books/monographs/edited volumes, over 700 identified papers in archival research and professional journals, 65 reviews/tutorial/industrial survey papers, about 150 Ph.D. theses, and several thousand industrial applications worldwide, (4) the area of what is known as "intelligent systems" has turned from fringe to mainstream in a large number of process engineering activities. These include monitoring and analysis of process operations, fault diagnosis, supervisory control, feedback control, scheduling and planning of process operations, simulation, and process and product design. The early emphasis on tools and methodologies, originated by research in artificial intelligence, has given place to more integrative approaches, which focus more on the engineering problem and its characteristics. So, today, one does not encounter as frequently as 10 years ago conference sessions with titles including terms such as "expert systems," "knowledge-based systems," or "artificial intelligence." Instead one sees many more mature contributions, from both the academic and industrial worlds, in mainstream engineering sessions, with significant components of what one would have earlier termed "intelligent systems." The evolving complementarity in the use of approaches from artificial intelligence, systems and control theory, mathematical programming, and statistics is a strong indication of the maturity that the area of intelligent systems is reaching.

A. THE CURRENT SETTING

The explosive growth of academic research and industrial practice in the synthesis, analysis, development, and deployment of intelligent systems is a natural phase in the saga of the Second Industrial Revolution. If the First Industrial Revolution in 18th century England ushered the world into an era characterized by machines that extended, multiplied, and leveraged human *physical capabilities*, the Second, currently in progress, is based on machines that extend, multiply, and leverage human *mental abilities*. (5) The thinking man, *Homo sapiens*, has returned to its Platonic roots where "all virtue is one thing, knowledge." Using the power and versatility of modern computer science and technology, software systems are continuously developed to preserve knowledge for it is perishable, clone it for it is scarce, make it precise for it is often vague, centralize it for it is dispersed, and make it portable for it is difficult to distribute. The implications are staggering and have already manifested themselves, reaching the most remote corners of the earth and the inner sancta of our private lives. In this expanding pervasiveness of computers, intelligent systems can affect and are affecting the way we educate, entertain, and govern ourselves, communicate with each other, overcome physical and mental disabilities, and produce material wealth. Computer-based deployment of "knowledge" has been thrust by modern sociologists into the center of our culture as the force most effective in resolving inequities in the distribution of biological, historical and material inheritance. But what is the tangible evidence? Software systems have been composed to do the following: (5–7) (i) harmonize chorales in the style of Johann Sebastian Bach and automate musical compositions into new territories; (ii) write original stanzas and poems with thematic uniformity, which could pass as human creations for about half of the polled readers; (iii) compose original drawings and "photographs" of nonexistent worlds; (iv) "author" complete books.

Equally impressive are the results in engineering and science. Characterized as "knowledgeable," "informed," "expert," "intelligent," or any other denotation, software systems have expanded tremendously the scope of automation in scientific and engineering activities. (4–13)

B. THE THEORETICAL SCOPE AND LIMITATIONS OF INTELLIGENT SYSTEMS

So what? a skeptic may ask. Are the above examples manifestations of the computer's long-awaited, human-like intelligence? No one familiar with Gödel's theorem of incompleteness would ask such a question, (14,15) for this theorem states that it is not possible to create a formal system that is both consistent and complete. As such, you cannot create a software system based on some sort of a formal system, i.e., a consistent set of axioms, which can reflect upon itself and discover (not invent) a new dimension of knowledge. (1)

Indeed, whenever you focus your attention on any of the so-called intelligent systems, and you take the time to learn the mechanisms they use to generate their marvelous and wondrous behavior, you come up with the anti-climactic realization that everything is quite ordinary and perfectly expectable with no surprises or mystical insights. Such reaction reminds us of how Sherlock Holmes reacted when a man questioned the brilliance of his deductive reasoning in solving one of his cases:

> Mr. Jabez Wilson laughed heavily. "Well, I never!" said he. "I thought at first that you had done something clever, but I see that there was nothing in it, after all." "Begin to think, Watson," said Holmes, "that I made a mistake in explaining. 'Omne ignotum promagnifico,' you know, and my poor little reputation, such as it is, will suffer shipwreck if I am so candid."

Similarly, Alan Turing, the father of the digital computer and creator of the Turing Test for checking the "intelligence" of a machine, put it this way:

> The extent to which we regard something as behaving in an intelligent manner is determined as much by our own state of mind and training as by the properties of the object under consideration. If we are able to explain and predict its behavior or if there seems to be little underlying plan, we have little temptation to imagine intelligence. With the same object, therefore, it is possible that one man would consider it as intelligent and another would not; the second man would have found out the rules of its behavior.

C. THE CHARACTER OF THE TEN PARADIGMS

All the paradigms of intelligent systems in this volume have plans and assume extensive amounts of knowledge. As such they are ordinary computer programs and they emulate a precise computational procedure, which uses a predefined set of data. In the Aristotelian form, "all instruction given or received (by the intelligent systems) by way of argument, proceeds from preexistent knowledge." Consequently, one should see all cases put forward by the individual chapters as nothing more than paradigms for new uses of the computer. Every one of them carries out deduction from a predefined set of knowledge, using explicit reasoning strategies. The reader should not search for inductive generation of new knowledge, even when the terms "induction" and "inductive reasoning" have been loosely employed in some chapters. Instead, the reader should see each chapter as a computer-based paradigm in *capturing, articulating,* and *utilizing* various forms of knowledge. As a result, the reader will notice that the ten chapters of this volume serve as a paradigm of an integrative attitude to the modeling and processing of knowledge. Nowhere in this volume will the reader find artificial debates on the superiority of a numerical over a symbolic approach or vice versa. On the contrary, the engineering methodologies advanced by the individual chapters indicate that *all available knowledge should be acquired, modeled, and used* within a framework that requires interaction and/or integration of processing methodologies from artificial intelligence, systems and control theory, operations research, statistics, and others.

It is this integrative attitude that today characterizes most of the work in the area of "intelligent systems for process engineering," as the editors of this volume have indicated in a recent review article. (4) It is this need for integrative approaches that has moved the applications of artificial intelligence into the mainstream of engineering activities. This is certainly the pivotal feature that characterizes the ten paradigms discussed in the subsequent chapters.

The ten chapters of this volume advance ten distinct paradigms for the use of ideas and methodologies from artificial intelligence in conjunction with techniques from various other areas. They represent the culmination of research efforts which started in 1986 at the *Laboratory for Intelligent Systems in Process Engineering* (LISPE) of the Chemical Engineering Department at MIT, and currently are spread over a half a dozen academic institutions. Each chapter, as the corresponding title indicates, is centered around two themes. The first theme (represented by the first part of a title) is drawn from the artificial intelligence techniques discussed in the specific chapter, while the second theme (represented by the second part of the title) focuses on a process engineering problem. It should be noted, though, that it is the process engineering problem, its formulation and characteristics, that sets the tone for every chapter. The various components of the corresponding intelligent systems serve specific needs. Nowhere will the reader find the "a technique in search for a problem to solve" attitude, which has led to the distortion of several engineering problems and the malignant proliferation of techniques. As a result, even if future developments suggest a change in the techniques used, the formulation of the engineering problems may retain the bulk of the essential features proposed by each of the ten chapters.

D. THEMES COVERED BY THE TEN CHAPTERS

Let us now give a brief synopsis of the themes advanced by each of the ten chapters. The five chapters of Volume 21 (Part I) advance paradigms which are related to product and process design, while the five chapters of Volume 22 (Part II) focus on aspects of process operations.

Volume 21: Product and Process Design

Chapter 1. MODELING LANGUAGES:
  Declarative and Imperative Descriptions of Chemical Reactions and Processing Systems

To model is to represent reality, and modeling as an essential task of any engineering activity is always *contextual*. Within the scope of differing engineering contexts, the same physical entity, e.g. molecule, chemical reaction, or process flowsheet, is represented with a broad variety of models. An enormous amount of effort is expended in the development and maintenance of a *Babel of models*, sporting different languages and being at cross purposes with each other, although like their biblical counterpart they share a common progenitor—in this case, the fundamentals of chemistry/physics and the principles of chemical engineering science. Creating a language that supports the expeditious generation of consistent models has become the key to unlocking the power of computer-aided tools, and unleashing the explosive synergism between human and computer. However, a modeling language is of little use if it only creates representations of physical entities as "things unto themselves" without meaningful semantic designation to what it purports to represent. Furthermore, the model of an entity should contain all knowledge that has some bearing on the representation of that entity, be that *declarative* or *imperative* (procedural) in character. Chapter 1 describes two modeling languages; LCR (*L*anguage for *C*hemical *R*easoning) to represent molecules and chemically reactive systems, and MODEL.LA. (*MODEL*ing *LA*nguage) for the representation of processing systems. Both are based on the same principles and have, to a large extent, a common structure. Both have been based on ideas and techniques which originated in artificial intelligence, and both have been implemented in a similar object-oriented programming environment.

Chapter 2. *AUTOMATION IN DESIGN:*
The Conceptual Synthesis of Chemical Processing Schemes

If you really know how to carry out an engineering task, then you can instruct a computer to do it automatically. This self-evident truism can be used as the litmus test of whether a human "really" knows how to, say, design an engineering artifact. Experience has shown that engineers have been able to automate the process of design in very few instances, thus demonstrating the presence of serious flaws in (a) their understanding of how to do design and/or (b) their ability to clearly articulate the design methodology, both of which can be traced to the inherent difficulty of making the "best" design decisions. The pivotal element in automating the design process is *modeling the design process itself*, which includes the following modeling tasks: (1) modeling the *structure of design tasks* that can take you from the initial design specifications to the final engineering artifact; (2) representing the *design decisions* involved in each task, along with the assumptions, simplifications, and methodologies needed to frame and make the design decisions; (3) modeling the *state of the evolving design*, along with the underlying rationale. Chapter 2 shows how one can use ideas and techniques from artificial intelligence, e.g., symbolic modeling, knowledge-based systems, and logic, to construct a computer-implemented model of the design process itself. Using Douglas' hierarchical approach as the conceptual model of the design process, this chapter shows how to generate models of the design tasks' structure, design decisions, and the state of design, thus leading to automation of large segments of the synthesis of chemical processing schemes. The result is a *human-aided, machine-based* design paradigm, with the computer "knowing" how the design is done, what the scope of design is, and how to provide explanations and the rationale for the design decisions and the resulting final design. Such a paradigm is in sharp contrast with the traditional *computer-aided, human-based* prototype, where the computer carries out numerical calculations and data fetching from files and databases, but has no notion of how the design is done, knowledge resting exclusively in the province of the individual human designer.

Chapter 3. SYMBOLIC AND QUANTITATIVE REASONING:
Design of Reaction Pathways through Recursive Satisfaction of Constraints

Given a fixed, predetermined set of elementary reactions, to compose reaction pathways (mechanisms) which satisfy given specifications in the transformation of available raw materials to desired products is a problem encountered quite frequently during research and development of chemical and biochemical processes. As in the assembly of a puzzle, the pieces (available reaction steps) must fit with each other (i.e., satisfy a set of constraints imposed by the precursor and successor reactions) and conform with the size and shape of the board (i.e., the specifications on the overall transformation of raw materials to products). Chapter 3 draws from *symbolic and quantitative reasoning* ideas of AI which allow the systematic synthesis of artifacts through a *recursive satisfaction of constraints* imposed on the artifact as a whole and on its components. The artifacts in this chapter are mechanisms of catalytic reactions and pathways of biochemical transformations. The former require the construction of *direct* mechanisms, without cycles or redundancies, to determine the basic legitimate chemical transformations in a reacting system. The latter are the chemical engines of living cells, and they represent legitimate routes for the biochemical conversion of substrates to products either desired from a bioprocess or essential for cell survival. The algorithms discussed in this chapter could be used in one of the following two settings: (a) Synthesize alternative pathways of chemical/biochemical reactions as a means to interpret overall transformations which are experimentally observed. (b) Synthesize reaction pathways in the course of exploring new, alternative production route. This chapter discusses examples in both directions. Although it is concerned only with constraints on the directionality and stoichiometry of elementary reactions, the ideas can be extended to include other types of constraints arising, for example, from kinetics or thermodynamics.

Chapter 4. INDUCTIVE AND DEDUCTIVE REASONING:
The Case of Identifying Potential Hazards in Chemical Processes

All reasoning carried out by computers is *deductive*; i.e., any software has all the necessary data, stored in various forms in a database, and possesses all the necessary algorithms to operate on the set of data and *deduce* some results. Many researchers in the area of cognitive psychology make similar claims on the reasoning mechanisms of human beings. The fact remains, though, that both humans and machines can use very simple "algorithms" on small sets of data and produce results which could not have been visible to the "naked eye" of direct reasoning. In such cases, we tend to talk about the *inductive* capabilities of either of the two. These ideas are nowhere more prominent than in the area of *hazards identification and analysis*. One often hears, "if I knew that the conversion of A to B could be catalyzed by the presence of C then I would have foreseen the last disaster, and have done something about it," with the speaker converting a problem of *inductive* identification (i.e., induce the possibility of a hazard from the list of chemicals) into an issue of deductive reasoning. Chapter 4 demonstrates that the identification of hazards is essentially an interplay between inductive and deductive reasoning. Through inductive reasoning one attempts to generate all potential hazardous top-level events which can be justified by the presence of a set of chemicals. The reasoning is called inductive because it has the potential to generate specific knowledge that was not "visible" ahead of time. Once the potentially harmful top-level events have been identified, deductive reasoning attempts to "walk" through the processing scheme, its unit operations, and their design or operating characteristics (assumptions, or decisions), and generate the preconditions which would enable the occurrence of a specific top-level event. The inductive reasoning procedures operate on a set of chemicals and create in an *exhaustive, bottom-up* manner many alternative reaction pathways, some of which could lead to a hazard, e.g., release of large amounts of energy over a short period of time. On the other hand, the deductive reasoning procedures are *goal-directed* and operate in a *top-down* manner. Chapter 4 develops the detailed framework for the implementation of these ideas which, among other benefits, offers the following advantages: (a) Formalizing the hazards identification problem and unifying the methodological approaches at any stage of the design activities and (b) systematizing the generation and evaluation of mechanisms for the prevention of hazards, or containment of their effects.

Chapter 5. *SEARCHING SPACES OF DISCRETE SOLUTIONS:*
   The Design of Molecules Possessing Desired Physical Properties

Strings of letters make words. From words to verses and stanzas, a poet composes a work with its own dynamic behavior, e.g., emotional impact on the reader, which transcends the character of its components. In an analogous manner, atoms form functional groups and these in turn yield molecules with distinct behavior, e.g., physical properties. It takes a Homeric or Shakespearean genius to convert letters to an epic with a predefined desired impact. It suffices to efficiently search a space of combinatorial alternatives in order to identify the molecules which satisfy the desired constraints on a set of physical properties. Often the requisite scientific knowledge is fragmented, dispersed, and nonformalized, making the deductive search for the desired molecules inefficient or impossible. The inductive "genius" of a scientist or engineer is needed to break the impasse in such cases. By evolution or revolution one needs to respond to tighter and shifting product specifications and identify new solvents, pharmaceuticals, imaging chemicals, herbicides and pesticides, refrigerants, polymeric materials, and many others. Chapter 5 sketches the characteristics of an intelligent, computer-aided tool to support the synthetic search for the desired molecules. With functional groups as the "letters" of an alphabet, automatic and interactive procedures compose and screen classes of potential molecules. The automatic synthesis algorithm defines and searches the space of discrete solutions (molecules) through a hierarchical sequence of the space's representations. However, one should never overestimate the effectiveness of search algorithms in locating the desired solutions. Quite frequently one needs to resort to human-driven, abductive jumps. Chapter 5 also describes how automatic search can become interwoven with effective man–machine interaction. Thus, the resulting computer-aided tool, the *Molecule Designer*, constitutes a paradigm of an intelligent system with two distinct but integrated and complementary capabilities. Examples of the synthesis of refrigerants, solvents, polymers, and pharmaceuticals illustrate the logic and features of the design procedures in the *Molecule Designer*.

Volume 22: Process Operations

Chapter 6: NONMONOTONIC REASONING:
The Synthesis of Operating Procedures in Chemical Plants

The inherent difficulty of planning a sequence of actions to take you from one point to another usually increases as more obstacles are placed in your way. The number of these obstacles (constraints) that you must circumvent determines the complexity of the task, because any time you run into one of them you must *backtrack* and try an alternative step or path of steps. Such *serial* (or *linear* or *monotonic*) construction of a plan is fraught with pitfalls and repeated backtracking. The more the constraints, the more inefficient the monotonic planning. If, on the other hand, an action-step (a Clobberer) leads to the violation of a constraint, then *do not backtrack*. Take another action-step (a White Knight) which, when it precedes a Clobberer, negates the impact of the Clobberer, and you never need to backtrack. So the more constraints the more efficient your planning process. Such *nonserial* (or *nonlinear*, or *nonmonotonic*) reasoning has become the essence of all modern and efficient planners, whether they are *logic-based* and *explicit*, or *implicit* enumerators of alternative plans. The purpose of Chapter 6 is twofold: (i) To introduce the ideas of *nonmonotonic reasoning* in the planning of process operations. (ii) To demonstrate how nonmonotonic planning can be used to synthesize operating procedures for chemical processes, either off-line for standard tasks (e.g., routine start-up or shut-down), or on-line for real-time response to large departures from desired conditions. It is shown that hierarchical modeling of process operations and operators is essential for the efficient deployment of nonmonotonic planning, and that the tractability of the resulting algorithms is strictly dependent on the form of the operators. In this regard, the modeling needs in this chapter draw heavily from the material of Chapter 1. Nonmonotonic planners handle with superb efficiency constraints on (a) the temporal ordering of operations, (b) avoidable mixtures of chemical species, and (c) bounding quantitative conditions on the state of a process. Consequently, they could be used to generate explicitly all feasible operating procedures, leaving a far smaller search space for the selection of the optimum procedure by a numerical optimizer.

Chapter 7. INDUCTIVE AND ANALOGICAL LEARNING:
  Data-Driven Improvement of Process Operations

Informed and systematic observation of naturally generated data can lead to the formulation of interesting and effective generalizations. While some statisticians believe that experimentation is the only safe and reliable way to "learn" and achieve operational improvements in a manufacturing system, other statisticians and all the empirical machine learning researchers contend that by looking at past historical records and sets of examples, it is possible to extract and generate important new knowledge. Chapter 7 draws from *inductive and analogical learning* ideas in an effort to develop systematic methodologies for the extraction of structured new knowledge from operational data of manufacturing systems. These methodologies do not require any a priori decisions/assumptions either on the character of the operating data (e.g., probability density distributions) or on the behavior of the manufacturing operations (e.g., linear or nonlinear structured quantitative models), and they make use of *instance-based learning* and *inductive symbolic learning* techniques developed in artificial intelligence. They are aimed to be complementary to the usual set of statistical tools that have been employed to solve analogous problems. Thus, one can see the material of Chapter 7 as an attempt to fuse statistics and machine learning in solving specific engineering problems. The framework developed in this chapter is quite generic and can be used to generate operational improvement opportunities for manufacturing systems (a) which are simple or complex (with internal structure), (b) whose performance is characterized by one or multiple objectives, and (c) whose performance metrics are categorical (qualitative) or continuous (real numbers). A series of industrial case studies illustrates the learning ideas and methodologies.

Chapter 8. EMPIRICAL LEARNING THROUGH NEURAL NETWORKS:
  The Wave-Net Solution

Empirical learning is an ever-lasting and ever-improving procedure. Although *neural networks* (NN) captured the imagination of many researchers as an outgrowth of activities in artificial intelligence, most of the progress was accomplished when empirical learning through NNs was cast within the rigorous analytical framework of the *functional estimation problem*, or *regression*, or *model realization*. Independently of the name, it has been long recognized that, due to the inductive nature of the learning problem, to achieve the desired accuracy and generalization (with respect to the available data) in a dynamic sense (as more data become available) one needs to seek the unknown approximating function(s) in functional spaces of varying structure. Consequently, a recursive construction of the approximating functions at multiple resolutions emerges as a central requirement and leads to the utilization of wavelets as the basis functions for the recursively expanding functional spaces. Chapter 8 fuses the most attractive features of a NN, i.e., representational simplicity, capacity for universal approximation, and ease in dynamic adaptation, with the theoretical soundness of a recursive functional estimation problem, using wavelets as basis functions. The result is the *Wave-Net* (*Wave*let *Net*work), a multiresolution hierarchical NN with localized learning. Within the framework of a Wave-Net, where adaptation of the approximating function is allowed, we have explored the use of the L^∞ error measure as the design criterion. One may cast any form of data-driven empirical learning within the framework of a Wave-Net to address a variety of modeling situations encountered in engineering problems, such as design of process controllers, diagnosis of process faults, and planning and scheduling of process operations. Chapter 8 discusses the properties of a Wave-Net and illustrates its use on a series of examples.

Chapter 9. *REASONING IN TIME:*
Modeling, Analysis, and Pattern Recognition of Temporal Process Trends

The plain record of a variable's numerical values over time does not invoke appreciable levels of cognitive activity in a human. Although it can cause a fervor of numerical computations by a computer, the levels of cognitive appreciation of the variable's temporal behavior remain low. On the other hand, if one presents the human with a graphical depiction of the variable's temporal behavior, the level of cognition increases and a wave of reasoning activities is unleashed. Nevertheless, when the human is presented with scores of graphs depicting the temporal behavior of interacting variables, his/her reasoning abilities are severely tested. In such a case, the computer will happily continue crunching numbers without ever rising above the fray and thus developing a "mental" model, interpreting correctly the temporal interactions among the many variables. Reasoning in time is very demanding, because time introduces a new dimension with significant levels of additional freedom and complexity. While the real-valued representation of variables in time is completely satisfactory for many engineering tasks (e.g., control, dynamic simulation, planning and scheduling of operations), it is very unsatisfactory for all those tasks which require decision-making via logical reasoning (e.g., diagnosis of process faults, recovery of operations from large unsolicited deviations, "supervised" execution of start-up or shut-down operating procedures). To improve the computer's ability to reason efficiently in time, we must first establish new forms for the representation of temporal behavior. It is the purpose of Chapter 9 to examine the engineering needs for temporal decision-making and to propose specific models which encapsulate the requisite temporal characteristics of individual variables and composite processes. Through a combination of analytical techniques, such as *scale-space filtering, wavelet-based, multiresolution decomposition of functions,* and modeling paradigms from artificial intelligence, Chapter 9 develops a concise framework that can be used to model, analyze, and synthesize the temporal trends of process operations. Within this framework, the modeling needs for logical reasoning in time can be fully satisfied, while maintaining consistency with the numerical tasks carried out at the same time. Thus, through the modeling paradigms of this chapter, one may put together intelligent systems which use consistent representations for their logical-reasoning and numerical tasks.

Chapter 10. INTELLIGENCE IN NUMERICAL COMPUTING:
Improving Batch Scheduling Algorithms through Explanation-Based Learning

Learning comes from reflection upon accumulated experience and the identification of patterns found among the elements of past experience. All numerical algorithms used in scientific and engineering computing are based on the same paradigm: *execute a predetermined sequence of calculation tasks and produce a numerical answer.* The implementation of the specific numerical algorithm is oblivious to the experience gained during the solution of a specific problem and, in the next encounter, a different, or even the same problem is solved through the execution of exactly the same sequence of calculation steps. The numerical algorithm makes no attempt to reflect upon the structure and patterns of the results it produced, or to reason about the structure of the calculations it has performed. Chapter 10 shows that this need not be the case. By allowing an algorithm to reflect upon and reason with aspects of the problems it solves and its *own structure of computational tasks, the algorithm can learn* how to carry out its tasks more efficiently. Such *intelligent numerical computing* represents a new paradigm, which will dominate the future of scientific and engineering computing. But, in order to unlock the computer's potential for the implementation of truly intelligent numerical algorithms, the *procedural* depiction of a numerical algorithm must be replaced by a *declarative* representation of the algorithmic logic. Such a requirement upsets an established tradition and imposes new educational challenges, which most educators and educational curricula have not, as yet, even recognized. This chapter shows how one can take a branch and bound algorithm, used to identify optimal schedules of batch operations, and endow it with the ability to learn to improve its own effectiveness in locating the optimal scheduling policies for flowshop problems. Given that most batch scheduling problems are NP-hard, it becomes clear how important it is to improve the effectiveness of algorithms for their solution. Using the Ibaraki framework, a branch and bound algorithm is declaratively modeled as a *discrete decision process*. Then explanation-based machine learning strategies can be employed to uncover patterns of generic value in the experience gained by the branch and bound algorithm from solving specific instances of scheduling problems. The logic of the uncovered patterns (i.e., new knowledge) can be incorporated into the control strategy of the branch and bound algorithm when the next problem is to be solved.

GEORGE STEPHANOPOULOS AND CHONGHUN HAN

REFERENCES

1. Stephanopoulos, G., Computers, Systems, Languages and Other Fragments. *CAST Newsletter*, Spring (1994).
2. Antsaklis, P. J. and Passino, K. M., eds., "An Introduction to Intelligent and Autonomous Control." Kluwer, Norwell, MA, 1993.
3. Åström, K. J. and McAvoy, T. J., Intelligent Control, *J. Proc. Cont.* **2**(3), 115 (1992).
4. Stephanopoulos, G. and Han, C., "Intelligent Systems in Process Engineering: A Review," *Proc. PSE*, Kyongju, Korea, 1994.
5. Kurtzweil, R., "The Age of Intelligent Machines." MIT Press, Cambridge, MA, 1990.
6. Mandelbrot, B. B., "The Fractal Geometry of Nature." W. H. Freeman, New York, NY, 1983.
7. Davis, P. J. and Hersh, R. "Descartes' Dream: The World According to Mathematics." Harcourt Brace Jovanovich, San Diego, CA, 1986.
8. Stephanopoulos, G. and Mavrovouniotis, M.L., eds., Artificial Intelligence in Chemical Engineering—Research and Development, *Comp. Chem. Eng.* **12**(9/10), (1988).
9. Stephanopoulos, G., Artificial Intelligence and Symbolic Computing in Process Engineering Design. *In* "Foundations of Computer-Aided Process Design" (J. J. Siirola, I. E. Grossmann, and G. Stephanopoulos, eds.), p. 21, Elsevier, New York, 1989.
10. Stephanopoulos, G., Artificial Intelligence: What Will Its Contributions Be to Process Control? *In* "The Second Shell Process Control Workshop" (D. M. Prett, C. E. Garcîa, and B. L. Ramaker, eds.), p. 591, Butterworths, Stoneham, MA, 1990.
11. Stephanopoulos, G., Brief Overview of AI and Its Role in Process Systems Engineering. *In* "Process Systems Engineering, Vol. I," CACHE, 1992.
12. Mavrovouniotis, M., ed., "Artificial Intelligence in Process Engineering." Academic Press, San Diego, CA, 1990.
13. Quantrille, T. E. and Liu, Y. A., "Artificial Intelligence in Chemical Engineering." Academic Press, San Diego, CA, 1991.
14. Hofstadter, D. R., "Gödel, Escher, Bach: An Eternal Golden Braid." Vintage Books, New York, 1980.
15. Penrose, R., "The Emperor's New Mind." Oxford University Press, Oxford, UK, 1989.

NONMONOTONIC REASONING: THE SYNTHESIS OF OPERATING PROCEDURES IN CHEMICAL PLANTS

Chonghun Han, Ramachandran Lakshmanan,[1] Bhavik Bakshi,[2] and George Stephanopoulos

Laboratory for Intelligent Systems in Process Engineering
Department of Chemical Engineering
Massachusetts Institute of Technology
Cambridge, Massachusetts 02139

I. Introduction	314
A. Previous Approaches to the Synthesis of Operating Procedures	316
B. The Components of a Planning Methodology	318
C. Overview of the Chapter's Structure	324
II. Hierarchical Modeling of Processes and Operations	324
A. Modeling of Operations	325
B. Modeling of Process Behavior	329
III. Nonmonotonic Planning	334
A. Operator Models and Complexity of Nonmonotonic Planning	336
B. Handling Constraints on the Temporal Ordering of Operational Goals	337
C. Handling Constraints on the Mixing of Chemicals	339
D. Handling Quantitative Constraints	343
E. Summary of Approach for Synthesis of Operating Procedures	348
IV. Illustrations of Modeling and Nonmonotonic Operations Planning	351
A. Construction of Hierarchical Models and the Definition of Operating States	351
B. Nonmonotonic Synthesis of a Switchover Procedure	359
V. Revamping Process Designs to Ensure Feasibility of Operating Procedures	368
A. Algorithms for Generating Design Modifications	369
VI. Summary and Conclusions	374
References	375

[1] Present address: Department of Chemical Engineering, University of Edinburgh, Scotland, UK.
[2] Present address: Department of Chemical Engineering, Ohio State University, Columbus, OH 43210, USA.

Planning a sequence of actions to take you from one point to another is usually a proposition whose inherent difficulty increases as more obstacles are placed in your way. The number of these obstacles (constraints), which you must skirt around, determines the complexity of the task, because any time you run into one of them you must *backtrack* and try an alternative step or path of steps. Such *serial* (or *linear* or *monotonic*) construction of a plan is fraught with pitfalls and repeated backtracking. The more the constraints, the more inefficient the monotonic planning. If, on the other hand, an action step (a Clobberer) leads to the violation of a constraint, then *do not backtrack*. Take another action step (a White Knight), which, when it precedes a Clobberer, negates the impact of the Clobberer, and you never need to backtrack. So, the more constraints, the more efficient your planning process. Such *nonserial* (or *nonlinear* or *nonmonotonic*) reasoning has become the essence of all modern and efficient planners, whether they are *logic-based* and *explicit*, or *implicit* enumerators of alternative plans. The purpose of this choice is twofold: (1) to introduce the ideas of *nonmonotonic reasoning* in the planning of process operations and (2) to demonstrate how nonmonotonic planning can be used to synthesize operating procedures for chemical processes, either off-line for standard tasks (e.g., routine startup or shutdown), or on-line for real-time response to large departures from desired conditions. It is shown that hierarchical modeling of process operations and operators is essential for the efficient deployment of nonmonotonic planning, and that the tractability of the resulting algorithms is strictly dependent on the form of the operators. In this regard, the ideas on modeling in this chapter draw heavily from the material of the first chapter in Volume 21 of this series. Nonmonotonic planners handle with superb efficiency constraints on (1) the temporal ordering of operations, (2) avoidable mixtures of chemical species, and (3) bounding quantitative conditions on the state of a process. Consequently, they could be used to generate explicitly all feasible operating procedures, leaving a far smaller search space for the selection of the optimum procedure by a numerical optimizer.

I. Introduction

The planning and scheduling of process operations, beyond the confines of single processing units, are fairly complex tasks. Thus, the synthesis of operating procedures for the routine startup or shutdown of a plant, equipment changeover, safe fallback from hazardous situations, changeover of the process to alternate products, and others, involves (1) a long list of noncommensurable objectives (operating economics, safety, environmental impact), (2) models describing the behavior of all units in a plant, (3) human supervision and intervention, and (4) the degree of the required automation. While mathematical programming techniques, such as branch and bound algorithms for mixed-integer linear or nonlinear (MILP or MINLP) formulations, have been used (Crooks and Macchietto, 1992) to locate the "optimal" plan, computer-based process control systems (Pavlik, 1984) "do not know" how to plan, schedule, and implement complex, nonserial schedules of process operations. The latter is considered to be largely a supervisory task left to the human operator. Nevertheless, there exists a natural tendency for increased automation in the control room (Garrison *et al.*, 1986), which takes one or both of the following forms: (1) a priori synthesis of operating procedures and subsequent implementation through programmed-logic control systems or (2) online, automatic planning and scheduling of operating procedures with real-time implementation through distributed control systems. The former approach is favored for the routine tasks such as startup, shutdown, and changeover, whereas the latter is essential for the optimization of operating performance, the intelligent response to large departures from operating norms, and the planning of emerging fallback operations in the presence of unsafe conditions.

The synthesis of operating procedures involves the specification of an ordered sequence of primitive operations, such as, opening or closing manual valves, turning on or off motors, and changing the values of setpoints, which when applied to a process will change the state of a plant from some given initial state and eventually bring it to a desired *goal* state (Lakshmanan and Stephanopoulos, 1988a). Typically, the transformation from the initial to the goal state is attained through a series of intermediate states (Fig. 1), each of which must satisfy a set of constraints, such as

1. *Physical constraints*, imposed by mass, energy, and momentum balances, chemical and phase equilibrium conditions, and rate phenomena.
2. *Temporal constraints*, imposed by various considerations; e.g., you cannot reach the state of a full tank if the state of the feeding pump has not been previously turned on.
3. *Logical, mixing constraints*, to avoid explosive mixtures, poisoning of catalysts, generation of toxic materials, deterioration of product quality, and other consequences resulting from the unintentional mixing of various chemicals.
4. *Quantitative inequality constraints*, requiring that the values of temperatures, pressures, flowrates, concentrations, and material or energy accumulations remain within a range defined by lower and upper bounds, in order to maintain the safety of personnel and

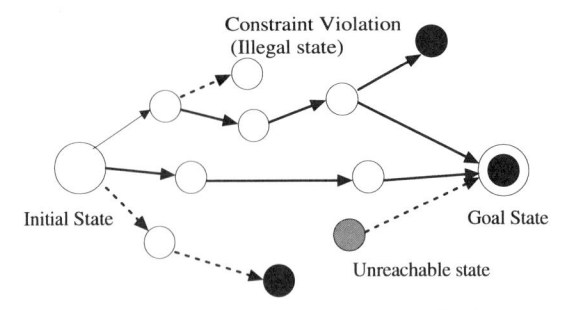

FIG. 1. Schematic of a typical planning problem (solid line signifies a feasible path and dotted line an infeasible path).

    equipment, environmental regulations, and production specifications.
5. *Performance constraints*, in terms of operating cost, product turnaround time, and others.

A. PREVIOUS APPROACHES TO THE SYNTHESIS OF OPERATING PROCEDURES

Domain-independent theories of planning have attempted to generate formulations and solution methodologies that do not depend on the characteristics of the particular area of application, whereas *domain-dependent* approaches have attempted to capitalize on the specific characteristics of a given problem domain, in order to construct special-purpose planning programs. Although a domain-independent planning theory would be preferable, there exists sufficient theoretical evidence to suggest that building a probably correct, complete, domain-independent planner that is versatile enough to solve real-world planning problems, is impossible (Chapman, 1985).

    Rivas and Rudd (1974) should be credited with the pioneering effort of synthesizing operating sequences that achieve certain operational goals, while ensuring that safety constraints were not violated. Chemical processes were modeled as networks of *valves* and *connectors* (e.g., pipes, vessels, unit operations), and the safety constraints were expressed as statements about the species that should never be present simultaneously in any location of the plant. The sequential logic algorithm that they proposed is equivalent to a restricted form of constraint propagation. Thus, high-level, abstractly expressed goals (input by the user) such as "start-a-heater" were propagated to lower-level goals of higher detail and were converted into a series of valve operations. Unfortunately, keeping track of the dependencies in order to generate *explainable* plans (a highly desirable feature) was not possible, as it was impossible to account for additional classes of constraints.

Kinoshita *et al.* (1982) divided a plant into sections around "key" processing equipment (including peripherals and attached pipes and valves) and identified a list of individual operations for each section. During the first phase of their methodology, they generate a tree of all possible sequences of operations for each section. Each tree described the *transition relationships among the states* of each section, and was endowed with the following information: (1) minimum and maximum time required for each operation, (2) the cost associated with each operation, and (3) constraints on the states of adjacent units. During the second phase of their work, Kinoshita *et al.* attempted to coordinate the sequences of the independent trees in an effort to achieve consistency among the operations of interacting units, by adjusting the timing and thus the temporal order of the various primitive operations.

In a similar spirit, Ivanov *et al.* (1980) had previously constructed *state transition networks* for the representation of operating states taken on by a plant as it goes from the initial state to the goal state. While the nodes of the network represent the states of a plant, the edges (arcs) signify the primitive operations that carry the processing units from one state to the next. To account for quantitative performance measures, Ivanov *et al.* assigned to each arc a weighting factor, which depended on the type of the performance measure. Within such framework, the synthesis of the "optimal" operating procedure becomes a search problem through a set of discrete alternatives. Ivanov *et al.* applied this formulation to the synthesis of optimal startup sequences of operations.

Tomita, Nagata, and O'Shima (1986), on the other hand, represented the topology of a plant through a directed graph with the nodes signifying the constituent equipment and the directed edges representing the pipelines. The valves were assigned to the appropriate edges of the digraph. Propositional logic was used to express the topology of the digraph and the temporal ordering of the various operations, thus leading to the generation of logically feasible operating plans. Subsequent work (Tomita *et al.*, 1989a, b) focused on the use of logic-based techniques and other ideas from artificial intelligence toward the development of an automatic synthesizer of operating procedures.

The work of Fusillo and Powers (1987) has attempted to define a formal theory for the synthesis of operating procedures. First, it introduced more expressive descriptions of the plants and their behavior, going beyond the

Boolean formulations favored by the pioneers in the field. Second, the planning methodology allowed the incorporation of global and local constraints, all of which were checked at each intermediate state to ensure that the evolving operating procedure was feasible. Third, they introduced the concept of a "stationary state," i.e., a state at which the plant itself does not change significantly with time (e.g., can wait until next action is taken), and where the operating goals are partially met. The "stationary state" was a breakthrough concept. It allows a natural breakoff point from a large schedule of operations, thus leading to an automatic generation of *intermediate operating goals*, which simplify the search for feasible operating procedures by decomposing the overall planning problem to smaller subproblems. In addition, a "stationary state" represents a convenient stopping-off point for recovery from operational errors, thus allowing the process to retreat back to an intermediate state thereby avoiding a complete shutdown. In subsequent publications (Fusillo and Powers, 1988a, b) the same authors introduced local quantitative models, and they also carried out detailed studies on the synthesis of purge operations.

Foulkes *et al.* (1988) have approached the synthesis of operating procedures from a more empirical angle. They have extended the work of Rivas and Rudd (1974) for the synthesis of complex pump and valve sequencing operations, relying on the use of logical propositions (implemented as rule-based expert systems), which capture the various types of constraints imposed on the states of a processing system.

B. The Components of a Planning Methodology

From Fig. 1 it is clear that the synthesis of an operating procedure could start from (1) the initial state and proceed forward to the goal state, (2) the goal state and proceed backward to the initial state, or (3) any known intermediate state(s) and proceed in both directions to bridge the gap with the specified initial and goal states. Furthermore, it should also be clear that there exist many paths connecting the initial to the goal state, which can be evaluated in terms of (1) their *feasibility*, i.e., whether their corresponding constituent states violate the specified constraints or not, and (2) a *performance index*, i.e., associated cost, total operational time, or other. To check the feasibility of a path we need an explicit *set of constraints* determining the allowable states of the process, and to evaluate performance we need a quantitative description of the desired performance index (or a vector of performance indices). Finally, in order to be able to evaluate the state of the process, we need *models* that describe the underlying physicochemical phenomena and the variety of primitive operations.

No planning methodology can be efficient or credible for the synthesis of process operating procedures, if it focuses only on the search for the "best" plan, delegating the modeling needs and the articulation of the requisite constraints to inferior tasks. The reverse is also true. Consequently, a concise methodology for the synthesis of operating procedures should provide a unified treatment of all three pivotal components, namely

1. Modeling of processing systems and primitive operations.
2. Explicit articulation of all constraints that an operating procedure should satisfy.
3. Search over the set of discrete plans and the identification of the feasible, and possibly of the "optimal" among them.

1. Formal Statement of the Operations Planning Problem

Consider a processing system composed of N subsystems, whose topological interconnections are determined by a set of streams, $S = \{s_{i,j}^{(p,q)}\}$; $i, j = 1, 2, \ldots, N$, with $s_{i,j}^{(p,q)} = 1$ if the pth output of the ith subsystem is the qth input to the jth subsystem, and zero otherwise. What is denoted as a subsystem could vary with the level of detail considered in the description of the processing system. At a high level of detail, a subsystem could be a processing equipment, an actuator, a safety device, or a controller. At a low level of detail, a subsystem could be a processing section, composed of several interconnected processing equipment with their actuators and controllers.

Let $\mathbf{x}_i$ denote the vector of state variables describing the behavior of the ith subsystem, and $\mathbf{u}_i^{(q)}, \mathbf{y}_i^{(p)}$ the vectors describing its qth input and pth output streams. The modeling relationships around each subsystem yield

$$\frac{d\mathbf{x}_i(t)}{dt} = \mathbf{f}_i[\mathbf{x}_i(t), \mathbf{u}_i^{(q)}(t); \quad q = 1, 2, \ldots, k_i]; \quad i = 1, 2, \ldots, N, \quad (1)$$

$$\mathbf{y}_i^{(p)}(t) = \mathbf{g}_i^{(p)}[\mathbf{x}_i(t), \mathbf{u}_i^{(q)}; \quad q = 1, 2, \ldots, k_i];$$

$$i = 1, 2, \ldots, N, \quad p = 1, 2, \ldots, m_i. \quad (2)$$

If the surrounding world is denoted as the zeroth subsystem, then the variables, $\mathbf{y}_0^{(p)}$; $p = 1, 2, \ldots, m_0$ represent the external inputs to the processing system, and in general they can be divided into two classes; the *real-valued*, $\mathbf{y}_{0,\text{RV}}^{(p)}$ and the *integer-valued*, $\mathbf{y}_{0,\text{IV}}^{(p)}$. The $\mathbf{y}_0^{(p)}$ variables can be used to represent the actions of human operators or supervisory control systems.

The values of the inputs, $\mathbf{u}_i^{(q)}$, $q = 1, 2, \ldots, k_i$; $i = 1, 2, \ldots, N$, are determined by a set of operations, such as opening or closing a manual valve, turning on or off a motor, or changing the value of a controller's setpoint. Each of these operations can be associated with a particular input and describes how the value of input, $\mathbf{u}_i^{(q)}$, is determined from the value of outputs, $y_i^{(p)}$;

$$\mathbf{u}_j^{(q)}(t) = Op_j^{(q)}\left[y_i^{(p)}(t); \quad p = 0, 1, 2, \ldots, m_i \text{ and} \right.$$
$$\left. \forall i: S_{ij}^{(p,q)} = 1; \quad \alpha_j^{(q)}(t)\right] \quad (3)$$

where $j = 1, 2, \ldots, N$.

Operations $Op_j^{(q)}$, can be seen as *operators* that can take on a *logical integer*, *analytic* (static or dynamic), or *hybrid* form. With each operation we have associated an integer-valued variable, $\alpha_j^{(q)}(t)$, which determines the discrete state that the operator $Op_j^{(q)}$, takes on with time.

Given an initial state of a processing system and the outputs of the surrounding world, one may solve the set of Eqs. (1), (2), and (3) and thus find the temporal evolution of the state describing the process behavior. As the various operators $Op_j^{(q)}$ take on different discrete states over time, the trajectory of the process' state goes through changes at discrete time points. Clearly, the order in which the various operations change state and the time points at which they change value, affect the temporal trajectory of the process evolution and determine the value of the performance index. Therefore, we are led to a *mixed-integer, nonlinear programming* (MINLP) problem with the following decision variables:

1. *Integer decisions*
   (a) $\alpha_j^{(q)}(t)$, i.e., the discrete state of each operator, $Op_j^{(q)}$.
   (b) The temporal ordering of the discrete-state changes of the various operators, e.g., $t_1 < t_2 < t_3$, where $\alpha_2^{(1)}(t_1) = 1$ (from 0), $\alpha_4^{(1)}(t_2) = 0$ (from 1), $\alpha_3^{(1)}(t_3) = 2$ (from 1).
   (c) $y_{0,IV}^{(p)}(t)$, i.e., integer-valued variables, which are inputs from the surrounding world.
   (d) The temporal ordering of the discrete-value changes of the integer-valued inputs $y_{0,IV}^{(p)}(t)$, from the surrounding world.
2. *Continuous decisions*. The time-dependent trajectories of the real-valued variables that are inputs from the surrounding world.

In addition to these above, the values that the states, x_i, $i = 1, 2, \ldots, N$; the outputs, $y_i^{(p)}$; $p = 1, 2, \ldots, m_i$; $i = 1, 2, \ldots, N$; and the inputs, $\mathbf{u}_i^{(q)}$; $q = 1, 2, \ldots, k_i$; $i = 1, 2, \ldots, N$ can take are limited by a set of engineering constraints, which are dictated by safety and operability considerations, physical limitations, or/ and production specifications. Finally, the selection of the "best" plan of operations requires the formulation of an objective function, or a vector of objective performance indices. Thus, for the general multiobjective synthesis of operating procedures we can state the following mixed-integer, dynamic optimization problem (termed *Problem* 1):

$$\underset{\mathbf{A};\, \mathbf{y}_0^{(p)}(t)}{\text{Minimize}} P = [P_1, P_2, \ldots, P_l]$$

subject to

$$\frac{d\mathbf{x}_i(t)}{dt} = \mathbf{f}_i[\mathbf{x}_i(t), \mathbf{u}_i^{(q)}(t)]; \quad q = 1, 2, \ldots, k_i]; \quad i = 1, 2, \ldots, N, \quad (1)$$

$$\mathbf{y}_i^{(p)}(t) = \mathbf{g}_i^{(p)}[\mathbf{x}_i(t), \mathbf{u}_i^{(q)}]; \quad q = 1, 2, \ldots, k_i];$$
$$p = 1, 2, \ldots, m_i, \quad i = 1, 2, \ldots, N. \quad (2)$$

$$\mathbf{u}_j^{(q)}(t) = Op_j^{(q)}[\mathbf{y}_i^{(p)}(t), \quad p = 1, 2, \ldots, m_i, \forall i: s_{ij}^{(p,q)} = 1; \quad \alpha_j^{(q)}(t)];$$
$$q = 1, 2, \ldots, k_j \text{ and } j = 1, 2, \ldots, N, \quad (3)$$

$$\mathbf{h}_i(\mathbf{x}_i; \mathbf{u}_i^{(q)}; \quad q = 1, 2, \ldots, k_i; \quad \mathbf{y}_i^{(p)};$$
$$p = 1, 2, \ldots, m_i) \leq 0; \quad i = 1, 2, \ldots, N, \quad (4)$$

where **A** is the vector of all $\alpha_j^{(q)}(t)$, for $q = 1, 2, \ldots, k_j$ and $j = 1, 2, \ldots, N$.

Problem-1 is a formidable challenge for mathematical programming. It is an NP-hard problem, and consequently all computational attempts to solve it cannot be guaranteed to provide a solution in polynomial time. It is not surprising then that all previous efforts have dealt with simplified versions of Problem-1. These simplifications have led to a variety of

(a) Representation models for the behavior of the subsystems, i.e., the form of the f_i and $\mathbf{g}_i^{(p)}$ functions.
(b) Descriptions for the operations, i.e., the form of the $Op_j^{(q)}$ operators.
(c) Search techniques for feasible or optimal plans.

2. The Character of Constraints and the Modeling Requirements

The character of the constraining functions, $\mathbf{h}_i$, given by Eq. (4), determines to a large extent the form of the required modeling functions, $\mathbf{f}_i$, $\mathbf{g}_i^{(p)}$, and $Op_i^{(p)}$. For example, Rivas and Rudd (1974) stipulated that the only constraints they were concerned were related to the avoidance of certain mixtures of chemicals, anywhere in the process. This being the case, all states, input and output variables can be characterized by Boolean values, and the required models are simple expressions of propositional calculus. Subsequently, the work of Fusillo and Powers (1987, 1988a,b) introduced qualitative and static quantitative constraints, which required more detailed $\mathbf{f}_i$ and $\mathbf{g}_i^{(p)}$ descriptions.

Another important consideration regarding the character of constraints, $\mathbf{h}_i$, is their *hierarchical articulation at multiple levels of detail*. For example, suppose that one is planning the routine startup procedure for a chemical plant. Safety considerations impose the following constraint on the temporal ordering of operations:

Start the reaction section last,

which is expressed at the abstract level of processing section. On the other hand, the constraint

in starting a heat exchanger, start the cold side first

is expressed at the more detailed level of processing equipment. Since both types of constraints must be accounted for during the planning of an operating procedure, it is clear that the computer-aided modeling of a processing system should allow for hierarchical representations at various levels of detail.

The final comment on the character of constraints is related to the available technology for equation solving. The general form of constraints (4) involves restrictions on the time-dependent trajectories of states, inputs, and outputs. Since the set of decisions of Problem-1 involves both integer and continuous variables, we need numerical methodologies that can solve large sets of differential algebraic equations [i.e., Eqs. (1), (2), and (3)] with continuous and discrete variables. The recent works in dynamic simulation (Pantelides and Barton, 1993; Barton, 1992; Marquardt, 1991; Holl *et al.*, 1988) are spearheading the developments in this area, but until these efforts produce efficient and robust solution algorithms, the formulations of Problem-1 will be forced to accept simple statements of constraints [Eq. (4)].

3. Search Procedures

The solution of Problem-1 requires extensive search over the set of potential sequences of operations. Prior work has tried either to identify all feasible operating sequences through *explicit* search techniques, or locate the "optimum" sequence (for single-objective problems) through the *implicit* enumeration of plans. The former have been used primarily to solve planning problems with Boolean or integer variables, whereas the latter have applied to problems with integer and continuous decisions.

Explicit search techniques for the identification of feasible operating procedures originated in the area of artificial intelligence. Typical examples involve the pioneering "means–end analysis" paradigm of Newell *et al.* (1960), and the various flavors of *monotonic planning* (e.g., Fikes and Nilsson, 1971), or *nonmonotonic planning* (e.g., Sacerdoti, 1975; Chapman, 1985). With the exception of the work of Lakshmanan and Stephanopoulos (1988a, b, 1990), who have used nonmonotonic planning, all other works on the synthesis of operating procedures for chemical processes have employed ideas of monotonic planning.

Starting either from the desired goal or the given initial state of a processing system, monotonic planning constructs an operating procedure step by step, moving "closer" all the time to the other end (i.e., initial or goal states, respectively). After each operating step the values for all state, input, and output variables are known by solving Eqs. (1), (2), and (3), respectively, and the satisfaction of constraints [Eq. (4)] is tested. The last operating step is revoked (or modified) if any of the constraints (4) is violated. Nonmonotonic planning, on the other hand, works by constructing and refining partial plans. A *partial plan* is a set of operating steps that leaves a certain amount of information unspecified. Perhaps the temporal order in which two operating steps are executed is not specified, or a particular step of the operating procedure may be assigned a *set* of potential feasible operations, rather than a *single* operation. Thus, at the end, *a single partial plan actually describes a large number of "completions" or total operating procedures*. This feature is extremely important for two reasons:

1. *Efficiency*. By conducting planning in terms of partial plans, nonmonotonic reasoning allows a single planning decision to represent a large number of plans, resulting in possibly exponential savings in efficiency (Chapman, 1985; Lakshmanan and Stephanopoulos, 1989).
2. *Integration*. The fact that nonmonotonic planning can capture large sets of feasible plans through single decisions implies that it can be integrated very naturally with the mixed-integer mathematical programming techniques if the identification of the "optimum" plan is required.

Implicit enumeration techniques for the identification of the optimum operating procedure originated in the area of operations research and are based largely on variants of the *branch-and-bound* paradigm. Through recent developments, they have become fairly efficient in locating the "optimum" solution, but they are still awkward in defining the set of feasible plans. Continued progress in the development of more efficient formulations of the planning problem and more efficient branch-and-bound strategies, to handle decision variables with many integer values, will provide a *practical tractability* for the solution of larger and richer versions of Problem-1.

C. OVERVIEW OF THE CHAPTER'S STRUCTURE

Section II addresses the most important facet of an efficient planning methodology, namely, "how to represent the behavior of processes and process operating steps." Three basic operators—simple propositional, conditional, and functional—are explored for their representational power to model process operations. In addition, the representation of process operational constraints dictates the need for a hierarchical modeling of processes and their subsystems, thus leading a modeling framework very similar to that of MODEL.LA. (see first chapter in Volume 21). In fact, we have used the formalism of MODEL.LA. to articulate all modeling needs, related to nonmonotonic planning of process operations.

Section III introduces the concept of nonmonotonic planning and outlines its basic features. It is shown that the tractability of nonmonotonic planning is directly related to the form of the operators employed; simple propositional operators lead to polynomial-time algorithms, whereas conditional and functional operators lead to NP-hard formulations. In addition, three specific subsections establish the theoretical foundation for the conversion of operational constraints on the plans into temporal orderings of primitive operations. The three classes of constraints considered are (1) temporal ordering of abstract operations, (2) avoidable mixtures of chemical species, and (3) quantitative bounding constraints on the state of processing systems.

In Section IV we provide illustrations of the modeling concepts presented in Section II and how the strategy of nonmonotonic planning has been used to synthesize the switchover operating strategy for a chemical process.

Section V extends the above ideas into a different problem, i.e., how to revamp the design of a process so that it can accommodate feasible operating procedures.

II. Hierarchical Modeling of Processes and Operations

In this section we will discuss a systematic approach for the representation of processes and operating steps. These representations, or *models*, conform to the general requirements presented in Section I,B; they can

(1) capture Boolean, integer (i.e., categorical, or qualitative), or real-valued variables of a process; and (2) emulate descriptions of the process or of the operations at multiple levels of detail with internal consistency.

A. MODELING OF OPERATIONS

The characterization of operating steps, i.e., the characterization of the operators, $Op_j^{(q)}$, depends on (1) the type of operation (i.e., Boolean, integer, analytic real-valued) and (2) the level of the desired detail (abstract operation over a section of a plant, primitive operations on valves, pumps, etc.). Let us see the classes of possible models, all of which could be available and used even within the same planning problem.

The STRIPS-Operator

Figure 2a shows the simplest possible model of an operating step, introduced by Fikes and Nilsson (1971) and known as the STRIPS operator. The *preconditions* are statements that must be true of the system and its surrounding world before the action can be taken. Similarly, the *postconditions* are statements that are guaranteed to be true after the actions corresponding to the specific operation have been carried out. The allowable forms of pre-, and postconditions should satisfy the following requirements:

(a) Pre- and postconditions must be expressed as propositions that have content, which is a tuple of elements, and can be negated.
(b) The elements of the preceding tuples can be variables or constraints and could be infinitely many of them.
(c) Functions, propositional operators, and quantifiers are not allowed.

The STRIPS operator is very simple and intuitively appealing. In attempting to use it for the synthesis of operating procedures for chemical processes, one finds immediately that it suffers from severe limitations, such as the following:

1. The temporal preconditions and postconditions in the STRIPS model vary from process to process, thus making impossible the creation of a generic set of operators a priori.
2. Given a set of preconditions [i.e., the values of $\mathbf{y}_i^{(p)}$ and $\alpha_j^{(q)}$; see Eq. (4)], the actions of process operation lead to new values for $\mathbf{u}_j^{(q)}$, which when propagated through Eq. (1) generate the new state, i.e., the action's postconditions. But Eq. (1) does include functional relationships, $\mathbf{f}_i$, thus leading to an overall functional operator, which is disallowed in a strict STRIPS model.

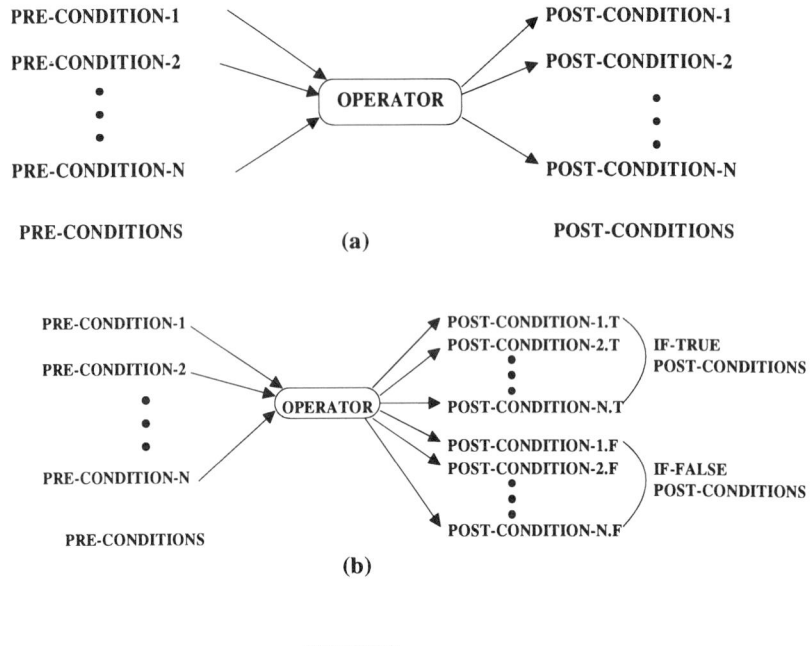

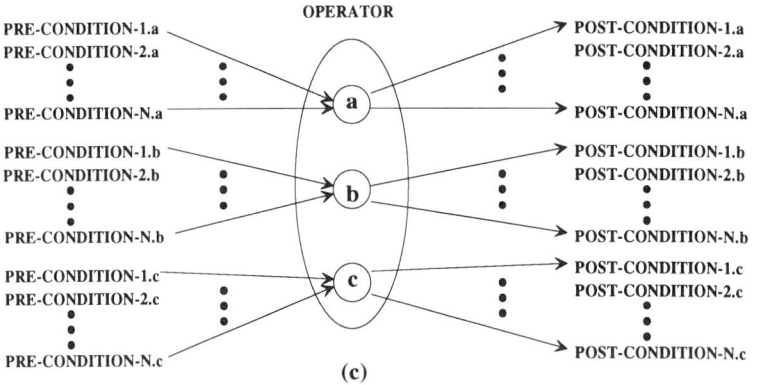

FIG. 2. The three planning operators: (a) STRIPS, (b) conditional, (c) functional.

Nevertheless, early work in the synthesis of operating procedures (Rivas and Rudd, 1974; Ivanov et al., 1980; Kinoshita et al., 1982; Fusillo and Powers, 1987) did employ various variants of the STRIPS operator on limited-scope problems with very useful results. Lakshmanan and Stephanopoulos (1987) also demonstrated the value of the STRIPS operators in ordering sequences of feasible operations during routine startup.

2. Conditional Operators

To overcome the limitations of the STRIPS operator, Chapman (1985) suggested the use of *conditional operators*, which produce two sets of postconditions (Fig. 2b), depending on whether all preconditions are true or any one of them is false. Conditional operators can describe a broader class of operations, but they do possess similar drawbacks when they are considered for the synthesis of operating procedures. In this regard, it is important to realize that the success of the operator-based approaches depend on the ability of the user to define, a priori, operator models for individual processing units, since such operators represent a fusion (i.e., simultaneous solution) of Eqs. (1)–(3) so that they can relate preconditions to the resulting postconditions. It is clear that their use is severely limited and cannot cover planning of operating procedures for (1) equipment/ product changeover, (2) optimizing control, and (3) certain types of safety fallback strategies.

3. Functional Operators

Instead of a conjunction of preconditions, as used by the STRIPS and conditional operators, the *functional* operator has a *set* of conjunctions of preconditions (Fig. 2c). Each element in the set describes some possible situation that might exist before the operator is applied. For each element of the set of preconditions, there is a corresponding element in the set of postconditions. The functional operator is a more flexible model than the STRIPS or conditional operators. It comes closer to the modeling needs for the synthesis of operating procedures for chemical processes, but as we will see in the next section, we need to introduce additional aspects in order to capture the network-like structure of chemical processes.

4. An Illustration

Consider the simple pipes-and-valves system shown in Fig. 3. Initially, oxygen is flowing into the system through inlet 1, and out through the outlet. We would like to develop an operating procedure to solve the following operational problem: "Route the flow of methane gas from the inlet 2 to the outlet without running the risk of explosion (i.e., avoid mixing oxygen and the hydrocarbon)."

Since the only conditions of interest relate to the presence or absence of materials in the various piping segments, the required models for the operations are very simple and can be given by the following three

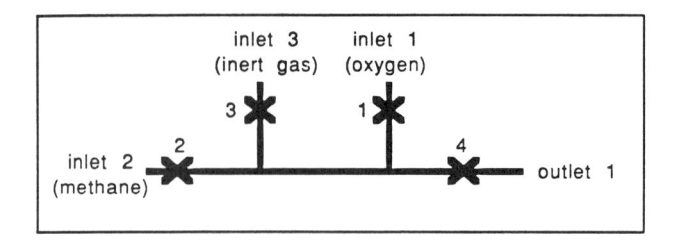

```
                inlet 3      inlet 1
               (inert gas)   (oxygen)
                    3 ✖      1 ✖
                      |        |
      inlet 2    2    |        |    4
      ─────────✖──────┴────────┴────✖────── outlet 1
      (methane)
```

INITIAL-STATE	GOAL-STATE
(flowing oxygen)	(flowing methane)
not(explosion)	not(explosion)

OPERATORS

(stop-flow x)   (establish-flow x)
pre-conditions: (flowing x)   pre-conditions: ()
post-conditions: not(flowing x)   post-conditions: (flowing x)

(purge x)   pre-conditions:
             not(flowing x);   not(equal inert-gas x)
             post-conditions: not(present x); (flowing inert-gas)

FRAME AXIOMS

(present methane) & (present oxygen) => (explosion!)

(flowing x) => (present x)

FIG. 3. Example operations planning problem. (Reprinted from *Comp. Chem. Eng.*, 12, Lakshmanan, R. and Stephanopoulos, G., Synthesis of operating procedures for complete chemical plants, Parts I, II, p. 985, 1003, Copyright 1988, with kind permission from Elsevier Science Ltd., The Boulevard, Langford Lane, Kidlington 0X5 1GB, UK.)

STRIPS-like operators:

1. Operator-1: (STOP_FLOW x)
            Preconditions: (flowing x)
            Postconditions: not(flowing x)
2. Operator-2: (ESTABLISH_FLOW x)
            Preconditions: ( )
            Postconditions: (flowing x)
3. Operator-3: (PURGE x)
            Preconditions: not(flowing x); not(equal inert-gas x)
            Postconditions: not(present x); (flowing inert-gas)

Using the initial and goal states as indicated in Fig. 3, it is easy to construct the following sequence of operations that satisfy the desired objectives:

(STOP_FLOW Oxygen),
(PURGE Oxygen), (ESTABLISH_FLOW Methane)

B. MODELING OF PROCESS BEHAVIOR

A processing facility is a network of unit operations, connected through material and energy flows. Through these connections, the effect of an operation is not confined to the operational behavior of the processing unit it directly affects and its immediate vicinity, but it may propagate and effect the state of all units in the plant. Thus, in addition to the complexity introduced by the functional dependence of process behavior on initial preconditions (see Section II,A), the conditions describing the state of a process resulting after the application of an operation must satisfy the constraints imposed by the network-like structure of the plant. Since the actual structure of the plant is not known at the time when we formulate the operator-models, we must rely on some other means for modeling the constraints characterizing the interactions among different units.

In the first chapter of Volume 2 (hereinafter referred to as **21**:1) we presented the general framework of MODEL.LA., a modeling language that can capture the hierarchical and distributed character of processing systems. We will employ all aspects of MODEL.LA. in order to develop a complete and consistent description of plants that will satisfy the modeling needs for the synthesis of operating procedures.

1. The Hierarchical Description of Process Topology

MODEL.LA.'s primitive modeling elements (see **21**:1), *Generic-Unit*, *Port*, and *Stream*, are capable of depicting any topological abstraction of a chemical process. In particular, the specific subclasses emanating from the *Generic-Unit* allow us to describe a process as (see **21**:1): (1) an overall *Plant-Sections*, where each section represents a grouping of units with a common operating framework, e.g., train of distillation columns; (2) a network of *Augmented-Units*, which encapsulate the structure of processing units along with their ancillary equipment, e.g., a distillation column with its feed preheater, condenser, and reboiler; and (3) a network of *Units*. Figure 4 shows a schematic of this topological hierarchy.

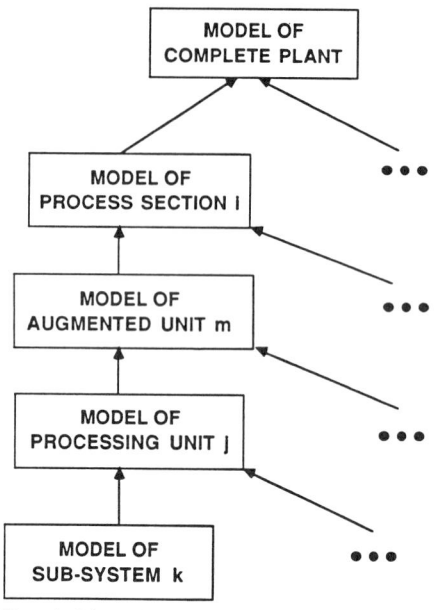

```
                    ┌──────────────┐
                    │   MODEL OF   │
                    │COMPLETE PLANT│
                    └──────────────┘
                      ↑          ↖
              ┌──────────────┐      • • •
              │   MODEL OF   │
              │PROCESS SECTION I│
              └──────────────┘
                  ↑         ↖
            ┌──────────────┐    • • •
            │   MODEL OF   │
            │AUGMENTED UNIT m│
            └──────────────┘
                ↑         ↖
          ┌──────────────┐   • • •
          │   MODEL OF   │
          │PROCESSING UNIT j│
          └──────────────┘
              ↑         ↖
        ┌──────────────┐  • • •
        │   MODEL OF   │
        │ SUB-SYSTEM k │
        └──────────────┘
```

FIG. 4. Hierarchical description of process flowsheets.

Figure 5 shows the sections of a plant producing gasoline from the dimerization of olefins. Using the explicit hierarchical descriptions given above, we can formulate the following series of abstract operators, which (as we will see in Section III) can provide significant help in reducing the number of alternative plans:

Plant level. (START_UP_PLANT);

Section level. (START_UP_FEED_PREPARATION), (START_UP_REACTION), (START_UP_RECOVERY), (START_UP_REFINING);

Augmented-Unit level. (START_UP_DEPROPANIZER), (START_UP_DEBUTANIZER), etc.

Any of these operators can be represented as a STRIPS, conditional, or functional operator, depending on the character of the constraints imposed on the particular operating procedure to be synthesized.

2. The Hierarchical Description of the Operating State

Whether we are dealing with an overall plant and its sections, augmented units, or individual units, we must view each system as a thermody-

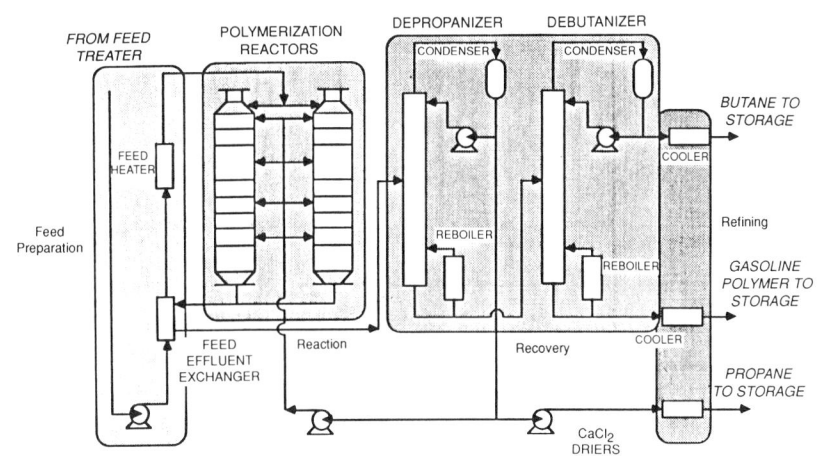

FIG. 5. Decomposition of gasoline polymerization plant.

namically "simple" system (Modell and Reid, 1983), which is characterized by a "state." From an operations planning point of view, this "state" can be viewed at different levels of detail, thus giving rise to a hierarchy of descriptions. MODEL.LA. provides the requisite modeling elements, which can be used to provide the necessary hierarchical descriptions of operating states:

a. Operational State of Variables. The modeling element, Generic-Variable (see **21**:1), allows the representation of the structured information that describes the behavior of "temperature," "pressure," "concentration," "flow," etc. Each of these instances of the *Generic-Variable* is an object with a series of attributes such as "current-value," "current-trend," "range-of-values," "record-over-x-minutes," and "average-over-x-minutes." Through these attributes we can capture all possible preconditions and postconditions that may be needed by the individual operators.

b. Operational State of "Terms". MODEL.LA.'s *Generic-Variable* possesses a series of subclasses (see **21**:1), whose sole purpose is to provide declarative description of certain compound variables, called *Terms*, such as "flow-of-component-x," "enthalpy-flow," "heat-flow," "diffusive-mass-flow," and "reaction rate." The declarative, rather than procedural, representation of compound variables enables the computer-aided planner to have an explicit description of the quantities involved in the preconditions and postconditions of an operation, thus replacing a numerical procedure by a series of explicit numerical or logical inferences, as the planning needs may be.

c. Operational State of Constraining Relationships. The operational state of "terms" is constrained by the laws of physics and chemistry and relationships that arise from engineering considerations. Typical examples are (1) the conservation principle (applied on mass, energy, and momentum); (2) phase or reaction equilibria; (3) reaction rates or transport rates for mass, energy, and momentum; (4) limits on pressure drops, heat flows, or work input, dictated by the capacity of processing units; and (5) experimental correlations between input and output "terms," etc. MODEL.LA.'s modeling element, *Constraint*, and its subclasses, *Equation*, *Inequality*, *Order-of-Magnitude*, *Qualitative*, and *Boolean* (see **21**:1) offer the necessary data structures to capture the information associated with any type of constraining relationships.

d. Operational State of a Process System. MODEL.LA's modeling element, *Modeling-Scope*, captures the consistent set of modeling constraints (described in the previous paragraph), which apply on a given process system. As the operational state of a system changes, the content of the *Modeling-Scope* may change; e.g., transition from a single to a two-phase content, transition from laminar to turbulent flow.

e. Example. The four-level hierarchy for the description of process operational states as shown in Fig. 6a offers a specific example. It offers a very rich and flexible modeling framework to capture any conditions (pre- or post-) associated with the definition of operators.

3. Maintaining Consistency among Hierarchical Descriptions of Process Topology or State

Operators, describing process operations, can be declared at any level of abstraction, but they should maintain consistent relations with each other, since they refer to the same process. For example, the top-level operator, (START_UP_PLANT), could be refined to the following sequence of operators (see also Fig. 5): (START_UP_RECOVERY), (START_UP_REFINING), (START_UP_FEED_PREPARATION), and (START_UP_REACTION). Clearly, the preconditions of (START_UP_PLANT) are distributed and represent a subset of the preconditions for all four more detailed operations. Similarly, the postconditions, derived from the startup operation of the four sections, should be

(a)

(b)

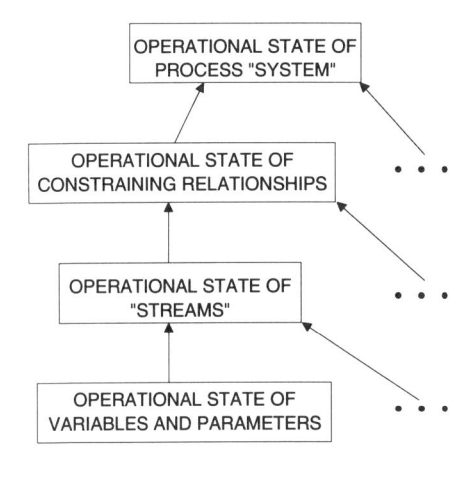

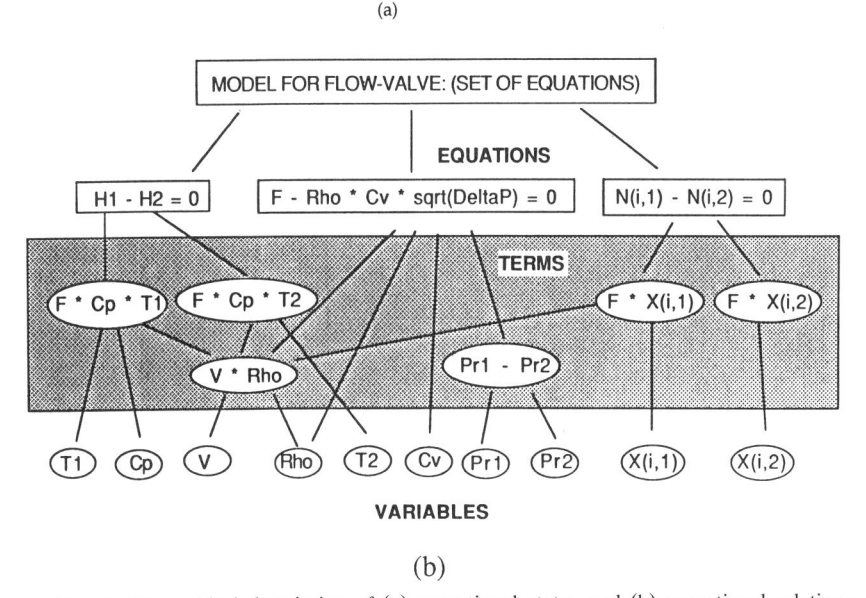

Fig. 6. Hierarchical description of (a) operational states and (b) operational relationships. (Reprinted from *Comp. Chem. Eng.*, 12, Lakshmanan, R. and Stephanopoulos, G., Synthesis of operating procedures for complete chemical plants. Parts I, II, p. 985, 1003, Copyright 1988, with kind permission from Elsevier Science Ltd., The Boulevard, Langford Lane, Kidlington 0X5 1GB, UK.)

consistent with the postulated postconditions of the overall (START_ UP_PLANT).

To achieve these consistencies, MODEL.LA. provides a series of semantic relationships among its modeling elements, which are defined at different levels of abstraction. For example, the semantic relationship (see **21**:1), **is-disaggregated-in**, triggers the generation of a series of relationships between the abstract entity (e.g., overall plant) and the entities (e.g., process sections) that it was decomposed to. The relationships establish the requisite consistency in the (1) topological structure and (2) the state (variables, terms, constraints) of the systems. For more detailed discussion on how MODEL.LA. maintains consistency among the various hierarchical descriptions of a plant, the reader should consult **21**:1.

III. Nonmonotonic Planning

Let us return to the simple operations planning problem of Fig. 3 (see also Section II,A,4). A non-monotonic planner faced with this problem proceeds as follows:

Step 1. Choose one of the desired goals (e.g., flowing methane).

Step 2. Select an operator that achieves the selected goal, e.g., (ESTABLISH_FLOW METHANE), which produces the desired goal as its postcondition.

Step 3. Propagate derived postconditions through the frame axioms (see Fig. 3) and take:

(flowing methane) $\Rightarrow$ (present methane)
(flowing oxygen) $\Rightarrow$ (present oxygen)

Step 4. Check for violation of constraints. In this case the constraint is also a frame axiom:

(present methane) AND (present oxygen) $\Rightarrow$ (explosion)

and is confirmed.

Step 5. Since the constraint was violated, the non-monotonic planner attempts to identify the proper operator among the set of available operators, which can alleviate the constraint violation, if this operator were "forced" to be applied before the constraint violation. In this example such an operator is (PURGE X). Thus, (PURGE X) must be applied before (ESTABLISH-FLOW METHANE). The variable X remains temporarily unbound to any particular value.

Step 6. Operation (PURGE X) stipulates the a priori satisfaction of the following two preconditions:

not(flowing X) AND not(equal inert-gas X)

Since the final goal state requires that (flowing methane) is true, then the value of X can be bound to oxygen. Then the following postconditions of (PURGE OXYGEN) are true:

not(present oxygen) (flowing inert-gas)

Step 7. The first precondition of (PURGE OXYGEN) becomes the new goal and requires that the following operation must take place before (PURGE OXYGEN):

(STOP_FLOW OXYGEN) producing not(flowing oxygen)

So the solution to the planning problem is now complete, and is given by the following sequence of elementary operations:

(STOP_FLOW OXYGEN) > (PURGE OXYGEN)
> (ESTABLISH_FLOW METHANE)

where the symbol > signifies the relative temporal ordering of two operations.

In summary, nonmonotonic planning of process operations has the following distinguishing features:

1. *Property 1: goal-driven approach*. This always starts from a desired goal state and tries to identify the requisite operations that will achieve it.

2. *Property 2: constraint-posting without backtracking*. Instead of backtracking, when a constraint violation is detected, nonmonotonic planning attempts to find an operator that would negate the preconditions leading to the constraint violation and forces this operator to be applied *before* the operation that causes the constraint violation. It leads to the posting of a new constraint determining the temporal ordering of operations. The operator that causes the violation of a constraint, negates, prevents, or undoes one or more of the planning goals, is called *Clobberer*. On the other hand, an operator that rectifies, or prevents any of the damage caused by a Clobberer will be called *nonmonotonic White Knight* (Chapman, 1985). Clearly, the success of any nonmonotonic planning methodology lies in its ability to identify these Clobberers and White Knights without having to resort to an exhaustive *generate-and-test* approach. Once these operators have been identified, the planner must

ensure that whenever a Clobberer is used in the plan, an accompanying White Knight is present.

3. *Property 3: generator of partial plans.* Nonmonotonic planning generates partially specified plans where a certain amount of information is not bound to specific values. A typical example was the value of X in the operation (PURGE X), that we saw in the illustration at the beginning of this section. Partial plans are abstractions, which can represent large sets of plans, and can lead to every efficient search strategies.

A. OPERATOR MODELS AND COMPLEXITY OF NONMONOTONIC PLANNING

The correctness of complete plans is encapsulated by the following theorem (Chapman, 1985).

Theorem 1 (Truth Criterion of Complete Plans). *In a complete plan, a proposition, p, is necessarily true in a situation, s, if and only if there exists a situation, t, previous or equal to s in which p is asserted, and there is no step between t and s that denies p.*

In this theorem, any proposition **p** can represent an operator (i.e., an operation step), whereas the situations **t** and **s** represent any intermediate state of the process. Although the validity of the theorem is general, its practical utility is confined to monotonic planning with STRIPS-like operators. For example, in nonmonotonic planning the plans are at any point when partially specified and a new mechanism is needed to guarantee that when the partial plan is completed, a given proposition (i.e., a given operation) is still true (i.e., consistent).

Chapman's work produced the following theorem, which provides the necessary and sufficient conditions for guaranteeing the truth of any given statement in a partial plan, if all operations are modeled by STRIPS-like operators.

Theorem 2 (Modal Truth Criterion). *A proposition p is necessarily true in a situation s if and only if the following two conditions hold:*

(a) *There is a situation t equal or necessarily previous to s in which p is necessarily asserted.*

(b) *For every step C, possibly before s, and every proposition, q, possibly codesignating with p, which C denies, there is a step W necessarily between C and s that asserts r, a proposition such that r and p codesignate whenever p and q codesignate.*

So, if **C** is the step in an operating procedure that introduces the

Clobberer, **q**, denying the validity of the postconditions of operation **p**, then **W** is the operating step (necessarily after **C**) at which the White Knight, **r**, comes to rescue and negates the effects of **q**. Theorem 2 also indicates that for each operation **p**, any preceding Clobberer must have the corresponding White Knight, if the nonmonotonic plan is to be correct.

For STRIPS-like operators, Chapman (1985) developed a polynomial-time algorithm, called TWEAK, around five actions that are necessary and sufficient for constructing a correct and complete plan. As soon as we try to extend these ideas to nonmonotonic planning with conditional operators, we realize that no polynomial-time algorithm can be constructed, as the following theorem explicitly prohibits (Chapman, 1985):

Theorem 3 (First Intractability Theorem). *The problem of determining whether a proposition is necessarily true in a nonmonotonic plan whose action representation is sufficiently strong to represent conditional actions is NP-hard.*

It is not surprising, then that, as we require functional operators to approximate more closely the majority of operations during the synthesis of process operating procedures, we must give up any expectation for a tractable algorithm (Lakshmanan, 1989):

Theorem 4 (Second Intractability Theorem). *The problem of determining whether a proposition is necessarily true in a nonmonotonic plan whose action representation employs functional operators is NP-hard.*

B. Handling Constraints on the Temporal Ordering of Operational Goals

This section and the following two will cover the treatment of three distinct classes of constraints within the general framework of nonmonotonic planning. Let us first consider constraints on the temporal ordering of operational goals. They are of the form "GOAL-A must be achieved before achieving GOAL-B" and appear at a certain level of process abstraction. Examples are

(a) (START_UP_FEED_TREATMENT_SECTION) before (START_UP_REACTOR).
(b) (START_FLOW_IN_COLD_SIDE) before (START_FLOW_IN_HOT_SIDE) of a heat exchanger.
(c) (OPEN_VALVE_1) before (OPEN_VALVE_2).

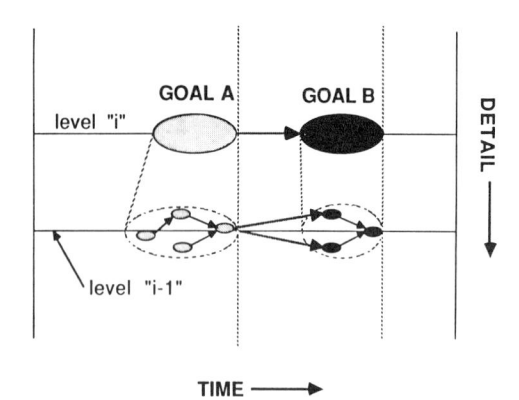

FIG. 7. Downward transformation of temporal constraints. (Reprinted from *Comp. Chem. Eng.*, 12, Lakshmanan, R. and Stephanopoulos, G., Synthesis of operating procedures for complete chemical plants. Parts I, II, p. 985, 1003, Copyright 1988, with kind permission from Elsevier Science Ltd., The Boulevard, Langford Lane, Kidlington 0X5 1GB, UK.)

A temporal constraint at a particular level of the abstraction hierarchy specifies a partial ordering on the goals in the level just below (more detailed), *causing the posting of new temporal constraints* on the goals of this level. Each of these new constraints specifies in turn a set of new temporal constraints at the level below it. In this manner, constraints at a high-level of abstraction are successively transmitted from level to level until they reach the most detailed level of individual processing units. All the newly generated constraints must be taken into account and satisfied by the non-monotonic planner. Figure 7 shows a schematic of the propagation of temporal constraints through two levels of abstraction. The directed path indicates the temporal ordering of goals.

Lakshmanan and Stephanopoulos (1988b) developed an algorithm, which automates the downward propagation of temporal constraints all the way to the level of processing units, where these constraints can be incorporated in planning of primitive operators (e.g., open or close valve, turnon or shutdown motor). This algorithm is provably correct having the following property.

1. Property 1: Correctness of the Algorithm for the Downward Propagation of Temporal Constraints

Let C_i be a temporal constraint, stated between two goals, A_i and B_i, at the ith level of the modeling hierarchy (see Section II,B), which indicates that A_i must be achieved before B_i. Let $A_{(i-k)}$ and $B_{(i-k)}$, with $k > 0$, be subgoals at the $(i-k)$th level of A_i and B_i, respectively. Then, the algorithm for the downward propagation of temporal constraints guarantees, on termination, that, for all $[A_{(i-k)}, B_{(i-k)}]$, a directed path from $A_{(i-k)}$ to $B_{(i-k)}$ exists in the constraint network $N_{(i-k)}$ and no corresponding path exists from $B_{(i-k)}$ to $A_{(i-k)}$.

The successive introduction of new constraints in the temporal ordering of operational steps at lower levels of the goal hierarchy, may create internal conflicts. For example, the propagation of constraint C_i may result in ordering A_i before B_i, while the propagation of constraint C_j may lead to ordering B_i before A_i. These conflicts manifest themselves in the form of directed circuits (cycles) at one or more levels in the goal hierarchy. A polynomial-time algorithm has been developed to detect the generation of directed cycles as new constraints are introduced (Lakshmanan and Stephanopoulos, 1988b). The user is informed and the offending constraint may be detracted.

a. Transformation of Integer–Goal Ordering Constraints onto Orderings of Primitive Operators. The temporal ordering of goals at higher levels of the goal hierarchy is successively transformed to temporal orderings of goals at lower levels. It now remains to be seen how the temporal ordering of goals at the lowest level of the hierarchy can be transformed into constraints on the ordering of primitive operators. Lakshmanan and Stephanopoulos (1988b) have developed such an algorithm which possesses the following property.

2. Property 2: Correctness of the Constraint Transformation Algorithm

The constraint transformation algorithm accepts a network of goals partially ordered by constraints, and generates a constraint network of primitive actions, such that, if there exists a directed path from goal A to goal B (i.e., A must be achieved before B) in the first network, and if OP-A is the primitive action that achieves goal A, and OP-B the action that accomplishes B, then OP-A and OP-B are labels on nodes in the generated network, and there exists a directed path from the node labeled with OP-A to the node labeled with OP-B.

This property guarantees that constraints on the temporal ordering of goals are correctly transformed into constraints on the temporal ordering of primitive operations that achieve these goals.

C. Handling Constraints on the Mixing of Chemicals

Safety, environmental, health, and performance considerations dictate that certain chemicals should not coexist anywhere in the plant. Rivas and Rudd (1974), O'Shima (1982), and Fusillo and Powers (1988a) have dealt with this important class of constraints, and have proposed approaches all of which are monotonic in character. In its most general form, a constraint on the mixing of chemicals involves the specification of a set of chemical species, and composition ranges for those species that ought to be avoided anywhere in the plant. Very often, though, the high cost of the consequences of an unsafe operation (e.g., an explosion) forces the operations planners to adopt a more conservative posture and require that the coexistence of a set of chemicals be avoided altogether. This class of constraints is expressed as an unordered set of chemical species $(x_1, x_2, \ldots, x_n)$ that should never be present at the same time anywhere in the plant. In this section we will examine how nonmonotonic planning handles such constraints and converts them into constraints on the temporal ordering of primitive operations through the following sequence of steps:

Step 1. *Construction of the influence graphs*, which represent the extent to which a given chemical species is present within the process in a given operational state.

Step 2. *Identification of potential constraint violations*, resulting from selected operations during the construction of partial plans.

Step 3. *Generation of temporal constraints on primitive operations*, to negate the violation of mixing constraints.

In the following paragraphs we will discuss the technical details involved in each of the preceding steps.

1. Construction of the Influence Graphs

Consider the directed graph representing the topology of a chemical process at a given operating state, i.e., the nodes of the graph represent process equipment and the directed edges of the graph the material flows. If the direction of the flow has not been computed or is uncertain, the corresponding edges are bidirectional. Each edge conveys information related to the type of chemical species flowing and the directionality of flow. All other state information (e.g., temperature, concentration) has been suppressed. Then, using the directed graph that represents the topology of the chemical process, the construction of the influence graphs (IGs) proceeds as follows (Lakshmanan and Stephanopoulos, 1990):

1. Locate all sources, s_i, of the chemical species, x_i.
2. Remove all edges that are immediately adjacent to closed valves.
3. Any node, v, which is on a directed path from any source, s_i, is labeled with x_i. Let N_i be the set of nodes labeled with x_i and let E_i be the set of associated edges. Nodes labeled with x_i and the attached edges contain the chemical species.

At the end of this three-step procedure we have generated a graph (i.e., a subgraph of the process graph) whose edges indicate the flow of species x_i and whose nodes contain the chemical x_i, at the given operating state.

2. Identification of Constraint Violations

Violations of the mixing constraints may occur at the *initial*, *goal*, or any *intermediate* state in the course of an operating procedure. Since the initial state is presumed to be a known, feasible state, it is generally not the case that it will contain potential mixing constraint violations. It is possible though that the human user or the planning program may specify a goal operating state that violates the mixing constraints.

The detection of potential mixing constraints violations at an intermediate operating state is based on a "worst-case scenario" with the maximum potential for the species to coexist, and which corresponds to the following situation:

(a) Keep open all the valves which are OPEN at the initial state.
(b) Keep open all the valves which must be OPEN at the goal state.

Construct the IGs for all species present in the various mixing constraints. Then:

> If a node, v, carries both labels, x_i and x_j, where (x_i, x_j) is an unacceptable pair of coexisting species, then the potential for a mixing constraint violation has been detected.

Clearly, if under the worst-case scenario no node can be found in the IGs that is labeled with all species present in a mixing constraint, then no potential violation of a mixing constraint exists. If, on the other hand, a potential violation has been detected, then we need to *generate additional constraints on the temporal ordering of primitive operations so that we can prevent or negate the preconditions of a mixing constraint*.

3. Generation of Constraints on the Temporal Ordering of Primitive Operations

The potential violation of a mixing constraint is equivalent to the presence of a potential *Clobberer*. Therefore, according to Theorem 2, we need to identify a *White Knight* and force him to proceed the action of the Clobberer. In terms of the synthesis of operating procedures, the "Clobberers" are the valves that are open under the worst-case scenarios, described in the previous paragraph. Obviously, the "White Knights" represent the set of valves that, when closed, will prevent the chemicals from mixing. We will call this set the *minimal separation valve set* (MSVS). Once the valves that should be closed have been identified, temporal ordering constraints are imposed, which specify that these valves be closed, *before* the valves that lead to a mixing constraint violation were opened. Lakshmanan and Stephanopoulos (1990) have developed a graph-based algorithm with polynomial-time complexity, which identifies the MSVS. However, the MSVS may not be an *acceptable separation valve set* (ASVS) for the given planning problem, since (1) some valve in the MSVS is required to be open in the goal state (in order to achieve a certain operating goal), or (2) the user may have some subjective preferences regarding the state of some valves. A concrete algorithm exists (Lakshmanan and Stephanopoulos, 1990) to convert the initial MSVS to an ASVS.

Once an ASVS has been found, the constraints on the temporal ordering of the valve operations in the ASVS must be established. This is systematically accomplished through an algorithm that undergoes through the following steps.

1. Let P be the set of valves that must be open under the worst-case scenario.
2. If p is an element of P and q is an element of ASVS, then, to negate the violation of a mixing constraint, the following temporal constraint must be generated:

   CLOSE(q) before OPEN(p)

4. Purging a Plant from Offending Chemicals

Once the ASVS has been identified and the temporal ordering of valve operations has been determined, no mixing constraints will be violated in the steady state. However, it is possible to violate the mixing constraints during the transient from one steady state to another. For example, during the changeover from oxygen to methane in the network of Fig. 3, it is possible that oxygen and methane will come into contact with each other even if the upstream oxygen inlet valve is closed, before the methane is allowed to enter. To avoid the contact between the offending chemicals, the first (e.g., oxygen) must be purged, before the second (e.g., methane) is allowed to flow in.

Lakshmanan (1989) has indicated that *"purging or evacuation operations are necessary, whenever the intersection of the IG of one component in the initial state with the IG of another (offending) component in the goal state is non-zero."* In such case, nonmonotonic planning generates a new intermediate goal, i.e., purge the plant of chemical x_i, and tries to achieve this goal first before admitting the offending chemical x_j in the process.

Consequently, the overall planning is broken into two phases and proceeds as follows (Lakshmanan and Stephanopoulos, 1990):

a. *Phase 1.* Find the path and destination of purge operations.

1. Identify all locations of a plant where a certain species must be purged. To accomplish this it is sufficient to scan the labels of the nodes in the IGs and identify those nodes that are labeled, e.g. x_i and x_j in two successive operating states, where (x_i, x_j) is a prohibited mixture of chemicals.
2. For each node v, where the offending chemicals (x_i, x_j) are present in the initial and goal states, respectively, find a directed path that leads the purgative material from its source to the sink through node v.

b. *Phase 2.* Generation of the purge planning island, i.e., intermediate planning goals.

1. Create an Intermediate operating State, i.e., introduce the intermediate goals requiring that *all valves along the purge path be open*.
2. Place the intermediate operating state between the initial and goal states and constrain the planner to achieve the goals in the intermediate state before opening the valves in the initial state, which were earlier identified as being open in the goal state (viz., worst-case scenario).

D. Handling Quantitative Constraints

During the synthesis of operating procedures, the operating state may be required to satisfy certain quantitative constraints. For example, during the operation of a reactor, the temperature may not be allowed to exceed a certain maximum, or the distillate from a distillation column may be required to have a minimum purity. Two chemicals may not be allowed to mix if the temperature is below a minimum. These constraints involve numerical values and may be stated as follows:

$\{$HCN should not mix with HCHO at $T \geq T_{max}\}$
$\{[$HCN$] \leq 0.1$ mol/L$\}$, in the reactor at all times.

Such constraints commonly arise from safety requirements, product specifications, environmental regulations, etc.

In this section we will present a formalized methodology that allows the transformation of quantitative bounding constraints into constraints on the temporal ordering of operators within the spirit of nonmonotonic planning.

1. Truth Criterion for Quantitative Constraints

The efficiency of nonmonotonic planning depends on its ability to identify Clobberers and White Knights efficiently. Theorems 3 and 4 (see Theorems 1–4 in Section III,A) have established the intractability of nonmonotonic planning for conditional and functional operators, which are rich enough to ensure adequate representation of process operations. Consequently, practical solutions can be derived and intractability can be avoided only through the use of domain-specific knowledge. A truth criterion, valid for the specific domain, should be identified and proved. Also, efficient ways of evaluating the domain-specific truth criterion should be given. Such a truth criterion has been identified and proved for synthesizing plans that are guaranteed to satisfy quantitative constraints in chemical plants. For posting quantitative constraints of the form $x.*.k$[2] the truth criterion is stated as follows:

Theorem 5: (Modal Truth Criterion for Quantitative Constraints). *A constraint $x.*.k$ is necessarily satisfied in a situation s iff there is a situation t equal to or necessarily previous to s in which $x.*.k$ is satisfied and for every step C possibly before s that possibly violates $x.*.k$, there is a step W necessarily before C that ensures $x.*.k$ in s.*

The statement and proof of the truth criterion (see Lakshmanan, 1989) for quantitative constraints is along the lines of Chapman's truth criterion for domain independent nonmonotonic planning. From this criterion, the plan modification operations that would ensure satisfaction of constraints are

1. If C is constrained to lie after s, its clobbering effect will not be felt at situation *s*. This process of constraining C to lie after s is called *demotion*.
2. A White Knight, W, may be selected and constrained to lie before C and therefore s, so as to ensure that the clobbering effect of C will not be felt.

The time required to evaluate the truth criterion depends on the complexity of the procedures for determining clobberers and white knights. The

---

[3]The notation '.*.' indicates relationships of the type $\langle, \rangle, =$, etc. between X and k.

algorithms for these operations are presented below. Clobbering of a constraint may often be caused by a set of valve operations that achieve an effect together. Such a set of valves is identified as an abstracted operator.

2. *Identification of Clobberers of Quantitative Constraints*

A goal is often achieved by multiple primitive (e.g., valve) operations. All the primitive operations that will together result in violation of a constraint will be Clobberers of that constraint. Consequently, the first step in the quantitative constraint-posting methodology is to identify abstracted operators as the sets of primitive operations. This is accomplished through the following procedure:

procedure ABSTRACT_OPERATOR
Input: Initial state, I; Final state, F; List of goals to be achieved,
    G and list of primitive valve operations required
    to achieve the goal state, V.
Output: Abstract operators, A consisting of the set of valve operations
    necessary to start or stop flow from each source to point
    where goal is specified
begin
for each g ∈ G do
    if RHS(g) = 0 **then** STATE := I
        {**comment**: if goal is to start/ stop flow, consider}
        else STATE := F   { final/ initial state as 'state' }
    remove closed valves from STATE
    j = 0
    for each point of goal specification, e_i **do**
    begin
        j = j + 1
        for each source, s_j **do**
        FIND_PATH(e_i, s_j)
        {**comment**: Search for paths from point where the goal
        is specified to sources of species and mark valves along it}
        AO-j := Intersection (open/closed valves in V,
        marked valves), for start/stop flow
    end
end

This algorithm provides a complete identification of all requisite operators, simple or abstract (i.e., sets of operators), as the following theorem guarantees (Lakshmanan, 1989).

Theorem 6: *Given the list of primitive operations to achieve a goal state, the algorithm* ABSTRACT_OPERATOR *will detect every simple or abstracted operator corresponding to the set of operations required for starting or stopping flow of materials from each source, s_j, to the point of constraint specification, e_i, necessary to achieve the goal state.*

The algorithm ABSTRACT_OPERATOR runs in time $O[g(V + s(V + E + V \ln V))]$, where g is the number of goals, s is the number of sources, V is the number of operators (e.g., valves) and E is the number of directed edges (i.e., connections) in the graph representing the topology of a process flowsheet. Any set of primitive operators that results in possible violation of a quantitative constraint will be a Clobberer of that constraint. For satisfaction of the truth criterion (Theorem 5), we need to identify those operators that could result in constraint violation. Clobberers of quantitative constraints may be identified efficiently only when the effect of just one operator is considered at a time. When multiple operators are considered simultaneously, the quantitative effect of each operator will depend on the magnitude of the others. Numerical simulation based on assumptions of the values of the variables may then be required to detect Clobberers. Since each variables could take any value within a range, such simulation would be intractable.

For the detection of Clobberers, it is necessary to propagate the quantitative effects of each abstracted operator through the plant to the point where the constraint is applicable. The algorithm for performing this task is given below.

```
procedure CLOBBERER_DETECTION
Input: Abstracted operators, A; final state, F;
       quantitative constraints, Q
Output: Abstracted operators that are Clobberers, C
begin
for each q ∈ Q do
    for each a ∈ A do
    begin
        QUANTITATIVE_PROPAGATE(q, eᵢ, F)
        {comment: propagate the quantitative values
          of the variables through the
          model equations}
        if constraint, q is violated then a ∈ C
    end
end
```

The following theorem guarantees the completeness of the above algorithm:

Theorem 7. *Every abstract operator, that will result in violation of a quantitative constraint, if implemented first and alone, will be detected as a Clobberer by the procedure* CLOBBERER_DETECTION.

3. Selection of Plan Modification Operations: The White Knight for Quantitative Constraints

Once the Clobberers that lead to violations of quantitative constraints have been identified, we must search for modifications of the operating procedure, which will negate the effects of the Clobberers on the procedure and thus restore its feasibility. There exist two general mechanisms for achieving this objective, and in the following paragraphs we will discuss each one of them.

a. Mechanism 1: Demotion of Clobberers. Demotion of the Clobberer involves constraining the clobberer to lie after the situation at which constraint satisfaction is being considered. Thus, if $\{AO - 1, AO - 2, AO - 3\}$ are three unordered operators, and $AO - 2$ is a Clobberer of a quantitative constraint, the plan may be modified by demotion of $AO - 2$ to give $\{AO - 1, AO - 3\} > \{AO - 2\}$, i.e., $AO - 1$ and $AO - 3$ may be applied at situation s, but $AO - 2$ is constrained to lie after s. Thus, posting quantitative constraints provide greater temporal order to the operators in the plan.

b. Mechanism 2: Identification of White Knights. If all operators for obtaining a feasible plan (i.e., satisfies both goals and constraints) have been correctly identified, then the quantitative constraints can be transformed into constraints on the temporal ordering of operations and demotion will always work. Nevertheless, if every operator identified for a single equipment is a potential Clobberer, then quantitative constraint violation cannot be avoided by the current set of operators and a White Knight has to be identified. The procedure for identifying White Knights is composed of two steps:

Step 1. It is necessary to identify the constraint being possibly violated and the qualitative change required in the variable to avoid the violation. For example, if a "$T < T_{\max}$" constraint is violated, then the qualitative change required in T to avoid the violation is "decrease T." Having identified the property to be affected and the direction of desired change, we now need to identify manipulations that may help us meet our objective.

Step 2. The qualitative value of the desired change is propagated through the steady-state model equations of the plant equipment, following the constraint propagation procedure of Steele (1980). Manipulations that cause the desired change and that are feasible are identified as White Knights and are constrained to lie before the situation of interest s, in accordance with the truth criterion.

The complete algorithm for the identification of White Knights is as follows:

procedure WHITE_KNIGHT
Input: Plant flowsheet in final state, F; constraint being clobbered, q.
Output: White Knight, W for the constraint q.
begin
for q **do**
    begin
    if q ≡ 'X(<, ≤)k' **then** L := 'decrease X' {**comment**: L − load}
    if q ≡ 'X(>, ≥)k' **then** L := 'increase X'
    if q ≡ 'X = k' and X < k **then** L := 'increase X'
    if q ≡ 'X = k' **then** X > k **then** L := 'decrease X'
    { **comment**: Such rules may be written for each type
     of constraint encountered }
    end
while v ≠ manipulatable valve **do**
    QUALITATIVE − PROPAGATE(L, F)
return v and required qualitative change
end

The following theorem (Lakshmanan, 1989) guarantees the completeness of the preceding algorithm.

Theorem 8. *The algorithm* WHITE-KNIGHT *will detect every manipulation, that will possibly avoid a quantitative constraint violation, as a White Knight.*

E. SUMMARY OF APPROACH FOR SYNTHESIS OF OPERATING PROCEDURES

The synthesis of operating procedures for a chemical plant, using nonmonotonic planning ideas, consists of two distinct phases: (1) formulation of the planning problem and (2) synthesis of operating plans. In the following two subsections we will examine the features of each phase and we will discuss the technical details for their implementation.

1. Phase 1: Formulation of the Planning Problem

The complete formulation of the planning problem implies the complete specification of the following four items.

a. Description of the Initial State. We must make sure that the specified starting operational state of a process is complete and satisfies the balances (mass, energy, and momentum), equilibrium, and rate phenomena in the process. This task is composed of the following steps:

1. Input of the initial operational state of the plant, using the linguistic primitives of MODEL.LA. (see **21**:1).
2. The specification of the initial state may be partial and lead to incomplete descriptions of various segments of the plant. The modeling facilities of MODEL.LA. contain a complete set of the balance equations, phase and chemical equilibrium, and rate relationships. These relationships are used to propagate the user-supplied specifications and thus complete the description of the initial state throughout the plant.
3. During the specification of the initial state, it is conceivable that the user-supplied information is not consistent with the physical constraining relationships. The planning program should possess procedures for detecting such conflicts and provide mechanisms for their resolution.

b. Description of the Final, Goal State. This proceeds in a manner similar to that for the description of the initial state.

1. Input the specifications of the desired goal state.
2. Check for potential conflicts in the specification of the goal state, and provide resolution for its consistent definition.
3. Generate concurrent goals, which usually come from the interaction among the various subsystems of the plant. This is accomplished through constraint propagation mechanisms, using the set of modeling relationships describing the process, analogous to the mechanism that ensures the completeness in the description of the initial state.

c. Specification and Identification of Operational Constraints in the Plan. In addition to the physical modeling constraints that govern the operation of the chemical process, we require that several "operational" constraints be met by an operating procedure, such as (1) temporal ordering in the execution of process operations, (2) avoiding the creation of undesirable mixtures of chemical species, and (3) maintaining the state variables within specific ranges of numerical values, e.g., bounding quantitative constraints.

d. Specification, Identification of Planning Islands. When an operating procedure is expected to be long and to involve a large number of intermediate states, it may be advantageous to explicitly identify intermediate goals, known as *planning islands* (Chakrabarti *et al.*, 1986), through which the operating procedure must pass, as it goes from the initial to the goal state. These planning islands are partial descriptions of intermediate states, which are known to lie on the paths of the most efficient, or feasible, plans, thus decomposing the overall operating plan into a series of subplans, which can be synthesized separately in a sequence. Lakshmanan (1989) has established specific procedures for the identification of planning islands.

2. Phase II: Synthesis of Operating Plans

Having defined the planning problem in a form that is "understood" by the computer, we can proceed to the synthesis of operating procedures through the following sequence of steps.

a. Identification of the Primitive Operators. The "means–ends analysis" of Newell *et al.* (1960) is employed to (1) identify the differences between the goal and initial states and (2) select the operators (i.e., operating actions) that would eliminate these differences.

b. Construction of Partial Plans. This consists of deriving a partial temporal ordering in the application of primitive operators. The partial ordering is driven by the constraints placed on the states of the desired operating plan, and proceeds as follows:

1. User-specified, temporal ordering of operational goals at higher levels of abstraction is propagated downwards in the hierarchy goals and is ultimately expressed as temporal ordering of primitive operations (see Section III,B).
2. Constraints on the disallowed mixtures of chemical species (a) are transformed into temporal orderings of primitive operations, or/ and

(b) introduce additional intermediate goals, e.g., planning islands for purge or/ and evacuation operations (see Section III,C).
3. Quantitative constrains (a) are transformed into temporal orderings of primitive operations (e.g., demotion of Clobberers) or (b) dictate the introduction of new operators (e.g., White Knights) to recover the feasibility of plans.

c. Synthesis of Complete Plans. At the present, nonmonotonic planning covers the three classes of constraints discussed above. For other classes of constraints, not amenable to direct transformation into temporal constraints between goals, the generate-and-test strategy of monotonic planning is employed. Thus, the partial plans generated by the constraint propagation of nonmonotonic planning are put together to develop feasible plans to solve the problem.

IV. Illustrations of Modeling and Nonmonotonic Operations Planning

In this section we will offer several illustrations of the various aspects of nonmonotonic operations planning (discussed in earlier sections) including the following: (1) development of hierarchical models for the process and its operations, (2) conversion of constraints to temporal orderings of primitive operations, and (3) synthesis of complete plans.

A. Construction of Hierarchical Models and Definition of Operating States

Consider the flowsheet shown in Fig. 8. It represents the power forming section of a refinery, along the ancillary storage tanks and piping network required for the catalyst regeneration. During the operation of the plant, the activity of the catalyst (chloroplatinic acid on alumina base) in the operating reactor, e.g., REACTOR-1, decays as a result of (1) coke formation, (2) reduction in the chloride content, and (3) size growth of the platinum crystals. We want to synthesize an operating procedure that will discontinue the operation of REACTOR-1, start the regeneration of its catalyst (i.e., remove coke, add chloride, redisperse the platinum crystals), and initiate the operation of the standby REACTOR-2. For a fairly detailed description of the plant, its modeling, and the synthesis of

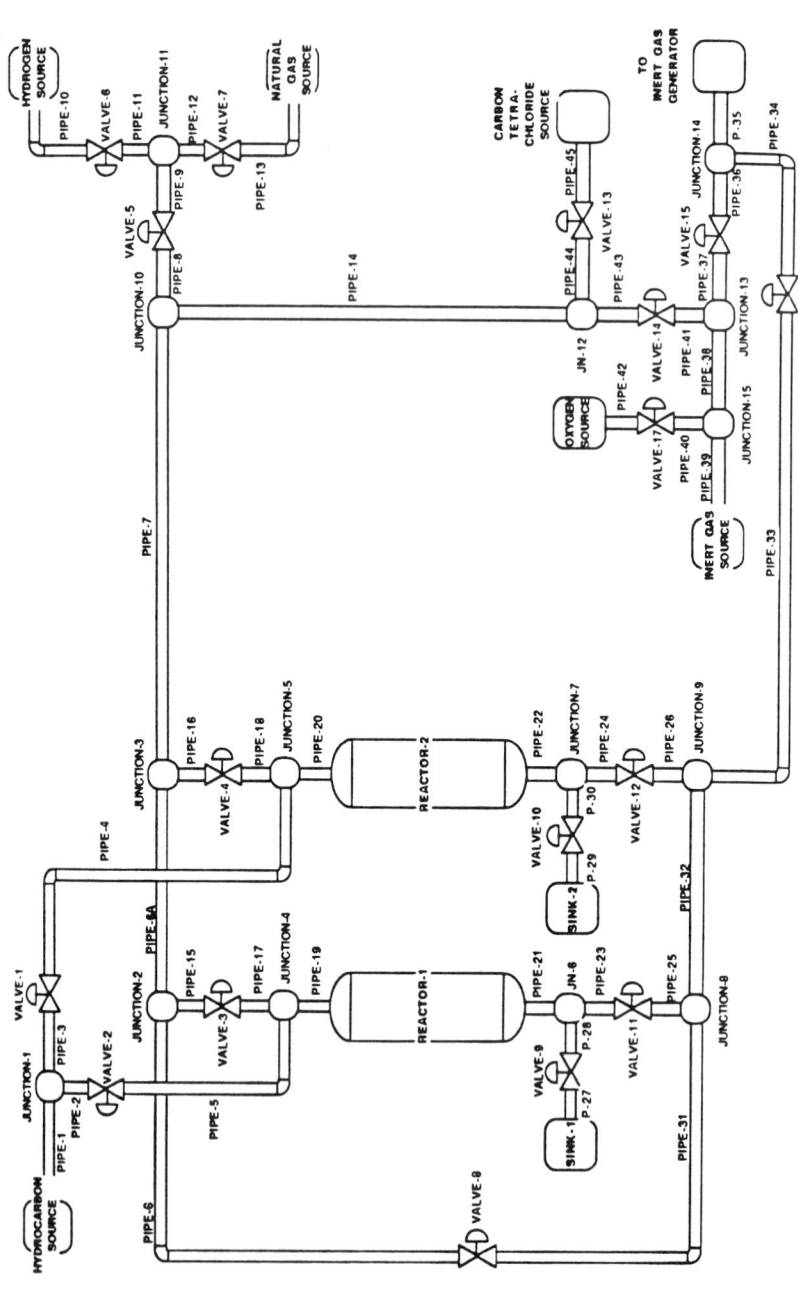

FIG. 8. Structure of the catalyst regeneration system. (Reprinted from *Comp. Chem. Eng.*, **12**, Lakshmanan, R. and Stephanopoulos, G., Synthesis of operating procedures for complete chemical plants, Parts I, II, p. 985, 1003, Copyright 1988, with kind permission from Elsevier Science Ltd., The Boulevard, Langford Lane, Kidlington OX5 1GB, UK.)

changeover operating procedures, the reader is referred to Lakshmanan (1989).

The flowsheet editing facilities of *Design-Kit* (Stephanopoulos *et al.*, 1987) allow the user to construct an iconic representation of the flowsheet shown in Fig. 8, while the system at the same time creates the requisite data structures to represent the topology of the process (i.e., units and their interconnections). These data structures represent instances of the basic modeling elements of MODEL.LA. (see **21**:1). For example, Fig. 9a shows the relevant attributes in the declarative description of "Power-Forming-Plant," which is an instance of the modeling element (see Section II,B), *Plant*. The latter is, in turn, a subclass of the generic modeling element, *Generic-Unit*. Every device in the plant of Fig. 9a is an instance of the modeling element, *Unit*. Figure 9b shows the relevant declarative description of the unit REACTOR-1. Since abstraction in the representation of a plant is essential for the synthesis of operating procedures, the user can represent the "Power-Forming-Plant" as composed of two instances of *Plant-Section* called "Power-Forming-Reaction-Section" and "Power-Forming-Catalyst-Regeneration-Section" (see Fig. 9c). Figure 9d shows the relevant declarative elements of the "Power-Forming-Reaction-Section," whose boundaries (i.e., component units and streams) have been defined by the user. MODEL.LA. allows the creation of mathematical models around any abstraction of the processing units, while maintaining internal consistency between the model of an abstract entity (e.g., Plant-Section) and those of its constituents. Figure 10 shows the classes modeling a flow-valve and their interrelationships. For more details on the generation of the mathematical models, see **21**:1.

1. Defining an Operating State and Ensuring Its Completeness and Consistency

Any operations planning problem requires the *complete* and *consistent* definition of the initial and the goal states. In addition, during the evolution of an operating procedure any intermediate state must be fully specified by the operating actions and the modeling relations. Let us focus our attention on one segment of the desired operating procedure, e.g., "take REACTOR-1 off-line and bring REACTOR-2 on-line."

In the *initial state*, REACTOR-1 is operating while REACTOR-2 is on standby. Consequently,

> Condition-1: Hydrocarbon-Flow-in-REACTOR-1 > 0
> Condition-2: Product-Flow-from-REACTOR-1 > 0
> Condition-3: Hydrocarbon-Flow-in-REACTOR-2 $= 0$
> Condition-4: Product-Flow-from-REACTOR-2 $= 0$

Instance: POWER-FORMING-PLANT
Is-a-member-of: PLANT
Is-composed-of: [REACTOR-1, REACTOR-2, VALVE-1, VALVE-2, ..., VALVE-17,
JUNCTION-1, ..., JUNCTION-15, PIPE-1, ..., PIPE-42,
HYDROCARBON-SOURCE, SINK-1, SINK-2, HYDROGEN-SOURCE,
NATURAL-GAS-SOURCE, CARBON-TETRACHLORIDE-SOURCE,
OXYGEN-SOURCE, INERT-GAS-SOURCE, TO-REGENERATOR]
⋮

Note: "X" is-a-member-of "Unit"
where "X" is any device in the above list.

(a)

Instance: REACTOR-1
Is-a-member-of: UNIT
Is-composed-of: []
Ports: [PORT-1, PORT-2]
⋮

Note: "PORT-1" is-a-member-of "PORT"
"PORT-2" is-a-member-of "PORT"

(b)

Instance: POWER-FORMING-PLANT
Is-a-member-of: PLANT
Is-composed-of: [POWER-FORMING-REACTION-SECTION,
POWER-FORMING-CATALYST-]-REGENERATION-SECTION]
⋮

(c)

Instance: POWER-FORMING-REACTION-SECTION
Is-a-member-of: PLANT-SECTION
Is-composed-of: [REACTOR-1, REACTOR-2, VALVE-1, ..., VALVE-4, VALVE-9, VALVE-12,
JUNCTION-1, ..., JUNCTION-9, PIPE-1, ..., PIPE-7, PIPE-15, ..., PIPE-26,
PIPE-31, ..., PIPE-33, HYDROCARBON-SOURCE, SINK-1, SINK-2]
⋮

(d)

FIG. 9. Object-oriented descriptions of the Power-Forming-Plant at two levels of abstraction: (a) in terms of processing units, and (c) in terms of processing sections. The descriptions of a processing unit (b) and a processing section (c).

```
┌─────────────────────────────────────┐         ┌──────────────────────────────┐
│ DEVICE :  FLOW-VALVE                │         │ DEVICE: PORT                 │
│                                     │         │ DEVICE NAME: P1              │
│ PORTS:  P1, P2 ────────────────────┼────────►│ ATTACHED TO: FLOW-VALVE      │
│ GENERAL RELATIONS: nil              │         └──────────────────────────────┘
│ REGIONS: OPEN, CLOSED,              │
│ OPERATIONS: OPEN-VALVE, CLOSE-VALVE │         ┌──────────────────────────────┐
│                                     │         │ DEVICE: PORT                 │
└─────────────────────────────────────┘         │ DEVICE NAME: P2              │
                                      └────────►│ ATTACHED TO: FLOW-VALVE      │
                                                └──────────────────────────────┘
```

REGION: OPEN

RELATIONS:

$$V = C_v \cdot (PRESSURE(P1) - PRESSURE(P2))^{0.5}$$

TEMPERATURE(P1) = TEMPERATURE(P2)

MOLE-FRACTION(species-i,P1) = MOLE-FRACTION(species-i,P2) [for all i]

REGION: CLOSED

RELATIONS:

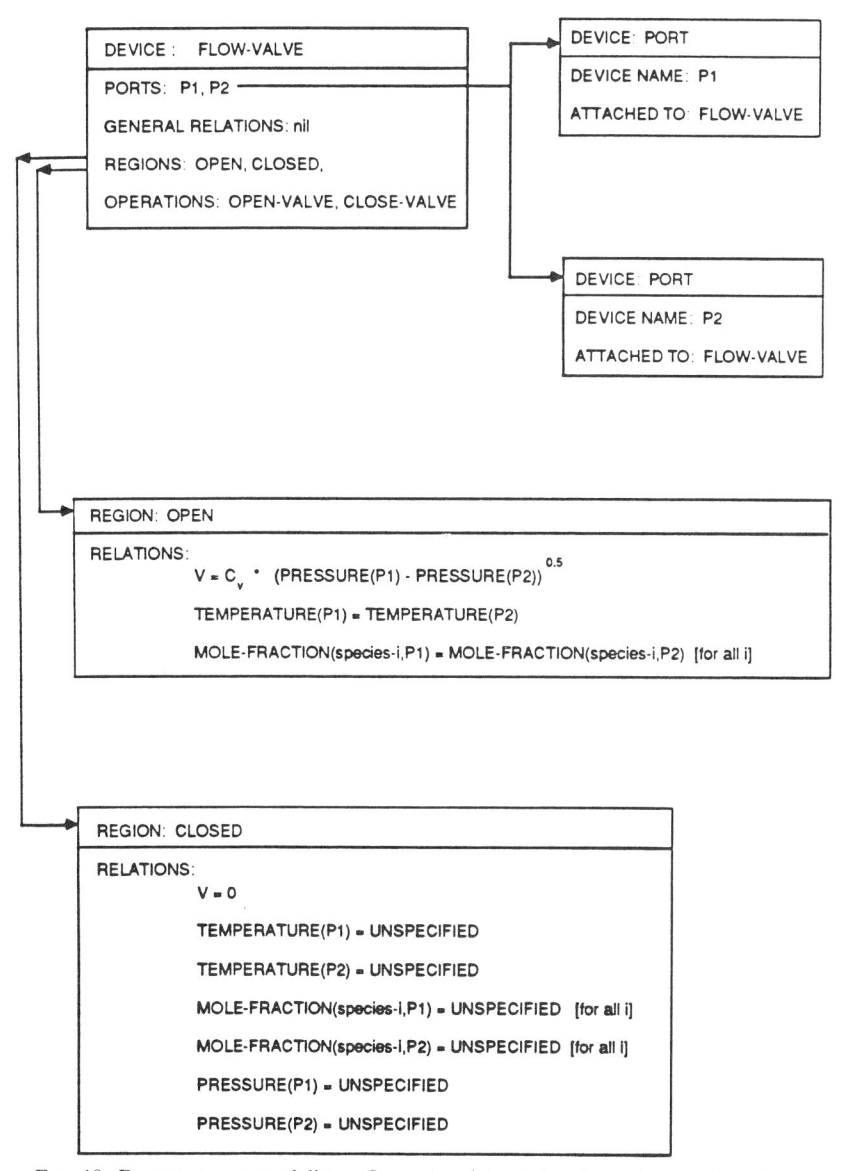

$V = 0$

TEMPERATURE(P1) = UNSPECIFIED

TEMPERATURE(P2) = UNSPECIFIED

MOLE-FRACTION(species-i,P1) = UNSPECIFIED [for all i]

MOLE-FRACTION(species-i,P2) = UNSPECIFIED [for all i]

PRESSURE(P1) = UNSPECIFIED

PRESSURE(P2) = UNSPECIFIED

FIG. 10. Data structure modeling a flow-valve. (Reprinted from *Comp. Chem. Eng.*, **12**, Lakshmanan, R. and Stephanopoulos, G., Synthesis of operating procedures for complete chemical plants, Parts I, II, p. 985, 1003, Copyright 1988, with kind permission from Elsevier Science Ltd., The Boulevard, Langford Lane, Kidlington 0X5 1GB, UK.)

Condition-1, when propagated through the modeling relationships, yields the following; VALVE-2 is OPEN and VALVE-3 is CLOSED. These values are then stored in the instances of the corresponding valves. Similarly, we take the following associations:

Condition-2 ⇒ [VALVE-9 is OPEN; VALVE-11 is CLOSED];
Condition-3 ⇒ [VALVE-1 is CLOSED; VALVE-4 is CLOSED];
Condition-4 ⇒ [VALVE-10 is CLOSED; VALVE-12 is CLOSED]

The propagation of these conditions through the modeling relationships is fairly straightforward, given the declarative richness of the data structures (e.g., see Fig. 10) representing the units of the plant, their interconnections and their physical behavior. Figure 11 shows the information flow through the modeling dependencies of variables, as Conditions 1–4 are propagated and produce the states of the associated operators (valves).

2. Ensuring the Completeness of the Initial State

The propagation of initial conditions through the modeling relationships is also used to ensure completeness in the definition of an initial state. For example, Condition-1 implies that the state of PIPE-1, PIPE-2, and PIPE-3 is characterized by the state of the flowing (or, stagnant, in PIPE-3) hydrocarbon. Similarly, Condition-2, when propagated backward, implies that the state of PIPE-27, PIPE-28, PIPE-21, and PIPE-23 is determined by the state of the flowing or stagnant product. The available initial conditions may not be able to define all facets of the initial state of a plant. For example, Conditions-1–3 cannot produce unambiguously the value of the state for VALVE-8, PIPE-6, and PIPE-31 (see Fig. 8). This is due to the fact that there is no directed path in the graph representing all modeling relationships, between the variables of the four conditions and the variables describing the state of VALVE-8, PIPE-6, and PIPE-31. In such cases the user must intervene and complete the specification for the initial state of the whole plant.

3. Checking the Consistency of the Initial State

Whereas the propagation of conditions through the modeling relationships always produces consistent completions in the definition of the initial state, this may not be true with the user-driven specifications. It is possible that a valve specified by the user to be OPEN, is "found" by the propagation of other conditions to be CLOSED. To detect such potential inconsistencies, we have developed a *dependency network* (Steele, 1980) to keep track of the flow of computations during constraint propagation. The dependency network is built on the undirected graph that represents the

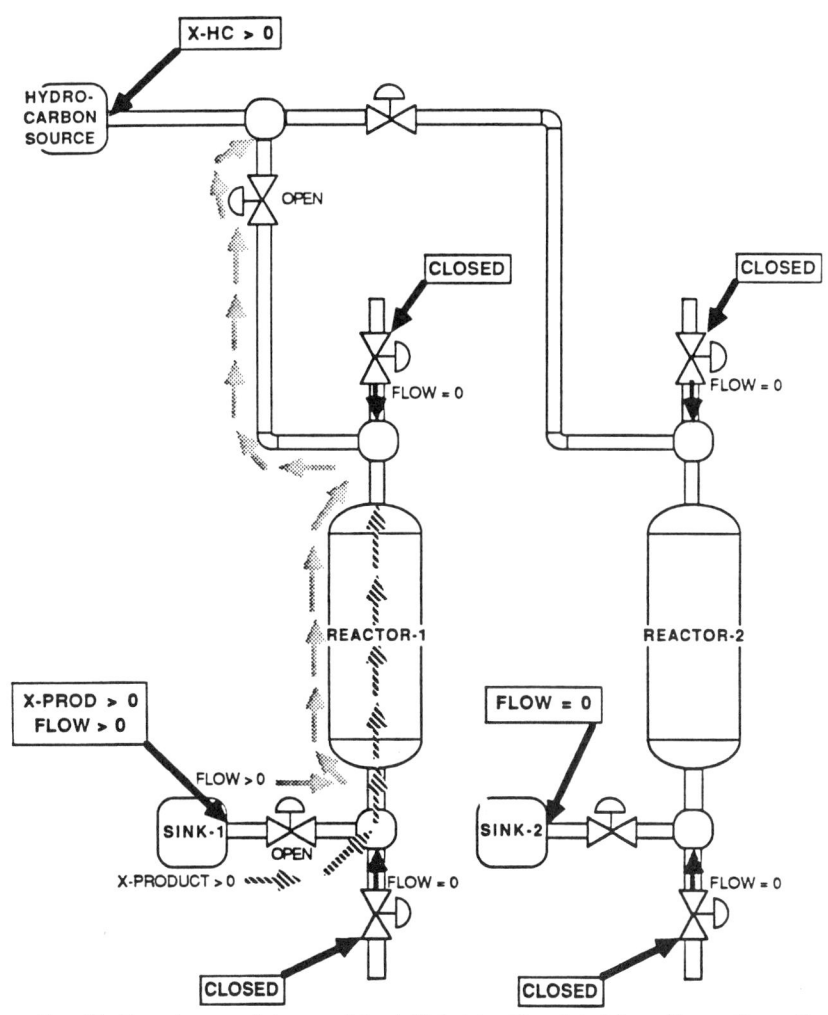

FIG. 11. Ensuring completeness of the initial state. (Reprinted from *Comp. Chem. Eng.*, **12**, Lakshmanan, R. and Stephanopoulos, G., Synthesis of operating procedures for complete chemical plants, Parts I, II, p. 985, 1003, Copyright 1988, with kind permission from Elsevier Science Ltd., The Boulevard, Langford Lane, Kidlington 0X5 1GB, UK.)

set of modeling relationships. With the given initial states as input variables, the undirected graph of modeling relationships is converted to a directed graph (dependency network), indicating the directionality of dependencies during the computation of initial states. If a conflict is encountered, backtracking through the dependency network reveals the source of conflict.

4. Complete and Consistent Definition of the Goal State

The mechanics for the definition of the goal state are fairly similar to those used for the definition of the initial state. Since the goal state is far more vague in the planners' mind than the initial state, the potential for conflicts is higher. Figure 12 shows the partial specification of the goal state, as well as the propagation of conditions to ensure completeness in the definition of the goal state.

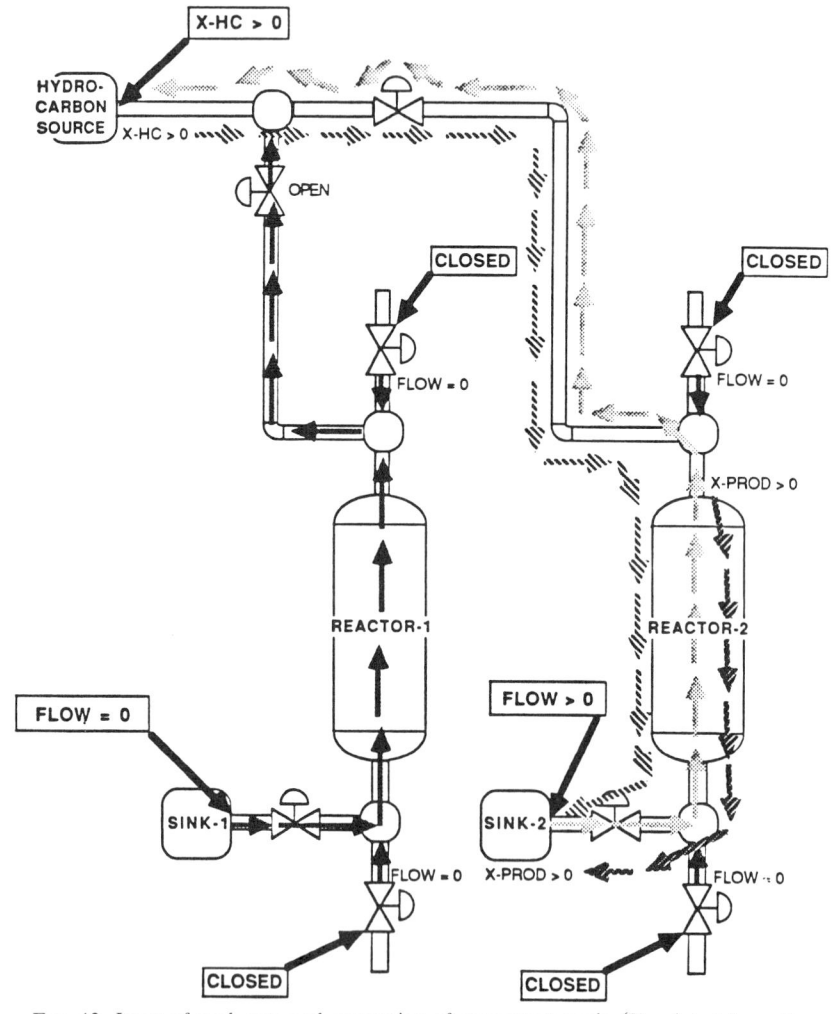

FIG. 12. Input of goal state and generation of concurrent goals. (Reprinted from *Comp. Chem. Eng.*, **12**, Lakshmanan, R. and Stephanopoulos, G., Synthesis of operating procedures for complete chemical plants, Parts I, II, p. 985, 1003, Copyright 1988, with kind permission from Elsevier Science Ltd., The Boulevard, Langford Lane, Kidlington 0X5 1GB, UK.)

B. NONMONOTONIC SYNTHESIS OF A SWITCHOVER PROCEDURE

Ethylenediaminetetraacetic acid (EDTA) is manufactured by the cyanomethylation process, whose main reactions are

$$HCHO + HCN \xrightarrow{H2SO4} H-C(H)(CN)-OH \quad \text{(Cyanohydrin)}$$

$$4\, H-C(H)(CN)-OH + H2N-C2H4-NH2 \xrightarrow{NAOH} \underset{HOOCH2C}{\overset{HOOCH2C}{>}}N-C2H4-N\underset{CH2COOH}{\overset{CH2COOH}{<}} + 4NH3$$

(EDA)   (EDTA)

Sulfuric acid, formaldehyde, and hydrogen cyanide are pumped into a glass-lined mixer (mixer 1, M1, of Fig. 13). Particular care is exercised so that the three charge operations are carried out in the order indicated above, to ensure the stability of the mixture at all times. In a separate segment of the plant, ethylenediamine (EDA) and dilute sodium hydroxide are charged and mixed in mixer 3 (M3 in Fig. 13). The solutions from mixer 1 and mixer 3 are pumped to the reactor (REACTOR, R1, in Fig. 13). When the reaction is complete, the reaction mixture is tested for traces of hydrogen cyanide. Dilute solution of formaldehyde is prepared in mixture 2 and is added to the reaction mixture, if there is any HCN present.

1. The Operating Situation Requiring a Switchover Procedure

The specific situation for which we want to synthesize an operating procedure is as follows. Hydrogen cyanide is handled by the top most line of piping, pumps, and valves (Fig. 13). Sulfuric acid and formaldehyde are handled by the next two flowlines in the flowsheet. The fourth flowline, which passes through pump PO4, has been shut down for routine maintenance. At some point during the operation of the plant, a leak is detected in the cooling jacket of mixer 1 and it has to be shut down in order to be repaired. Mixer 2 can be used in emergencies for limited periods of time only, because it is not a glass-lined vessel. As a result of this situation, we need to develop an operating procedure that will (1) shut mixer 1 down and disconnect it from the rest of the plant and (2) divert the flow of these materials, i.e., hydrogen cyanide, sulfuric acid, and formaldehyde, to mixer 2.

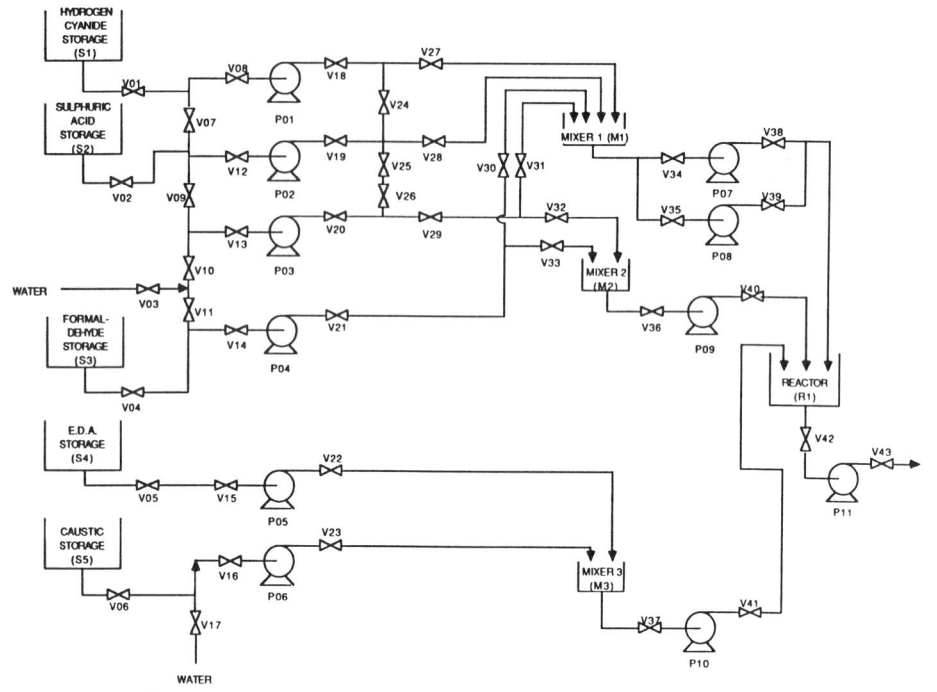

FIG. 13. Flowsheet of the ethylenediaminetetraacetic acid (EDTA) plant.

2. Constraints on the Operating Procedure

As we switch the operation from mixer 1 to mixer 2, certain constraints must be kept satisfied at all times. These constraints are as follows:

Constraint-1. Nowhere in the plant is HCN allowed to mix with formaldehyde.

Constraint-2, 3. The concentrations of HCN and H_2SO_4 in the mixer should satisfy the following constraints at all times:

$$[\text{HCN}] \leq 0.1 \text{ mol/L}$$

$$[H_2SO_4] \geq 1 \text{ mol/L}$$

Constraint-4. The temperature in the mixer should not exceed 30°C:

$$T_{M2} \leq 30°C$$

3. Initial and Goal States

The starting situation is partially defined by the state of the following valves:

{CLOSED-VALVES} = {V03, V07, V09, V14, V17, V24, V25, V26, V30, V32}
{OPEN-VALVES} = {V22, V23, V27, V28, V29}

Propagation of the above states through the modeling relationships completes the definition of the initial state of the plant. The results are shown in Fig. 14a. (Note that valves V34 through V43, which are downstream of the mixers, are considered closed, and they do not affect the synthesis of the switchover procedure.) The desired goal state is operationally defined by the following conditions:

Mixer-1-input-flow-i = 0   i = HCN, H_2SO_4, HCHO
Mixer-1-output-flow = 0
Mixer-2-input-flow-i = positive   i = HCN, H_2SO_4, HCHO

Propagating these values through the modeling equations, we can complete the definition of the goal state (see Fig. 14b). Comparing initial and

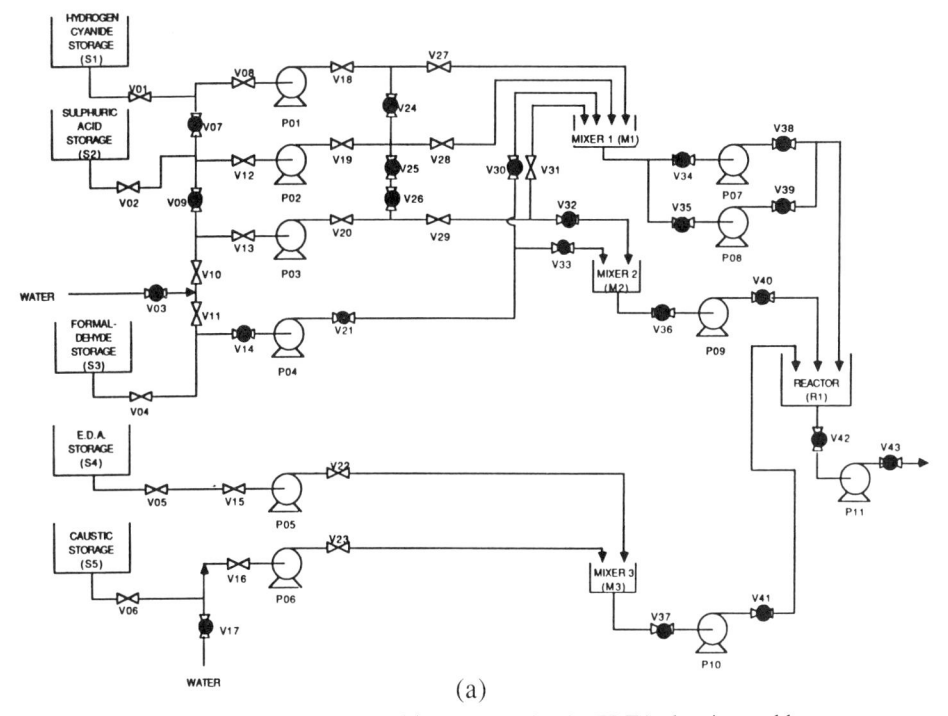

FIG. 14. The (a) initial and (b) goal states for the EDTA planning problem.

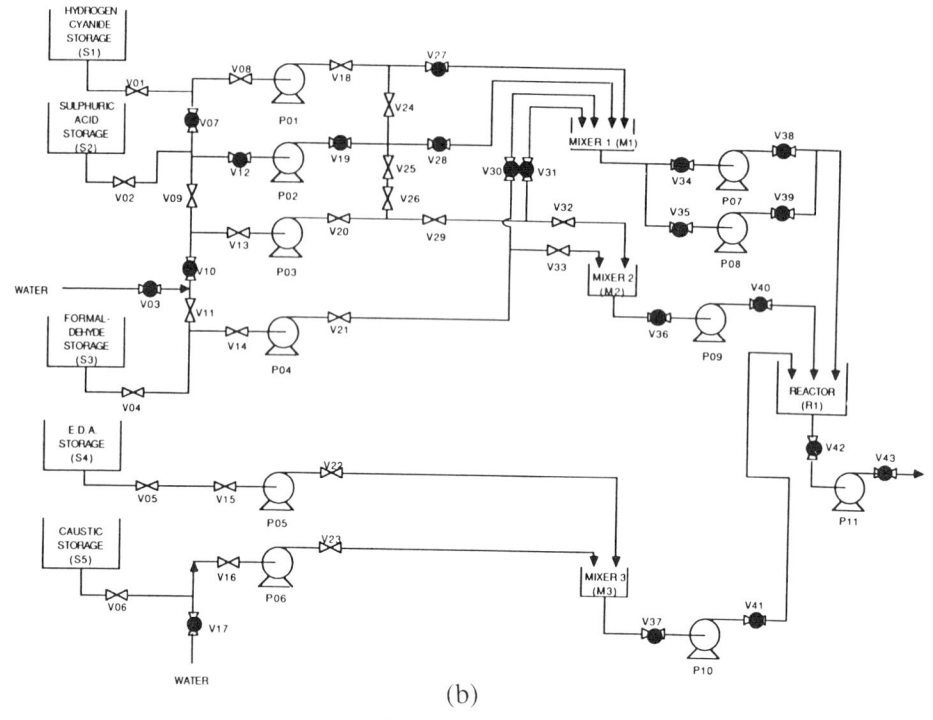

FIG. 14. Continued.

goal states, we conclude that the means–ends analysis requires the availability of the following operations to achieve the desired goal state:

V = {CLOSE V27, CLOSE V28, CLOSE V31, CLOSE V10, OPEN V9, CLOSE V12, CLOSE V19, OPEN V24, OPEN V25, OPEN V26, OPEN V32, OPEN V14, OPEN V21, OPEN V33}.

4. Abstraction of Operators

In Section III,D we presented an algorithm for the aggregation of primitive valve actions into abstract operators. Applying this algorithm over the set V of actions to achieve the goal state, we take:

AO-1: {CLOSE V27} ⇒ Stop-Flow-HCN-M1
AO-2: {CLOSE V12, CLOSE V19, CLOSE V28} ⇒ Stop-Flow-H_2SO_4-M1
AO-3: {CLOSE V10, CLOSE V31} ⇒ Stop-Flow-HCHO-M1
AO-4: {OPEN V24, OPEN V25, OPEN V26, OPEN V32}
 ⇒ Start-Flow-HCN-M2
AO-5: {OPEN V09, OPEN V32} ⇒ Start-Flow-H_2SO_4-M2
AO-6: {OPEN V14, OPEN V21, OPEN V33} ⇒ Start-Flow-HCHO-M2.

Then, the set V takes the following form:

$$V = \{AO\text{-}1, AO\text{-}2, AO\text{-}3, AO\text{-}4, AO\text{-}5, AO\text{-}6\}.$$

5. Transformation of Constraint-1 (Mixing Constraint)

In Section III,C we argued that one should identify the worst-case scenario for the violation of Constraint-1 and take action to ensure that a White Knight operation is carried out before any offending operation (i.e., Clobberer), in order to negate its impact (i.e., violation of mixing constraint). The worst-case scenario occurs when the plant has the following two sets of valves OPEN at the same time:

{Valves that are OPEN at the initial state}.
{Valves that are characterized as OPEN at the goal state}.

Then, we construct the influence graphs (IGs) of HCN and HCHO for the initial and goal states (Fig. 15), and from these we can identify the

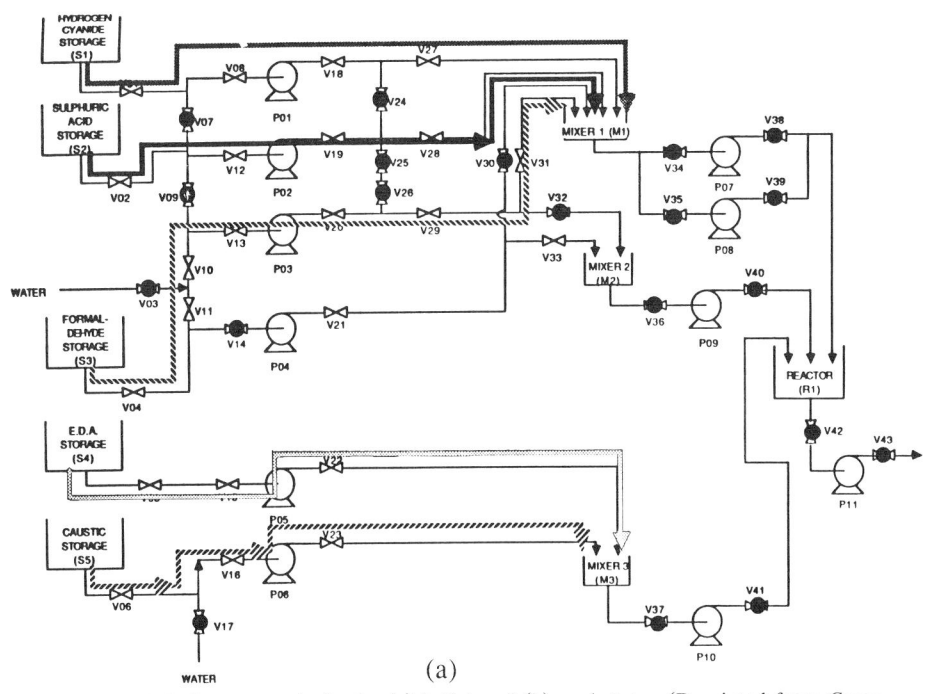

FIG. 15. Influence graphs in the (a) initial and (b) goal states. (Reprinted from *Comp. Chem. Eng.*, 14, Lakshmanan, R. and Stephanopoulos, G., Synthesis of operating procedures for complete chemical plants, Parts III, p. 301, Copyright 1990, with kind permission from Elsevier Science Ltd., The Boulevard, Langford Lane, Kidlington OX5 1GB, UK.)

FIG. 15. Continued.

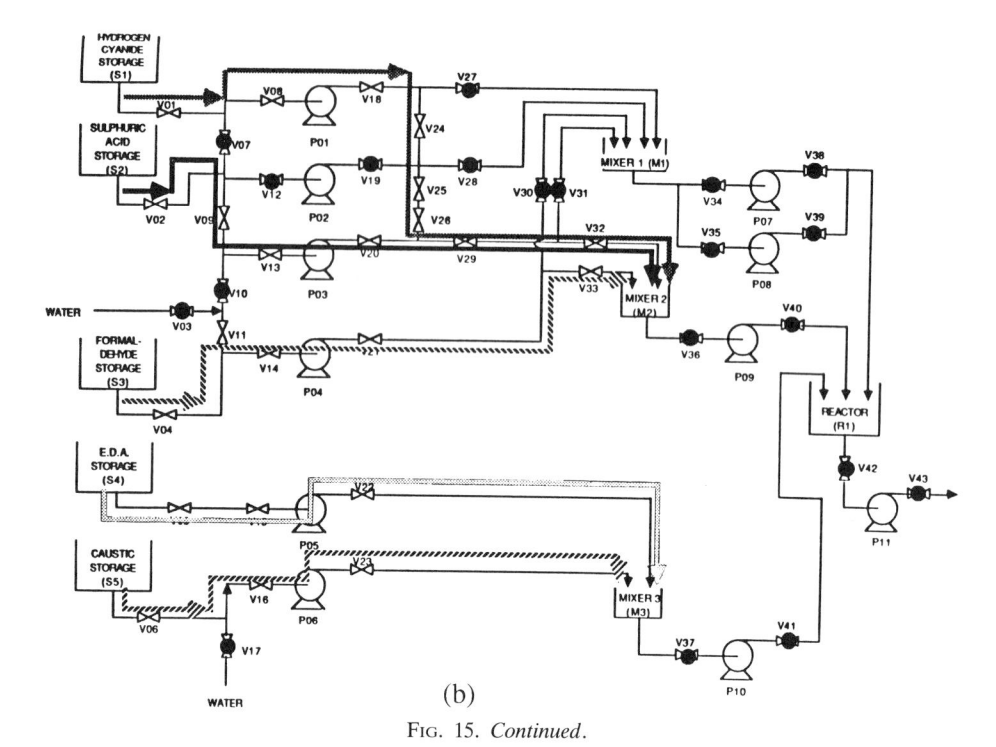

region of potential mixing between HCN and HCHO (intersection of IGs for HCN and HCHO) at the worst-case scenario (Fig. 16). From Fig. 16 we see that the IGs intersect at the valves V29 and V32. It is also clear that {V10} is the minimal separation valve set (MSVS; see Section III,C), i.e., the set of valves that, if closed, would prevent the mixing of HCN and HCHO. Thus, V10 must be closed before any valve in the initial state is opened. Looking at the elements of the set V in the previous paragraph, we derive the following temporal ordering of valve operations:

(CLOSE V10) before {OPEN V9, OPEN V24, OPEN V25, OPEN V26, OPEN V32, OPEN V14, OPEN V21, OPEN V33]   (Ordering-1)

So the mixing constraint has been transformed into the above constraint on the temporal ordering of primitive operations.

6. *Generation of a Purge/Evacuation Procedure*

Valves V29 and V32 and the associated pipes contain leftovers of HCHO and must be purged with H_2SO_4 before the other chemical,

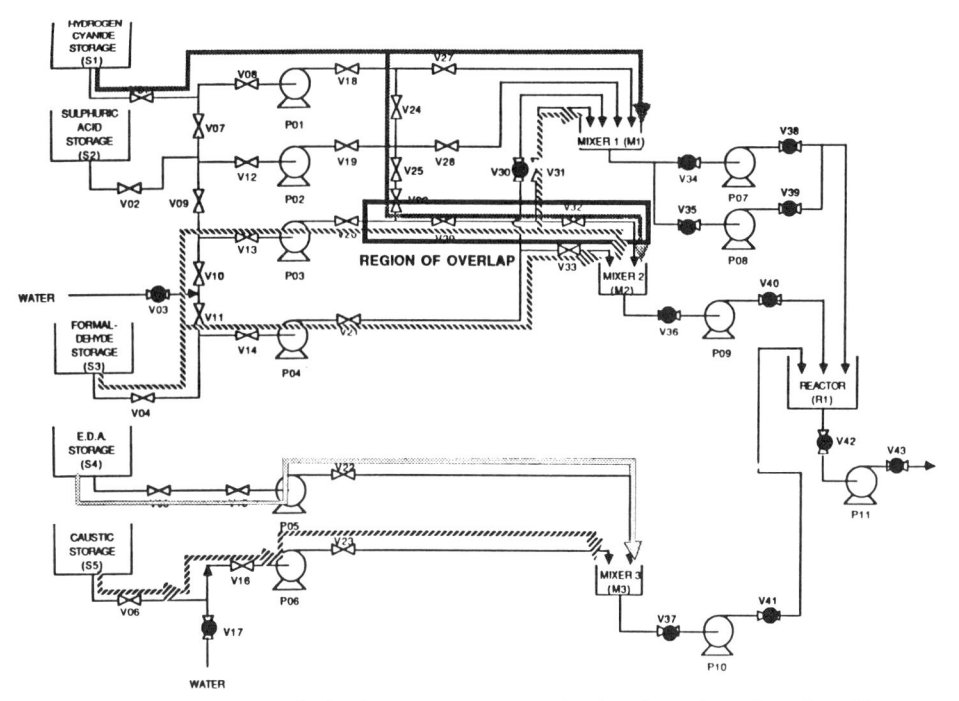

FIG. 16. Influence graphs in the worst-case state, showing the region of overlap. (Reprinted from *Comp. Chem. Eng.*, **14**, Lakshmanan, R. and Stephanopoulos, G., Synthesis of operating procedures for complete chemical plants, Parts III, p. 301, Copyright 1990, with kind permission from Elsevier Science Ltd., The Boulevard, Langford Lane, Kidlington 0X5 1GB, UK.)

i.e., HCN, is allowed to flow through. (Remember the required order of addition: H_2SO_4 before HCHO before HCN.) The path-finding algorithm (see Section III,C) identifies the following path from the storage of the purgative material (i.e., H_2SO_4) to mixer 2:

Purge Path = {PUMP PO3; VALVE V20; VALVE V29; VALVE V32}

This path represents a "planning island," i.e., the goals of an operating subprocedure, which must be completed after the MSVS has been closed, and before the operations that allow HCN in the line. Therefore, we have the following temporal constraints:

(CLOSE V10) before (PURGE OPERATIONS) before (OPEN V24)
(Ordering-2)

where (PURGE OPERATIONS) is the sequence of operating steps required for the purging of HCHO and stands on its own as an independent operating procedure. The synthesis of primitive steps leads to

(PURGE OPERATIONS) = {OPEN V09, START PO3, OPEN V32, CLOSE V32, STOP PO3, CLOSE V09}

7. Transformation of Constraints-2 and -3

Suppose that we were to load mixer M2 with the following sequence of abstract operators (see earlier paragraph in this section for definition of abstract operators):

{(Start-Flow-HCN-M2), (Start-Flow-H_2SO_4-M2), (Start-Flow-HCHO-M2)}.

If the initial concentrations of the raw material were (in moles per liter), $[HCN]_0 = 1$, $[H_2SO_4]_0 = 1.5$, and $[HCHO]_0 = 2$, then the preceding sequence of operations would immediately have violated Constraint-2, since $[HCN] = 1 > 0.1$. Similarly, Constraint-3 is violated by the preceding sequence, since $[HSO_4] = 0 < 1.5$. In Section III,D, we discuss the notion of "demotion" of Clobberers. Let us see how it works here and leads to temporal ordering of operations:

(a) (Start-Flow-HCN-M2) is a Clobberer of $[HCN] \le 0.1$. Demoting this Clobberer, we require that

(Start-Flow-HCN-M2) after (Start-Flow-H_2SO_4-M2; Start-Flow-HCHO-M2)

(b) (Start-Flow-HCN-M2) and (Start-Flow-HCHO-M2) are Clobberers of $[H_2SO_4] \ge 1$. Demoting both Clobberers, we establish the following temporal ordering constraints:

(Start-Flow-H_2SO_4-M2) before
(Start-Flow-HCN-M2; Start-Flow-HCHO-M2)

Taking (a) and (b) together, we see that Constraints-2 and -3 have been transformed to the following temporal ordering:

(Start-Flow-H_2SO_4-M2) before (Start-Flow-HCHO-M2)
before (Start-Flow-HCN-M2)

(Ordering-3)

8. Transformation of Constraint-4

Let the temperature of the three materials in the storage tanks be 35°C. Then, all three operations, (Start-Flow-HCN-M2), (Start-Flow-H_2SO_4-M2) and (Start-Flow-HCHO-M2), are Clobberers of the Constraint-4, since $T_{M2} = 35°C > 30°C$. No demotion of any of the three operators would restore feasibility, and we are led to the need for a White Knight. Searching the plant for relevant operations we identify the following White Knight:

(OPEN VALVE-CW) ⇒ Start-Cooling-Water-through-Jacket

and we constrain the White Knight to be carried out before any of three Clobberers, i.e.,

(OPEN VALVE-CW) before
[(Start-Flow-HCN-M2), (Start-Flow-H$_2$SO$_4$-M2), (Start-Flow-HCHO-M2)]
(Ordering-4)

9. Synthesis of the Complete Switchover Operating Procedure

In the previous paragraphs we saw how the ideas of nonmonotonic planning have transformed the required operational constraints into constraints ordering the temporal sequencing of operations. These temporal constraints define partial segments of the overall operating procedure. Now, let us see how they can be merged into a complete operating procedure. Let us recall the set V of operations, which was defined earlier through the means–ends analysis. Any complete plan must carry out the abstract operators AO-1 through AO-6. From each temporal ordering we take the following implications:

Ordering-1 ⇒ (AO-3: Stop-Flow-HCHO-M1) before (AO-4; AO-5; AO-6).
Ordering-2 ⇒ (AO-3) before (PURGE OPERATIONS) before (AO-4).
Ordering-3 ⇒ (AO-5) before (AO-6) before (AO-4).
Ordering-4 ⇒ (OPEN VALVE-CW) before (AO-4, AO-5, AO-6).

The following three sequences satisfy all the preceding constraints on the temporal ordering of operations:

Procedure-A. (AO-3) > (PURGE OPERATIONS) > (OPEN VALVE-CW) > (AO-5) > (AO-6) > (AO-4).
Procedure-B. (AO-3) > (OPEN VALVE-CW) > (PURGE OPERATIONS) > (AO-5) > (AO-6) > (AO-4).
Procedure-C. (OPEN VALVE-CW) > (AO-3) > (PURGE OPERATIONS) > (AO-5) > (AO-6) > (AO-4).

Each of the abstract operators, AO-3, AO-4, AO-5, AO-6, is composed of more than one operation. No constraints have been stated or generated to determine their relative order. For example, (AO-3) could be implemented as (CLOSE V10) > (CLOSE V31) or (CLOSE V31) > (CLOSE 10).

It is also important to note that abstract operators (AO-1) and (AO-2) are not participating in any temporal ordering constraint. Consequently, they could be placed anywhere in the chain of operations of a procedure. For example, *Procedure-A* can be completed with (AO-1) and (AO-2) in any of the following ways:

Procedure A-1. (AO-1) > (AO-2) > (AO-3) > (OPEN VALVE-CW) > ....
Procedure A-2. (AO-1) > (AO-3) > (AO-2) > (OPEN VALVE-CW) > ....
Procedure A-3. (AO-2) > (AO-1) > (AO-3) > (OPEN VALVE-CW) > ....
Procedure A-4. (AO-2) > (AO-3) > (AO-1) > (OPEN VALVE-CW) > ....
Procedure A-5. (AO-3) > ··· > (AO-5) > (AO-6) > (AO-1) > (AO-2).
Procedure A-6. (AO-3) > ··· > (AO-5) > (AO-6) > (AO-2) > (AO-1) etc.

V. Revamping Process Designs to Ensure Feasibility of Operating Procedures

During the planning of operating procedures for complete chemical plants, a very important stage is the one in which the computer tries to identify whether there is a possibility that one or more of the operating constraints will be violated. This is analogous to the detection of Clobberers in Chapman's program TWEAK. As is the case in TWEAK, when Clobberers are detected, they must be countered by a suitable White Knight unless they are subdued by demotion or promotion. In the planning methodology presented in this chapter, the worst-case scenario lies between the initial and goal states, and promotion or demotion after or before these two states is clearly not possible. Thus the only alternative is for the program to locate a White Knight to counter the Clobberer.

It should be obvious that there is no guarantee that a suitable White Knight can be found to counter at least one of the Clobberers. In these instances a modification must be made to the process flowsheets so as to introduce structures that will serve as White Knights. In this section we describe algorithms that allow a computer-based methodology to automatically generate the necessary process modifications.

Let us look more closely at the situation where a Clobberer has been detected. In the case of qualitative mixing constraints, this involves an overlap of influence graphs in the worst-case scenario. The White Knights that are needed to counter this Clobberer are (a) an acceptable separation valve set (ASVS), and (b) a purgative and a purge route. In this analysis, we shall assume that the absence of White Knights is due to one of these two situations. Thus the modifications to the flowsheet must provide either an ASVS or a purgative and purge route (or both) depending on the needs of the situation at hand. The remainder of this section provides a methodology for accomplishing both these tasks.

A. ALGORITHMS FOR GENERATING DESIGN MODIFICATIONS

In this subsection, we present the algorithms that are to be used by a computer program that attempts to make design modifications to a flowsheet structure so as to allow the system to generate feasible operating plans where none existed before. The algorithms are classified into two types: those that are used to ensure the existence of an ASVS, and those that are used for generating a purge source and a purge route.

1. Flowsheet Modifications for Generating an ASVS

The methodology for proposing flowsheet modifications to ensure the existence of an ASVS is described in this section.

a. Stage 1. Determine a Minimal Separation Pipe Set

1. Construct the reduced pipe network corresponding to the topological graph of the worst-case state.
   (a) If the current node represents a PIPE or a "supersource," do nothing.
   (b) If the current node represents any other type of processing equipment, do the following:
      (i) Delete the current node from the topological graph.
      (ii) Let U be the set of nodes incident to the in-edges of the current node.
      (iii) Let V be the set of nodes incident to the out-edges of the current node.
      (iv) For every pair (u, v), where u is in U and v is in V, create a directed edge (u, v) and add it to the topological graph.

In the transformed graph, the only nodes are pipes and "supersources" and an edge leads from one node to another iff there is a path connecting the corresponding nodes in the topological graph, and no other pipe or source lies on this path.

2. Now, replace all directed edges with undirected edges. The problem of finding a minimal separation pipe set (MSPS) now reduces to the problem of determining a *vertex separator* in the reduced pipe network for the pair of supersources for the dangerous species under consideration. A polynomial–time algorithm exists for determining a vertex separator (see Even, 1979).

The MSPS may not be an acceptable separation pipe set (ASPS) for the given planning problem. This is because it is possible that some pipe in the MSPS is required to have flow through it in the goal state, or the user may simply have some preference regarding the state of flow through some of the pipes. However, from a heuristic standpoint, the minimal set is a good starting point for evolving into an ASPS, since disturbing the flow in these pipes upsets the overall flow only minimally. The procedure for evolving from an MSPS to an ASPS is described in stage 2.

b. Stage 2. Determine an Acceptable Separation Pipe Set

1. Let the current separation pipe set (SPS) of edges be the MSPS already determined.
2. Identify the pipes (if any) in the current SPS that are constrained to have positive or negative flows in the goal state. If there are no such pipes, then the current SPS is feasible. Go to step 3. If, on the other hand, there are some such pipes, mark them as unsatisfactory and proceed to step 4.
3. Present the MSPS graphically to the user (by highlighting the pipes on the flowsheet, for example). Ask the user to specify any pipes that he or she prefers to keep open. Mark all such pipes as "unsatisfactory." If there are no unsatisfactory pipes, return the current SPS as the ASPS.
4. Remove all "unsatisfactory" pipes from the reduced pipe network. Go to stage 1, step 2.

Once an ASPS has been found, the final stage of the methodology can be started.

c. Stage 3. Generation of an ASVS. For each pipe in the ASVS, locate a corresponding pipe in the topological graph. For each pipe located in the topological graph, place a valve on the pipe. The set of valves that are added to the flowsheet in this step will form an ASVS for the planning problem under consideration.

Thus, if the planner fails to come up with a feasible plan, and the reason for failure is that no ASVS was found, the preceding algorithm may be used to propose flowsheet modifications (in the form of extra valves) to create an appropriate ASVS.

2. Flowsheet Modifications for Purge Sources and Routes

Another reason why the planner may fail to find a feasible plan is that no suitable purgative or purge route can be found. The modification of the flowsheet structure to allow for this purging is described in this section. The algorithms provided here will, however, address only the structural modifications that need to be performed, and will not address the chemistry-related issues of deciding on a suitable chemical species to introduce as a purgative. This aspect of the problem can be addressed by encoding chemistry knowledge in the form of a rule-based system or by asking the user to choose an inert component. We also make the assumption that all of the flows in the worst-case scenario are defined. This may not necessarily be the case, but as yet no algorithm is known that will propose structural modifications to the flowsheet in the absence of such flow information.

ALGORITHM **CREATE-PURGE-ROUTE**

1. Construct a copy of the initial state topological graph. In this graph close all the valves that are members of the ASVS. This will partition the topological graph into two subgraphs G_A and G_B, each of which will contain the source of only one of the two dangerous components A and B, respectively.
2. Locate all the nodes in G_A where B was present in the initial state. Let the subgraph that consists of these nodes be called G'. Similarly, locate all the nodes in G_B where A was present in the initial state and create a subgraph G" that consists of these nodes and their incident edges. For each of the graphs G' and G", do the following:
    (a) Find all the minimal and maximal nodes in G' (or G") (see Lakshmanan, 1989).
    (b) Create a supersource of the chemical species that was identified (by either the user or a rule-based system) as a suitable purgative. For each minimal node in G' introduce a pipe node and appropriate connecting edges leading from the supersource to the minimal node.
    (c) Create a *sink* for the chemical species used as a purgative. For each maximal node in G' introduce a pipe node and the appropriate connecting edges leading from the maximal nodes to the sink.

372   C. HAN, R. LAKSHMANAN, B. BAKSHI, AND G. STEPHANOPOULOS

At the end of this construction process, the flowsheet will have been suitably modified so as to allow the purging of one dangerous species before admitting the second.

3. Case Studies

In this subsection we examine two modifications of the EDTA plant discussed earlier, in Section IV. The purpose of the illustration is to demonstrate the working of the algorithms that generate flowsheet modifications when (1) no MSVS or ASVS can be found and (2) no purge source or purge route is available.

In this example the flowsheet of the EDTA process has been modified by removing the following valves: v10, v11, v13, and v20. The result of the modifications is shown in Fig. 17.

When the planning algorithms are run on this plant no MSVS is found since, in the worst-case state, there are no valves that will prevent hydrogen cyanide and formaldehyde from mixing in the region of overlap. Thus the computer will try to use the methodology described in the

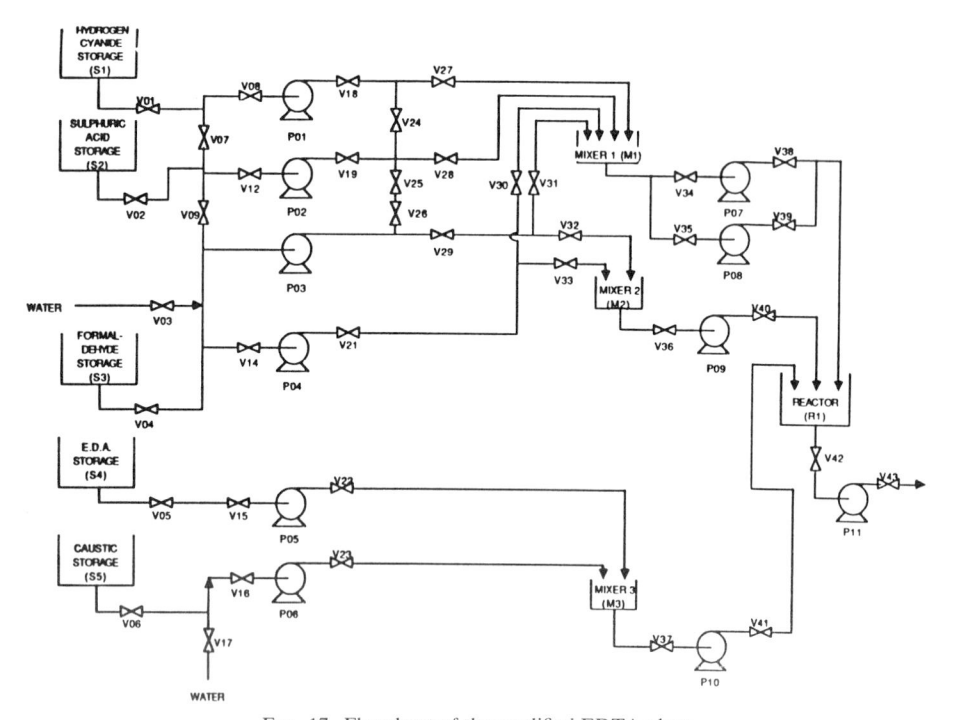

FIG. 17. Flowsheet of the modified EDTA plant.

previous paragraphs to come up with flowsheet modifications that will allow the computer to find an MSVS.

The algorithms described earlier are run on the modified plant, and the computer suggests that a valve be placed between the water inlet and the junction just above it. This modification is shown in Fig. 18, and this valve serves as the MSVS in the planning problem.

The second example that is used to illustrate the design methodologies is a modification to the EDTA problem as follows. The structure of the flowsheet is exactly the same as the one presented in Section IV. The only change to the problem is the statement that sulfuric acid and formaldehyde should not be allowed to come into contact with each other.

It may be recalled that in the initial analysis sulfuric acid was used as a purge species. It is obvious that this is not possible anymore since one of the dangerous components, namely formaldehyde, cannot be brought into contact with sulfuric acid. Thus the computer must generate new piping and source and sink structure to allow the computer to find a purge route.

The algorithm CREATE-PURGE-ROUTE described earlier is run under the worst-case scenario of the original EDTA plant with the added

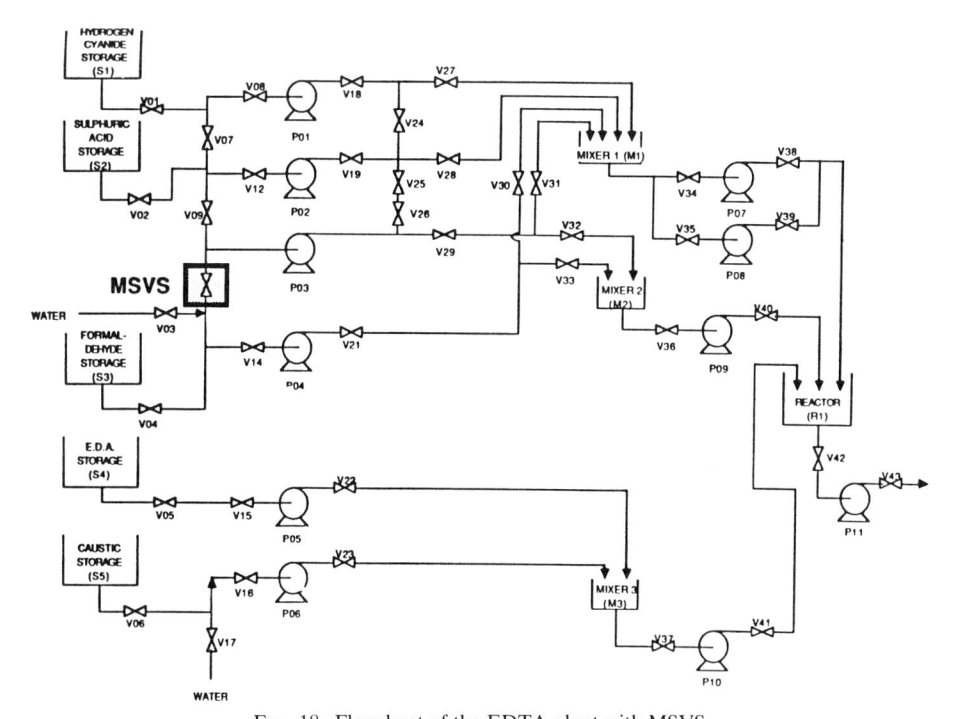

FIG. 18. Flowsheet of the EDTA plant with MSVS.

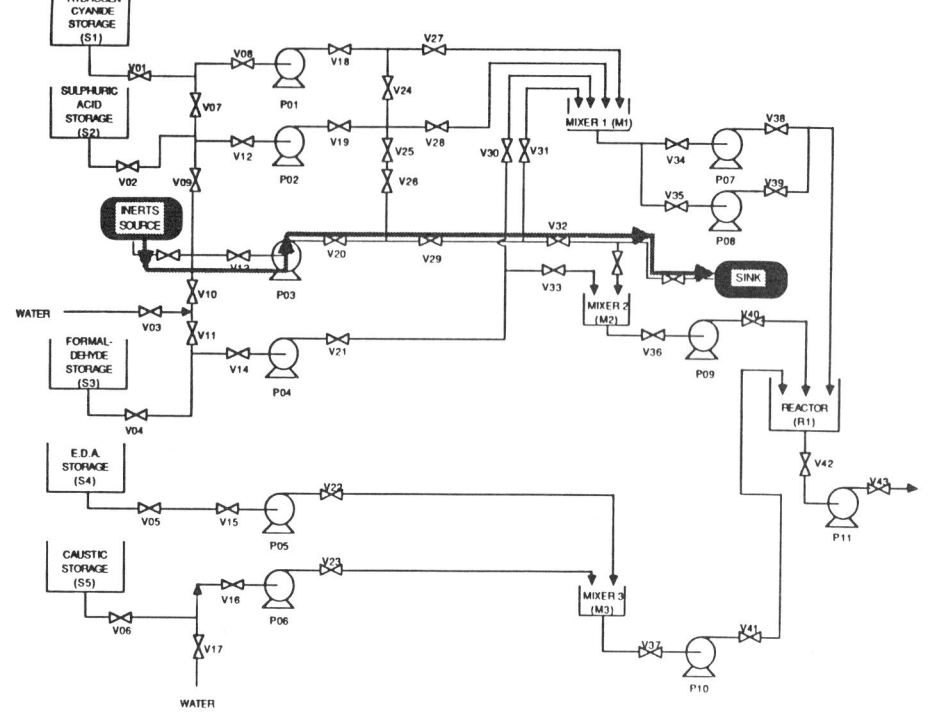

FIG. 19. Flowsheet of the EDTA plant showing purge route.

mixing constraint, and piping modifications are generated in the flowsheet. The final flowsheet with the purge route is shown in Fig. 19.

Note that, as mentioned earlier, the program does not attempt to specify what component is actually used as the inert purgative. A rule-based system with extensive knowledge of explosive mixtures is needed to accomplish this task. The construction of such a knowledge base, while useful, is assumed to be outside the scope of the planning problem. The program is designed only to propose the piping and valving modifications to the flowsheet.

VI. Summary and Conclusions

The synthesis of operating procedures for continuous chemical plants can be represented as a mixed-integer nonlinear programming problem, and it has been addressed as such by other researchers. In this chapter we have attempted to present a unifying theoretical framework, which addressed the logical (or integer) components of the problem, while allowing a smooth integration of optimization algorithms for the optimal selection of the values of the continuous variables. Nonmonotonic reasoning has been found to be superior to the traditional monotonic reasoning approaches, and capable to account for all integer decision variables. Critical in its success is the modeling approach, which is used to represent process operational tasks at various levels of abstraction.

Nevertheless, a series of interesting issues arise, which must be resolved before these ideas find their way to full industrial implementation:

1. *Dynamic simulation with discrete-time events and constraints.* In an effort to go beyond the integer (logical) states of process variables and include quantitative descriptions of temporal profiles of process variables one must develop robust numerical algorithms for the simulation of dynamic systems in the presence of discrete-time events. Research in this area is presently in full bloom and the results would significantly expand the capabilities of the approaches, discussed in this chapter.

2. *Design modifications for the accommodation of feasible operating procedures.* Section V introduced some early ideas on how the ideas of planning operating procedures could be used to identify modifications to a process flowsheet, which are necessary to render feasible operating procedures. More work is needed in this direction. Clearly, any advances in dynamic simulation with discrete-time events would have beneficial effects on this problem.

3. *Real-time synthesis of operating procedures.* Most of the ideas and methodologies, presented in this chapter, are applicable to the a priori, off-line, synthesis of operating procedures. There is a need though to address similar problems during the operation of a chemical plant. Typical examples are the synthesis of operational response (i.e., operating procedure) to process upsets, real-time recovery from a fallback position, and supervisory control for constrained optimum operation.

References

Barton, P. I., The modeling and simulation of combined discrete/continuous processes. Ph.D. Thesis, University of London (1992).

Chakrabarti, P. P., Ghose, S., and DeSarkar, S. C., Heuristic search through islands. *Artif. Intell.* **29**, 339 (1986).

Chapman, D., "Planning for Conjunctive Goals," MIT AI Lab. Tech. Rep. AI-TR-802. Massachusetts Institute of Technology, Cambridge, MA, 1985.

Crooks, C. A., and Macchietto, S., A combined MILP and logic-based approach to the synthesis of operating procedures for batch plants. *Chem. Eng. Commun.* **114**, 117 (1992).

Even, S., "Graph Algorithms." Computer Science Press, Rockville, MD, 1979.

Fikes, R. E., and Nilsson, N. J., STRIPS: A new approach to the application of theorem proving to problem solving. *Artif. Intell.* **2**, 198 (1971).

Foulkes, N. R., Walton, M. J., Andow, P. K., and Galluzzo, M., Computer-aided synthesis of complex pump and valve operations. *Comput. Chem. Eng.* **12**, 1035 (1988).

Fusillo, R. H., and Powers, G. J., A synthesis method for chemical plant operating procedures. *Comput. Chem. Eng.* **11**, 369 (1987).

Fusillo, R. H., and Powers, G. J., Operating procedure synthesis using local models and distributed goals. *Comput. Chem. Eng.* **12**, 1023 (1988a).

Fusillo, R. H., and Powers, G. J., Computer aided planning of purge operations. *AIChE J.* **34**, 558 (1988b).

Garrison, D. B., Prett, M., and Steacy, P. E., Expert systems in process control. A perspective. *Proc. AAAI Workshop Control*, Philadelphia (1986).

Holl, P., Marquardt, W., and Gilles, E. D., Diva—a powerful tool for dynamic process simulation. *Comput. Chem. Eng.* **12**, 421 (1988).

Ivanov, V. A., Kafarov, V. V., Kafarov, V. L., and Reznichenko, A. A., On algorithmization of the startup of chemical productions. *Eng. Cybern. (Engl. Transl.)* **18**, 104 (1980).

Kinoshita, A., Umeda, T., and O'Shima, E., An approach for determination of operational procedure of chemical plants. *Proc. PSE'82, Kyoto* 114 (1982).

Lakshmanan, R., and Stephanopoulos, G., "Planning and Scheduling Plant-Wide Startup Operations," LISPE Tech. Rep. (1987).

Lakshmanan, R., and Stephanopoulos, G., Synthesis of operating procedures for complete chemical plants. I. Hierarchical, structured modeling for nonlinear planning. *Comput. Chem. Eng.* **12**, 985 (1988a).

Lakshmanan, R., and Stephanopoulos, G., Synthesis of operating procedures for complete chemical plants. II. A nonlinear planning methodology. *Comput. Chem. Eng.* **12**, 1003 (1988b).

Lakshmanan, R., Synthesis of operating procedures for complete chemical plants. Ph.D. Thesis. Massachusetts Institute of Technology, Dept. Chem. Eng., Cambridge, MA (1989).

Lakshmanan, R., and Stephanopoulos, G., Synthesis of operating procedures for complete chemical plants. III. Planning in the presence of qualitative, mixing constraints. *Comput. Chem. Eng.* **14**, 301 (1990).

Marquardt, W., Dynamic process: Simulation-recent progress and future challenges. *In* "Chemical Process Control, CPC-IV" (Y. Arkun, and W. H. Ray, eds.) CACHE, AIChE Publishers, New York, 1991.

Modell, M., and Reid, R. C., "Thermodynamics and its Applications," 2nd ed., Prentice-Hall, Englewood Cliffs, NJ, 1983.

Newell, A., Shaw, J. C., and Simon, H. A., Report on a general problem solving program. *Proc. ICIP*, 256 (1960).

Pantelides, C. C., and Barton, P. I., Equation-oriented dynamic simulation. Current status and future perspectives. *Comput. Chem. Eng.* **17**, S263 (1993).

Pavlik, E., Structures and criteria of distributed process automation systems. *Comput. Chem. Eng.* **8**, 295 (1984).

Rivas, J. R., and Rudd, D. F., Synthesis of failure-safe operations. *AIChE J.* **20**, 311 (1974).

Sacerdoti, E. D., The non-linear nature of plans. *Adv. Pap. IJCAI, 4th*, 206, (1975).

Steele, G. L., "The Definition and Implementation of a Computer Programming Language Based On Constraints," AI Lab. Tech. Rep. AI-TR-595. Massachusetts Institute of Technology, Cambridge, MA, 1980.

Tomita, S., Hwang, K., and O'Shima, E., On the development of an AI-based system for synthesizing plant operating procedures. *Inst. Chem. Eng. Symp. Ser.*, 114 (1989a).

Tomita, S., Hwang, K., O'Shima, E., and McGreavy, C., Automatic synthesizer of operating procedures for chemical plants by use of fragmentary knowledge. *J. Chem. Eng. Jpn.* **22**, 364 (1989b).

Tomita, S., Nagata, M., and O'Shima, E., Preprints IFAC Workshop, Kyoto, 66 (1986).

INDUCTIVE AND ANALOGICAL LEARNING: DATA-DRIVEN IMPROVEMENT OF PROCESS OPERATIONS[1]

Pedro M. Saraiva

Department of Chemical Engineering
University of Coimbra
3000 Coimbra, Portugal

I. Introduction	378
II. General Problem Statement and Scope of the Learning Task	381
III. A Generic Framework to Describe Learning Procedures	384
A. Generic Formalism	385
B. Major Departures from Previous Approaches	385
IV. Learning with Categorical Performance Metrics	389
A. Problem Statement	389
B. Search Procedure, S	391
C. Case Study: Operating Strategies for Desired Octane Number	394
V. Continuous Performance Metrics	396
A. Problem Statement	396
B. Alternative Problem Statements and Solutions	398
C. Taguchi Loss Functions as Continuous Quality Cost Models	401
D. Learning Methodology and Search Procedure, S	403
E. Case Study: Pulp Digester	405
VI. Systems with Multiple Operational Objectives	408
A. Continuous Performance Variables	409
B. Categorical Performance Variables	409
C. Case Study: Operational Analysis of a Plasma Etching Unit	413
VII. Complex Systems with Internal Structure	417
A. Problem Statement and Key Features	417
B. Search Procedures	424
C. Case Study: Operational Analysis of a Pulp Plant	426
VIII. Summary and Conclusions	431
References	432

---

[1] The work reported in this chapter was performed while the author was on leave as a Ph.D. student in the Department of Chemical Engineering, Massachusetts Institute of Technology, Cambridge, MA, 02139.

Informed and systematic observation of data, naturally generated, can lead to the formulation of interesting and effective generalizations. Whereas some statisticians believe that experimentation is the only safe and reliable way to "learn" and achieve operational improvements in a manufacturing system, other statisticians and all the empirical machine learning researchers contend that by looking at past records and sets of examples, it is possible to extract and generate important new knowledge. This chapter draws from *inductive and analogical learning* ideas in an effort to develop systematic methodologies for the extraction of structured new knowledge from operational data of manufacturing systems. These methodologies do not require any a priori decisions and assumptions either on the character of the operating data (e.g., probability density distributions) or the behavior of the manufacturing operations (e.g., linear or nonlinear structured quantitative models), and make use of *instance-based learning* and *inductive symbolic learning* techniques, developed by artificial intelligence (AI). They are aimed to be complementary to the usual set of statistical tools that have been employed to solve analogous problems. Thus, one can see the material of this chapter as an attempt to fuse statistics and machine learning in solving specific engineering problems. The framework developed in this chapter is quite generic and, as the subsequent sections illustrate, it can be used to generate operational improvement opportunities for manufacturing systems (1) that are simple or complex (with internal structure), (2) whose performance is characterized by one or multiple objectives, and (3) whose performance metrics are categorical (qualitative) or continuous (real numbers). A series of industrial case studies illustrates the learning ideas and methodologies.

I. Introduction

With rapid advances in hardware, database management and information processing systems, efficient and competitive manufacturing has become an information-intensive activity. The amounts of data presently collected in the field on a routine basis are staggering, and it is not unusual to find plants where as many as 20,000 variables are continuously monitored and stored (Taylor, 1989).

It has also been recognized that the ability of manufacturing companies to become *learning* organizations (Senge, 1990; Shiba *et al.*, 1993; Hayes *et al.*, 1988), achieving high rates of continuous and never-ending improvements (with a special focus on *total manufacturing quality*), is a critical factor determining the survival and growth of the company in an increasingly competitive global economy (Deming, 1986; Juran, 1964).

As a result, there has been a significant and increasing awareness and interest on how the accumulation of data can lead to a better management of operational quality, which involves two complementary steps (Saraiva and Stephanopoulos, 1992a): (a) *control within prespecified limits* and (b) *continuous improvement of process performance*. The aim of the first is to detect "abnormal" situations, identify and eliminate the *special causes* that produced them. But, bringing the process under bounded, statistical control is not enough. The final level of variability thus achieved is the result of *common and sustained causes*, present within the process itself, and must not be considered as unavoidable. Therefore, the second step dictates that one should move from process control to process improvement, i.e., continuously search for common causes, ways of reducing their impact and challenge the current levels of performance (Juran, 1964).

The bulk chemical commodity producing companies (e.g., refineries, petrochemicals) have been practicing this philosophy for some time, using dynamic models to contain operational variability through feedback controllers, and employing static models to determine the optimal levels of operating conditions (Lasdon and Baker, 1986; Garcia and Prett, 1986).

On the other hand, plants involving solids processing (e.g., pulp and paper) with poorly understood physicochemical phenomena, not lending themselves to description through first-principles models, have not been using effectively the mountains of accumulated operational data. This lack of exploration and analysis of data is particularly severe for records of data collected while the processes were kept under "normal" operating contexts. Thus, given that less than 5% of the points contained in a conventional 6σ control chart are likely to attract any attention, the bulk of the acquired operating records is simply "stored," i.e., wasted. For such manufacturing plants there is a strong need and incentives to develop theoretical frameworks and practical tools that are able to extract useful and operational knowledge from existing records of data, leading to continuous improvement of process operational performance. This chapter represents a specific effort that attempts to fulfill that need.

Several statistical, quality management, and optimization data analysis tools, aimed at exploring records of measurements and uncover useful information from them, have been available for some time. However, all of them require from the user a significant number of assumptions and a priori decisions, which determine in a very strict manner the validity of the final results obtained. Furthermore, these classical tools are guided essentially by an academic attitude that emphasizes numerical accuracy over the extraction of substantive knowledge, whereas the formats chosen to express the solutions are often difficult to be interpreted and understood by human operators or to be implemented by them in manufacturing operations.

Rather than trying to replace any of the above traditional techniques, this chapter presents the development of complementary frameworks and methodologies, supported by symbolic empirical machine learning algorithms (Kodratoff and Michalski, 1990; Shavlik and Dietterich, 1990; Shapiro and Frawley, 1991). These ideas from machine learning try to overcome some of the weaknesses of the traditional techniques in terms of both (1) the number and type of a priori decisions and assumptions that they require and (2) the knowledge representation formats they choose to express final solutions.

One can thus state the primary goal of the approaches summarized in this chapter as follows:

To develop and apply assumption-free learning frameworks and methodologies, aimed at uncovering and expressing in adequate solution formats performance improvement opportunities, extracted from existing data which were acquired from plants that cannot be described effectively through first-principles quantitative models. (1)

The remaining material of this chapter has the following structure: Section II provides a more specific definition of the problems that are addressed, and it identifies the scope of applications and the type of manufacturing systems that the presented methodologies aim to cover. Then, Section III introduces a generic framework for the development and description of the learning algorithms that will allow us to present the several methodologies on a common basis. In the same section we will also enumerate the most critical characteristics and features that differentiate the approaches of this chapter from previous ones. Sections IV and V illustrate the problem statements and solution methodologies employed when the manufacturing system's performance is measured by a *categorical* (qualitative) or *continuous* (real-valued) variable. Section VI extends the ideas presented in Sections IV and V to manufacturing systems whose performance is characterized by multiple, possibly noncommensurable objectives, and Section VII makes extensions to complex manufacturing systems with internal structure. Concrete applications are presented at the end of these sections. Some final conclusions together with a critical summary of the main results are presented in Section VIII.

II. General Problem Statement and Scope of the Learning Task

In its most general form, the problem that we address in this chapter can be stated as follows:

$$\begin{aligned} &\textbf{\textit{Given}} \text{ a set of existing } (\mathbf{x}, y) \text{ data records,} \\ \text{where} \quad &\mathbf{x} \text{ is a vector of operating or decision variables, which} \\ &\text{are believed to influence the values taken on by } y; \\ &y \text{ is a performance metric, usually assumed to be a quality} \\ &\text{characteristic of the product or process under analysis;} \\ &\textbf{\textit{Learn}} \text{ how to improve the system performance above its current levels} \end{aligned} \quad (2)$$

The work described in this chapter revisits this old problem by adopting a new perspective, exploring alternative formats for the presentation of solutions and problem-solving strategies. Such alternative problem-solving strategies are particularly useful in addressing systems where the weaknesses of traditional approaches become particularly severe. It is thus important to clarify the scope of application and the type of situations within the broad area covered by the problem statement (2) that the methodologies of this chapter aim to cover. That's what we will try to do in the following paragraphs.

1. *Supervisory control layer of decisionmaking*. Although learning capabilities and data analysis activities should be present across all levels of decisionmaking (Fig. 1), countless studies and authors (National Research Council, 1991; Latour, 1976, 1979; Launks *et al.*, 1992; Klein, 1990; Sargent, 1984; Ellingsen, 1976; Moore, 1990) have shown that the greatest

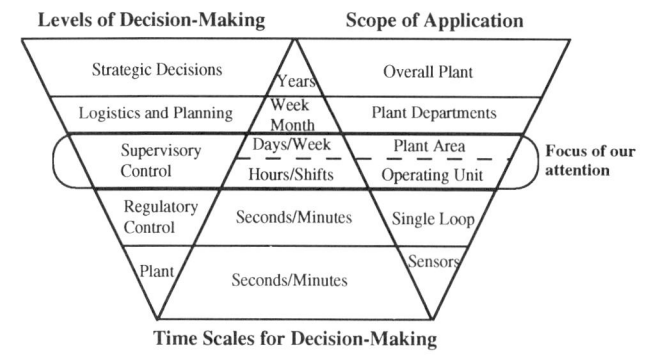

FIG. 1. Levels, time scales, and application scopes of decisionmaking activities.

opportunities and benefits are associated with improvements at the steady-state optimization or supervisory control layer. Shinnar (1986) makes clear this point by stating that in the chemical industry 80% of all practical control problems and 90% of the financial gains must be found and addressed at this level. Accordingly, we will be performing a steady-state analysis of data that has been collected and defined according to the supervisory control layer of decisionmaking.

2. *Plants lacking credible first-principles models.* The learning methodologies of this chapter require very few assumptions and decisions to be made a priori by the user, and rely on an exploratory analysis of data (Tukey, 1977) to uncover operational knowledge. Thus, they are particularly useful to analyze systems poorly understood and for which no first-principles models with acceptable accuracy are available. In such plants supervisory control decisions are quite often taken by human operators, rather than by computers. Reflecting on this fact, across the scale of automation in the spectrum of human–machine interaction (Sheridan, 1985) the approaches of this chapter fall within the scope of decision support systems (Turban, 1988); i.e., they do not intend to replace human intervention, but instead to interact with the users, and provide them as much useful information as possible. Yet, it is still the human operator who is responsible for the choice of a particular final solution, among a number of promising alternatives presented to him or her, and for selecting the particular course of action to follow. That being the case, it is critical for the knowledge extracted from the data to be represented in operational, explicit, and easy-to-understand formats, so that better use of that knowledge can be made by the human operators, who are not necessarily experts in pattern recognition, optimization, or statistics. Thus, the learning methodologies must reflect these concerns. Indeed, a strong emphasis is placed on the language used by the techniques of this chapter to express final solutions.

3. *Considerable amounts of data available on routine basis.* In learning activities, there is a fundamental tradeoff between the a priori existing knowledge and data intensity (Gaines, 1991). The more one already knows, the less data are needed to reach a given conclusion. As was said earlier, we are especially interested in studying complex systems for which no accurate quantitative models of behavior are available. Therefore, the learning methodologies must not rely or depend on the validity of any strong assumptions made about the system under study. A price has to be paid, as a result of this strategic choice; to reach the same level of accuracy and resolution in the final solutions found, more data are needed than if stronger assumptions could be made a priori and they happened to

be valid ones. This data-intensive nature of the approaches is not however likely to be problematic, since today in most manufacturing organizations there is no real shortage of data, but rather large amounts of unexplored measurement records. Furthermore, the case studies that were conducted show that even with moderate amounts of data (always less than 1000 records), it is possible to find solutions that result in quite significant improvements over the current levels of performance.

4. *Flexibility in problem definition.* Within the scope of application problems, outlined in the previous paragraphs, the learning methodologies are quite flexible and can handle a number of different situations and problem formulations. Both types of systems, where the performance metric used to evaluate performance, y, is continuous and categorical (i.e., can assume only one among a discrete number of possible values, such as "good," "bad," and "excellent"), can be analyzed. In systems of nontrivial size several performance objectives must often be taken into account. Extensions of the basic learning methodologies to handle such multiobjective problem formulations have been developed and tested. Finally, besides simple systems without internal structure, assumed to be isolated from the remaining world and self-sufficient for decisionmaking purposes, one has frequently to consider more complex systems, such as a complete plant, composed of a number of interconnected subsystems. A learning architecture for these complex systems was conceived and applied to a specific case study. It is based on the same procedures applied for simple systems, on top of which we have added articulation, coordination and propagation procedures. Thus, the learning methodologies of this chapter cover a broad range of possible industrial applications, and are able to handle simple or complex systems, with single or multiple objectives, and categorical or continuous performance evaluation variables.

5. *Prototypical application examples.* To provide a more concrete notion of the type of systems where our approaches are expected to be particularly helpful and useful, we conclude this section with a sample of prototypical examples of what the performance metric, y, in the problem statement (2) may represent, together with a definition of the corresponding systems:

Example 1: kappa (κ) index of the unbleached pulp produced by a Kraft digester

Example 2: activity, yield, and total production cost of proteins from fermentation units

Example 3: composition, molecular-weight distribution, and structure of complex copolymers

Example 4: selectivity and uniformity of etching rates at a chemical vapor deposition unit that produces silicon wafers

Example 5: analysis of the operation of an activated sludge wastewater treatment unit or of an entire pulp plant

As can be inferred from this sample of examples, industrial sectors where the learning methodologies of this chapter look particularly promising are pulp and paper, biotechnology, cement and ceramics, microelectronics, and discrete-parts manufacturing, i.e., complex plants or pieces of equipment that include solids processing, whose behavior is poorly understood and for which no accurate first-principles models exist. It should not be construed, though, that the methodologies of this chapter cannot also be applied to the analysis of other well-understood systems. However, as we move to such well-known systems, for which reliable quantitative models are already available, the suggested problem-solving approaches gradually loose their competitive advantages over traditional ones.

III. A Generic Framework to Describe Learning Procedures

In this section we present a formal framework to describe learning procedures that is sufficiently rich and general to allow us to concisely express (1) all the previous approaches that have been proposed for the formulation and solution of the problems stated in statement (2), and (2) all the different alternative definitions and methodologies that will be discussed in later sections of the chapter. This description language provides a common basis to introduce our different application contexts (categorical or continuous y, single or multiple objectives, simple or complex systems) and corresponding algorithms. It also allows us to make a clear statement of their most important departures from previous conventional approaches, thus facilitating a comparative analysis.

Subsequently, we use the common description framework in order to identify and analyze the following three most important distinguishing features that differentiate the learning frameworks of this chapter from traditional optimization and statistical techniques:

(a) Formats chosen to express the solutions
(b) Criteria used to evaluate the solutions
(c) Procedures used to estimate the solutions' evaluation criteria

A. A Generic Formalism

Any learning procedure, aimed to address and solve problems given by the problem statement (2) at the supervisory control level of decisionmaking, can be expressed by the following quartuple:

$$L = (\xi, \psi, f, S), \qquad (3)$$

where

(a) $\xi \in \Xi$ represents a generic solution, ξ, defined in the solution space, Ξ.
(b) $\psi \in \Psi$ is the performance criterion, ψ, defined in the performance space, Ψ, that one chooses to evaluate the merit of a generic solution ξ.
(c) f is the model or procedure that maps the solution into the performance space, i.e., it allows one to compute an estimate of the performance criterion, ψ^{est}, for any potential solution ξ, $\psi^{est} = f(\xi)$.
(d) S is a search procedure that explores f in order to identify specific solutions, $\xi^* \in \Xi$, that look particularly promising according to their estimated performances, $f(\xi^*)$. This final fourth element is optional and it is absent from conventional approaches whose goal is strict estimation (prediction of ψ for a given particular choice of ξ).

B. Major Departures from Previous Approaches

An examination of previous classical learning procedures reveals that they differ from each other only with respect to the choices of ψ, f, and S. All of them share the same basic format for ξ and the corresponding solution space, Ξ. Let's assume that each $(\mathbf{x}, y)$ pair in the problem statement (2) contains a total of M decision variables:

$$\mathbf{x} \equiv [x_1, \ldots, x_m, \ldots, x_M]^T \in \Xi_{decision}, \qquad (4)$$

where $\Xi_{decision}$ stands for the decision space, composed of all feasible $\mathbf{x}$ vectors, a subspace of $\mathscr{R}^M$.

Traditional approaches adopt a solution space Ξ that coincides with $\Xi_{decision}$, and thus any final solution (ξ^* or $\mathbf{x}^*$) has the same format as $\mathbf{x}$, consisting of a real vector that defines a single point in the decision space.

By considering such a language to express solutions, one ignores the fact that in the type of problems we want to study decision variables behave as random variables, and there is always some variability associated with them. No matter how good control systems happen to be, in reality we will always have to live with ranges of values for the decision variables (concentrations, pressures, flows, etc.), eventually bounded within a narrow, but not null, operation window. As a consequence of not taking into account that one has to operate within a given zone of the decision space, rather than at a single point, the final solutions found by conventional approaches may be suboptimal even when perfect $f(\mathbf{x})$ models are available, since their evaluation criterion, ψ, reflects only the performance achieved at a particular point in the decision space, and completely ignores the system behavior around that point. The zone of the decision space that surrounds the $\mathbf{x}^*$ solution with the best $\psi(\mathbf{x}^*)$ does not correspond in general to the zone of the decision space, $\mathbf{Z}^*$, where best average performance, $\psi(\mathbf{Z}^*)$, can be achieved.

1. Hyperrectangles as a Convenient Solution Format

As a result of the above observations, once one chooses to adopt solution formats where each ξ represents a region rather than a point in the decision space, it has yet to be decided what type of zones and shapes should be considered. Since our goal is not to achieve full automation, but to provide support to human operators, it is critical for the new language, used to express solutions, to be understood by them and lead to results that are easy to implement. In that regard, after observing how people tend to articulate their reasoning activities in the control rooms we concluded that they essentially follow an "orthogonal thinking" paradigm: *human operators express themselves by means of conjunctions of statements about individual decision variables* (e.g., the concentration is high and the pressure low), x_m, *and not through some more or less intricate linear or nonlinear combination of them* (as is the case with multivariate statistical techniques such as principal-components analysis, partial least squares, factor analysis, or neural networks). Combining the need to identify zones in the decision space, rather than points, with the need to preserve the individuality of each decision variable, rather than losing it in linear or nonlinear combinations, hyperrectangles (conjunctions of ranges of x_m values) in the decision space appear as a very convenient and the natural format to express solutions.

Thus, a critical departure from previous approaches, common to all our learning methodologies, is the adoption of a solution format that consists of hyperrectangles (not points) defined in the decision space.

2. Interval Analysis Nomenclature

Thus, interval analysis (Moore, 1979; Alefeld and Herzberger, 1983) provides the adequate support and notation formalism to express solutions. To distinguish real numbers from intervals, we will use capital letters for intervals. Also, bold typing is employed to represent both real variable and real interval vectors. A real interval X is a subset of $\mathcal{R}$ of the form

$$X \in \mathbf{I} \equiv \{x \in \mathcal{R} \mid i(X) \leq x \leq s(X)\}, \quad (5)$$

where

(a) $\mathbf{I}$ is the space of all closed real intervals.
(b) $i(X)$ is the lower bound of X.
(c) $s(X)$ is the upper bound of X.
(d) $w(X) = s(X) - i(X)$ is the width of X.
(e) $m(X) = [i(X) + s(X)]/2$ is the midpoint of X.

Extending the notation to hyperrectangles in $\mathcal{R}^M$, an M-dimensional interval vector, $\mathbf{X}$, has as its components real intervals, X_m, defined by ranges of x_m:

$$\mathbf{X} \equiv [X_1, \ldots, X_m, \ldots, X_M]^T \in \mathbf{I}^M;$$
$$\mathbf{m}(\mathbf{X}) = [m(X_1), \ldots, m(X_m), \ldots, m(X_M)]^T \in \mathcal{R}^M. \quad (6)$$

Since a real vector is a degenerate interval vector whose components are null width intervals, previous conventional pointwise solution formats can be considered a particular case of the suggested alternative and more general solution space, obtained when the minimum allowed region size is reduced to zero, thus converting hyperrectangles into single points.

3. Major Differences

The notation introduced above allows us to make now a more explicit and condensed enumeration of the major characteristics and differences, with respect to the (ξ, ψ, f, S) key components, that separate our learning methodologies from other approaches.

Solution format, ξ. The solution space consists of hyperrectangles ($\xi = \mathbf{X} \in \mathbf{I}^M$), instead of points ($\xi = \mathbf{x} \in \mathcal{R}^M$), defined in the decision space.

Performance criterion, ψ. The quality of any potential solution, $\mathbf{X}$, is determined by the average system performance achieved within the zone of the decision space identified by $\mathbf{X}, \psi(\mathbf{X})$, not by the individual performance obtained at any particular point $\mathbf{x}, \psi(\mathbf{x})$.

Mapping procedure, f. The models that perform the mapping from the solution to the performance space have as argument a given hyperrectangle, $\mathbf{X}$, rather than a point, $\mathbf{x}$. They compute an estimate of the average performance that is expected within $\mathbf{X}$, $\psi^{\mathrm{est}}(\mathbf{X}) = f(\mathbf{X})$, not a single pointwise prediction, $\psi^{\mathrm{est}}(\mathbf{x}) = f(\mathbf{x})$.

Search procedure, S. The search procedures explore the modified mapping models, $\psi^{\mathrm{est}}(\mathbf{X}) = f(\mathbf{X})$, in order to generate and identify a set of final solutions, $\mathbf{X}^*$, that look particularly promising according to the corresponding estimated performance scores, $\psi^{\mathrm{est}}(\mathbf{X}^*)$.

As we will see in subsequent sections, the mapping procedures, f, adopted in our learning methodologies, are based on direct sampling approaches:

For any given $\mathbf{X}$ and ψ, we find those $(\mathbf{x}, y)$ pairs for which $\mathbf{x} \in \mathbf{X}$, and use this random sample to compute $\psi^{\mathrm{est}}(\mathbf{X})$ and build confidence intervals for $\psi(\mathbf{X})$.

These direct sampling strategies provide $\psi^{\mathrm{est}}(\mathbf{X})$ estimators that are consistent, unbiased, and do not require a priori assumptions about the system behavior, thus remaining consistent and unbiased regardless of the validity of any assumptions. In addition to a point estimate of performance, they also provide a probabilistic bound on the uncertainty associated with it, through the construction of confidence intervals for $\psi(\mathbf{X})$. Furthermore, the accuracy of the estimates obtained is limited only by the amounts of data that are available, not by any of the structural or functional form choices that have to be made with other mapping models. Finally, they are also computationally efficient, since all the effort involved consists of a search for those pairs of data falling inside $\mathbf{X}$, followed by a computation of the average and standard deviation among these records.

The preceding set of characteristics and properties of the $\psi^{\mathrm{est}}(\mathbf{X})$ estimators makes our type of mapping procedures, f, particularly appealing for the kinds of systems that we are especially interested to study, i.e., manufacturing systems where considerable amounts of data records are available, with poorly understood behavior, and for which neither accurate first-principles quantitative models exist nor adequate functional form choices for empirical models can be made a priori. In other situations and application contexts that are substantially different from the above, while much can still be gained by adopting the same problem statements, solution formats and performance criteria, other mapping and search procedures (statistical, optimization theory) may be more efficient.

A solution space, Ξ, consisting of hyperrectangles defined in the decision space, $\mathbf{X}$, is a basic characteristic common to all the learning methodologies that will be described in subsequent sections. The same does not happen with the specific performance criteria ψ, mapping models f, and search procedures S, which obviously depend on the particular nature of the systems under analysis, and the type of the corresponding performance metric, y.

IV. Learning with Categorical Performance Metrics

The nature of the performance metric, y, is determined by the characteristics of the specific process under analysis. Since we are particularly interested in analyzing situations where y is related to product or process quality, it is quite common to find systems where a categorical variable y is chosen to classify and evaluate their performance. This may happen due to the intrinsic nature of y (e.g., it can only be measured and assume qualitative values, such as "good," "high," and "low"), or because y is derived from a quantization of the values of a surrogate continuous measure of performance (e.g., $y =$ "good" if some characteristic z of the product has value within the range of its specifications, and $y =$ "bad," otherwise).

In this section we will introduce the problem statements adopted for this type of performance metric, briefly describe the learning methodology employed to address it [for a more complete presentation, see Saraiva and Stephanopoulos (1992a)], and show a specific application case study.

A. Problem Statement

When y is a categorical variable, which can assume only one among a total of K discrete possible values, pattern recognition (Duda and Hart, 1973; James, 1985) provides an adequate context for the introduction of the learning methodologies. The most important features that separate these learning methodologies from existing classification techniques (such as linear discriminant functions, nearest-neighbor and other nonparametric classifiers, neural networks, etc.), are summarized in Table I, and briefly discussed below.

TABLE I
CONVENTIONAL PATTERN RECOGNITION AND SUGGESTED ALTERNATIVE

	Conventional classification	Suggested alternative
ξ	$\mathbf{x} \in \mathcal{R}^M$	$\mathbf{X} \in \mathbf{I}^M$
ψ	$y(\mathbf{x})$ or $p(y=j\|\mathbf{x})$	$y(\mathbf{X})$ or $p(y=j\|\mathbf{X})$
f	Technique-dependent	$\dfrac{n_j(\mathbf{X})}{n(\mathbf{X})}$
S	Nonexistent	Induction of decision trees

1. Conventional Classification Techniques

All conventional classification procedures are aimed at answering one of the following questions:

$$\text{Given a generic vector of values } \mathbf{x} \equiv [x_1, \ldots, x_m, \ldots, x_M]^T \in \mathcal{R}^M, \text{ what is the corresponding } y \text{ value?} \quad (7a)$$

and/or

$$\text{Given a generic vector of values } \mathbf{x} \equiv [x_1, \ldots, x_m, \ldots, x_M]^T \in \mathcal{R}^M, \text{ what are reasonable estimates for the conditional probabilities } p(y=j|\mathbf{x}), j = 1, \ldots, K? \quad (7b)$$

Thus, they share exactly the same solution (Ξ) and performance criteria (Ψ) spaces. Furthermore, since their role is simply to estimate y for a given $\mathbf{x}$, no search procedures S are attached to classical pattern recognition techniques. Consequently, the only element that differs from one classification procedure to another is the particular mapping procedure f that is used to estimate $y(\mathbf{x})$ and/or $p(y=j|\mathbf{x})$. The available set of $(\mathbf{x}, y)$ data records is used to build f, either through the construction of approximations to the decision boundaries that separate zones in the decision space leading to different y values (Fig. 2a), or through the construction of approximations to the conditional probability functions, $p(y=j|\mathbf{x})$.

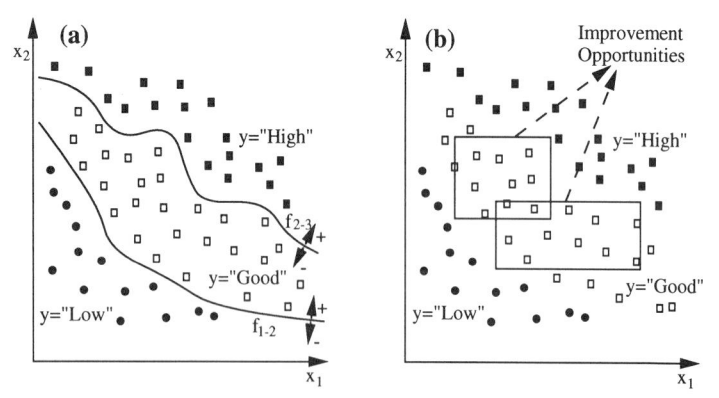

FIG. 2. (a) Conventional pattern recognition; (b) alternative problem statement and solution format.

2. The Learning Methodology

On the other hand, the question that we want to see answered is the following one (Fig. 2b):

What are hyperrectangles in the decision space, $\mathbf{X} \in \mathbf{I}^M$, inside which one gets only a desired y value, or at least a large fraction of that y value? (8)

The solution space thus consists of hyperrectangles in the decision space, $\mathbf{X} \in \mathbf{I}^M$, and the corresponding performance criteria are the conditional probabilities of getting any given y value inside $\mathbf{X}$, $p(y=j|\mathbf{X})$, $j = 1, \ldots, K$, or the single most likely y value within $\mathbf{X}$.

The mapping procedure, f, that allows us to compute $p(y=j|\mathbf{X})$ estimates, starts with a search performed over all the available $(\mathbf{x}, y)$ pairs that leads to the identification of the $n(\mathbf{X})$ cases for which $\mathbf{x} \in \mathbf{X}$. If we designate as $n_j(\mathbf{X})$ the number of such records with $y = j$, the desired estimates, $p^{\text{est}}(y=j|\mathbf{X})$, are given by

$$p^{\text{est}}(y=j|\mathbf{X}) = \frac{n_j(\mathbf{X})}{n(\mathbf{X})}, \quad j = 1, \ldots, K. \tag{9}$$

Using the normal approximation to a binomial distribution, confidence intervals (CIs) for $p(y=j|\mathbf{X})$ can be established for a specific significance level, α:

$$\text{CI} = \left[p^{\text{est}}(y=j|\mathbf{X}) \pm t_{(\alpha/2, n(\mathbf{X})-1)} \cdot \sqrt{\frac{p^{\text{est}}(y=j|\mathbf{X}) \cdot (1 - p^{\text{est}}(y=j|\mathbf{X}))}{n(\mathbf{X})}} \right], \tag{10}$$

where t stands for the critical value of the Student's distribution.

The search procedure, S, used to uncover promising hyperrectangles in the decision space, $\mathbf{X}^*$, associated with a desired y value (e.g., $y =$ "good"), is based on *symbolic inductive learning algorithms*, and leads to the identification of a final number of promising solutions, $\mathbf{X}^*$, such as the ones in Fig. 2b. It is described in the following subsection.

B. SEARCH PROCEDURE, S

In order to introduce the search procedure, S, we start by showing how classification decision trees lead to the definition of a set of hyperrectangles, and how they can be constructed from a set of $(\mathbf{x}, y)$ data records.

Then we describe the conversion of the knowledge captured by the induced decision trees into a set of final solutions, **X***.

1. Classification Decision Trees and Their Inductive Construction

A classification decision tree allows one to predict in a sequential way the y value (or corresponding conditional probabilities) that is associated with a particular **x** vector of values. At the top node of the tree (A in Fig. 3), a first test is performed, based on the value assumed by a particular decision variable (x_3). Depending on the outcome of this test, vector **x** is sent to one of the branches emanating from node A. A second test follows, being carried out at another node (B), and over the values of the same or a different decision variable (e.g., x_6). This procedure is

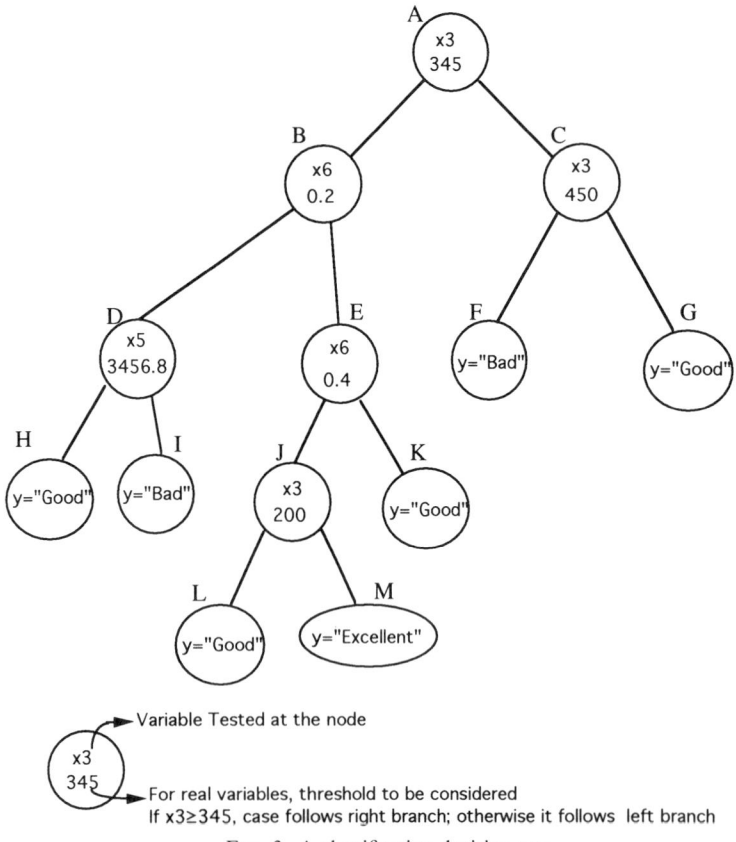

FIG. 3. A classification decision tree.

repeated until, after a last test, vector **x** follows a branch that leads to a terminal node (or leaf), labeled with a particular y value and/or set of conditional probabilities, which provides the $y(\mathbf{x})$ and/or $p(y=j|\mathbf{x})$ estimates.

Although decision trees contain a number of attractive features, including competitive accuracy, when considered strictly as classification devices (Saraiva and Stephanopoulos, 1992a), the most important point for our purposes is that each of the tree's terminal nodes identifies a particular hyperrectangle in the decision space, **X**, associated with a given y value. For example, node M defines a $y =$ "excellent" rectangle that corresponds to the following rule:

$$\text{If } 200 \leq x_3 < 345 \text{ and } 0.2 < x_6 \leq 0.4, \text{ then } y = \text{"Excellent,"} \quad (11)$$

Once a decision tree such as this has been constructed from the available set of $(\mathbf{x}, y)$ pairs, by picking up those leaves labeled with the desired y value, one gets an initial collection of promising hyperrectangles. Some additional transformations, summarized in the next paragraph, convert this initial collection into a final set of solutions, $\mathbf{X}^*$.

The algorithm that we employ to build a classification decision tree from $(\mathbf{x}, y)$ data records belongs to a group of techniques known as *top-down induction of decision trees* (TDIDT) (Sonquist et al., 1971; Fu, 1968; Hunt, 1962; Quinlan, 1986, 1987, 1993; Breiman et al., 1984).

The construction starts at the root node of the tree, where all the available $(\mathbf{x}, y)$ pairs are initially placed. One identifies the particular split or test, s, that maximizes a given measure of information gain (Shannon and Weaver, 1964), $\Phi(s)$. The definition of a split, s, involves both the choice of the decision variable and the threshold to be used. Then, the $(\mathbf{x}, y)$ root node pairs are divided according to the best split found, and assigned to one of the children nodes emanating from it. The information gain measure, $\Phi(s)$, for a particular parent node t, is

$$\Phi(s) = -\sum_{k=1}^{K} P_k(t)\log_2 P_k(t) + \sum_{c=1}^{R} \frac{N(t_c)}{N(t)} \sum_{k=1}^{K} P_k(t_c)\log_2 P_k(t_c), \quad (12)$$

where

(a) $N(t)$ is the number of $(\mathbf{x}, y)$ pairs included in t.
(b) R is the total number of children nodes, t_c, $c = 1, \ldots, R$, emanating from t.
(c) $P_k(t) = N_k(t)/N(t)$ is an estimate of the $p(y=k|t)$ conditional probability.
(d) $N_k(t)$ is the number of $(\mathbf{x}, y)$ pairs assigned to node t for which $y = k$.

This splitting procedure is now applied recursively to each of the children nodes just created. The successive expansion process continues until terminal nodes or leaves, over which no further partitions are performed, can be identified.

The preceding strategy for the construction of decision trees provides an efficient way for inducing compact classification decision trees from a set of (**x**, *y*) pairs (Moret, 1982; Utgoff, 1988; Goodman and Smyth, 1990). Furthermore, tests based on the values of irrelevant x_m variables are not likely to be present in the final decision tree. Thus, the problem dimensionality is automatically reduced to a subset of decision variables that convey critical information and influence decisively the system performance.

*2. From Decision Trees to Final Solutions, **X****

Once an induced decision tree has been found, it is subjected to an additional pruning and simplification treatment, whose details can be found in Saraiva and Stephanopoulos (1992a). Pruning and simplification lead to the definition of a revised and reasonably sized final decision tree, with greater statistical reliability. Each of the terminal nodes of this final decision tree corresponds to a particular hyperrectangle in the decision space, **X**, which is labeled either with the *y* value that is most likely to be obtained within that hyperrectangle, or with the corresponding conditional probabilities estimates, $p^{est}(y=j|\mathbf{X})$, $j = 1, \ldots, K$. These terminal nodes are further refined through an expansion process aimed at enlarging the zone of the decision space that they cover, followed by statistical significance tests that may introduce additional simplifications (Saraiva and Stephanopoulos, 1992a). By the end of this refinement stage, we get a group of improved, statistically significant, simplified, and partially overlapped hyperrectangles. Those that lead predominantly to the desired *y* value constitute the final group of solutions, **X***, and are presented to the user together with a number of auxiliary evaluation scores (Saraiva and Stephanopoulos, 1992a). It is the user's responsibility to analyze this set of hyperrectangles, **X***, make a selection among them, and define the course of action to follow.

C. Case Study: Operating Strategies for Desired Octane Number

To illustrate the potential practical capabilities of the learning methodology, we will now present the results obtained through its application to records of operating data collected from a refinery unit (Daniel and Wood, 1980). Additional industrial case studies can be found in Saraiva and Stephanopoulos (1992a).

The y variable that we will consider derives from a quantization of the octane numbers of the gasoline product, z, assuming one out of three possible values:

(a) $y = 1$ ("very low") for $z \leq 91$
(b) $y = 2$ ("low") when $91 < z < 92$
(c) $y = 3$ ("good") for $z \geq 92$

In this particular problem, one wishes to achieve values of z as high as possible, and thus to identify zones in the decision space where one gets mostly $y = 3$.

There are four decision variables: three different measures of the feed composition (x_1, x_2, x_3) and the value of an unspecified operating condition (x_4).

In Fig. 4 we present the final induced decision tree, as well as the partition of the (x_1, x_4) plane defined by its leaves, together with a projection of all the available $(\mathbf{x}, y)$ pairs on the same plane. These two decision variables are clearly influencing the current performance of the refinery unit, and the decision tree leaves perform a reasonable partition of the plane. To achieve better performance, we must look for operating zones that will result in obtaining mostly $y = 3$ values. Terminal nodes 2

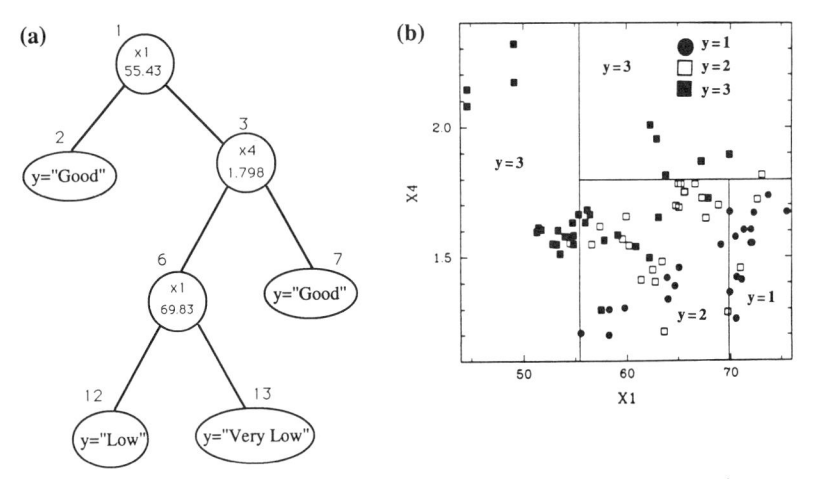

FIG. 4. (a) Induced decision tree; (b) partition of the plane defined by its leaves.

and 7 identify two such zones. The corresponding final solutions, $\mathbf{X}^*$, found after going through all the additional steps of refinement and validation, are

$$\mathbf{X}_1^* = \{x_1 \in [44.6, 55.4]\}, \text{ with } p^{\text{est}}(y = 3|\mathbf{X}_1^*) = 0.9,$$

$$\mathbf{X}_2^* = \{x_4 \in [1.8, 2.3]\}, \text{ with } p^{\text{est}}(y = 3|\mathbf{X}_2^*) = 1.0$$

It can be noticed from the conditional probability estimates that one should expect to get almost only "good" y values while operating inside these zones of the decision space, as opposed to the current operating conditions, which lead to just 40% of "good" y values.

V. Continuous Performance Metrics

In Section IV we considered a categorical performance metric y. Although that represents a common practice, especially when y defines the quality of a product or process operation, there are many instances where system performance is measured by a continuous variable. Even when y is quality-related, it is becoming increasingly clear that explicit continuous quality cost models should be adopted and replace evaluations of performance based on categorical variables.

This Section addresses cases with a continuous performance metric, y. We identify the corresponding problem statements and results, which are compared with conventional formulations and solutions. Then Taguchi loss functions are introduced as quality cost models that allow one to express a quality-related y on a continuous basis. Next we present the learning methodology used to solve the alternative problem statements and uncover a set of final solutions. The section ends with an application case study.

A. Problem Statement

A number of different techniques have been suggested and applied to address situations where y is a continuous variable. Table II summarizes the most important characteristics of our approach and major features that differentiate it from conventional procedures.

TABLE II
CONVENTIONAL APPROACHES AND SUGGESTED ALTERNATIVE

	Conventional approaches	Suggested alternative
ξ	$\mathbf{x} \in \mathscr{R}^M$	$\mathbf{X} \in \mathbf{I}^M$
ψ	$y(\mathbf{x})$ or $E(y\|\mathbf{x})$	$E(y\|\mathbf{X})$
f	Technique-dependent	$\dfrac{1}{n(\mathbf{X})} \sum_{j=1}^{n(\mathbf{X})} y_j$
S	Optimization	Exploratory search

1. Conventional Procedures

All conventional approaches (mathematical and stochastic programming, parametric and nonparametric regression analysis) adopt as a common solution format real vectors, $\mathbf{x} \in \mathscr{R}^M$, and as performance criterion, ψ, the expected value of y, $E(y|\mathbf{x})$, given $\mathbf{x}$, or the single y value that corresponds to a specific $\mathbf{x}$, $y(\mathbf{x})$, if one assumes a fully deterministic relationship between y and $\mathbf{x}$. Just as in Section IV, the element that essentially distinguishes the several techniques is the mapping procedure, f, used to compute $\psi^{\text{est}} = f(\mathbf{x})$. In order to find a final solution $\mathbf{x}^*$, optimization algorithms form the search procedure, S, that leads to the identification of the particular point in the decision space that maximizes or minimizes $\psi^{\text{est}} = f(\mathbf{x})$.

These conventional approaches usually follow a two-step sequential process:

Step 1. The $(\mathbf{x}, y)$ records of data are employed to build the f mapping.

Step 2. S does not make direct use of the original data, but rather employs f to find the final solution, $\mathbf{x}^*$, that optimizes $\psi^{\text{est}} = f(\mathbf{x})$.

2. The Learning Methodology

For the reasons already discussed in Section III, our solution space consists of hyperrectangles in the decision space, $\mathbf{X} \in \mathbf{I}^M$, not single points, $\mathbf{x}$. The corresponding performance criterion used to evaluate solutions, ψ, is the expected y value within $\mathbf{X}$:

$$\psi(\mathbf{X}) = E(y|\mathbf{X}). \tag{13}$$

These conceptual changes in both solution formats and performance criteria are independent of the particular procedures chosen to estimate $\psi(\mathbf{X})$ and search for a set of final solutions, $\mathbf{X}^*$. However, as was also discussed in Section III, for the types of systems that we are specially interested in analyzing, direct sampling strategies to estimate $\psi(\mathbf{X})$ offer a number of advantages. The mapping model that we employ, f, is similar to the one adopted for categorical y variables. A search is performed over all the available $(\mathbf{x}, y)$ data records, leading to the identification of a total of $n(\mathbf{X})$ pairs for which $\mathbf{x} \in \mathbf{X}$. The performance criterion estimate, $\psi^{\text{est}}(\mathbf{X})$, is the sample y average among these $n(\mathbf{X})$ pairs:

$$\psi^{\text{est}}(\mathbf{X}) = f(\mathbf{X}) = \frac{1}{n(\mathbf{X})} \sum_{j=1}^{n(\mathbf{X})} y_j \qquad (14)$$

The corresponding confidence interval, CI, for $E(y|\mathbf{X})$, at a given significance level, α, is

$$\text{CI} = \left[f(\mathbf{X}) \pm t_{(\alpha/2, n(\mathbf{X})-1)} \cdot \frac{s_y(\mathbf{X})}{\sqrt{n(\mathbf{X})}} \right], \qquad (15)$$

where $s_y(\mathbf{X})$ stands for the sample standard deviation:

$$s_y(\mathbf{X}) = \left\{ \frac{1}{n(\mathbf{X}) - 1} \sum_{j=1}^{n(\mathbf{X})} [y_j - f(\mathbf{X})]^2 \right\}^{0.5}. \qquad (16)$$

If one is interested only in finding the single feasible hyperrectangle (i.e., respecting minimum width constraints imposed due to control limitations) that minimizes $\psi^{\text{est}}(\mathbf{X})$, to find that hyperrectangle one may choose as search procedure, S, any optimization routine. However, our primary goal is to conduct an exploratory analysis of the decision space, leading to the identification of a set of particularly promising solutions, $\mathbf{X}^*$, that are presented to the decisionmaker, who is responsible for a final selection and a choice of the course of action to follow. The search procedure adopted in our learning methodology, S, reflects this goal, and will be described in a subsequent paragraph.

B. ALTERNATIVE PROBLEM STATEMENTS AND SOLUTIONS

Recognizing that, due to unavoidable variability in the decision variables, one has to operate within a zone of the decision space, and not at a single point, we might still believe that finding the optimal pointwise solution, $\mathbf{x}^*$, as usual, would be enough. The assumption behind such a belief is that centering the operation around $\mathbf{x}^*$ will correspond in practice to the adoption of a zone in the decision space to conduct the operation, $\mathbf{X}$ [with $\mathbf{m}(\mathbf{X}) = \mathbf{x}^*$], that is equivalent or close to the best possible zone, $\mathbf{X}^*$. However, because the evaluation of performance at a single point in the decision space, $\mathbf{x}$, completely ignores the system behavior around that point, the preceding assumption in general does not hold: *searching for an optimal hyperrectangle leads to a final solution*, $\mathbf{X}^*$, *that is likely to lie in a region of the decision space quite distant from* $\mathbf{x}^*$, *and* $\mathbf{m}(\mathbf{X}^*) \neq \mathbf{x}^*$. This observation emphasizes how critical it is to adopt the modified problem statements described at the beginning of Section V, where a direct and explicit search for the best zone to operate replaces the classical optimization paradigm, which ignores variability in the decision variables and seeks to identify as precisely as possible the optimal $\mathbf{x}^*$. No matter how accurately $\mathbf{x}^*$ is determined, targeting the operation around it can represent a quite suboptimal answer when the random nature of the decision variables is taken into account.

To illustrate how different $\mathbf{m}(\mathbf{X}^*)$ and $\mathbf{x}^*$ may happen to be, let's consider as a specific example (others can be found in Saraiva and Stephanopoulos, 1992c) a Kraft pulp digester. The performance metric y, that one wishes to minimize, is determined by the kappa index of the pulp produced and the cooking yield. Two decision variables are considered: *H-factor* (x_1), and *alkali charge* (x_2). Furthermore, we will assume as perfect an available deterministic empirical model (Saraiva and Stephanopoulos, 1992c), f, which expresses y as function of $\mathbf{x}$, i.e., that $y = f(x_1, x_2)$ is perfectly known.

If one follows the conventional optimization paradigm, adopting pointwise solution formats, the best feasible answer, $\mathbf{x}^*$, which minimizes $f(\mathbf{x})$, is $\mathbf{x}^* = (200; 17.9)$, as can be confirmed by examining the contour plots of f shown in Fig. 5a.

On the other hand, when the unavoidable variability in the decision space is considered explicitly, the goal of the search becomes the identification of the optimal hyperrectangle, $\mathbf{X}^*$, which solves the following problem:

$$\min_{\mathbf{X} \in \mathbf{I}^2} f(\mathbf{X}), \qquad (17)$$

subject to

$$w(X_1) \geq \Delta x_1 = 300,$$

$$w(X_2) \geq \Delta x_2 = 1.0,$$

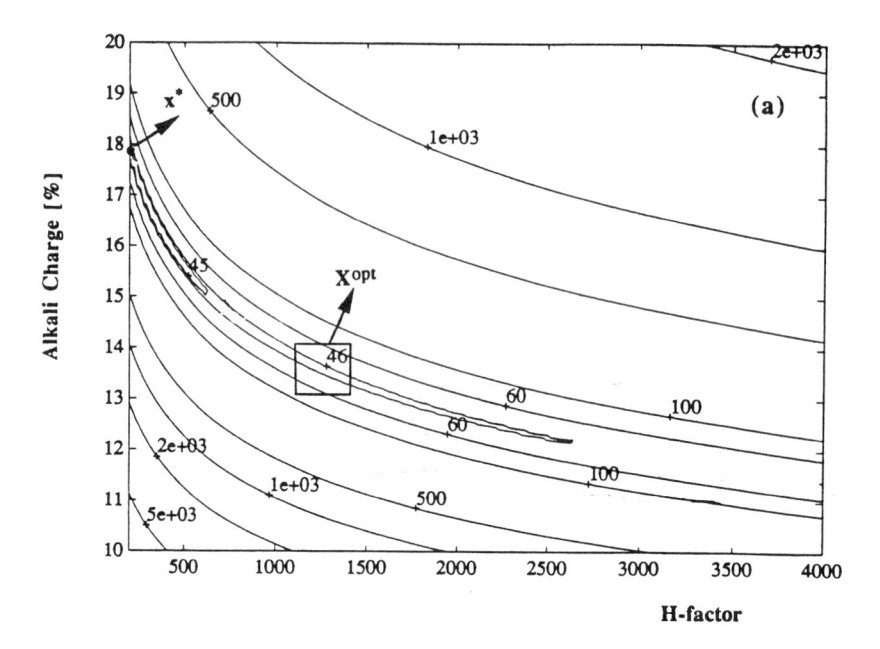

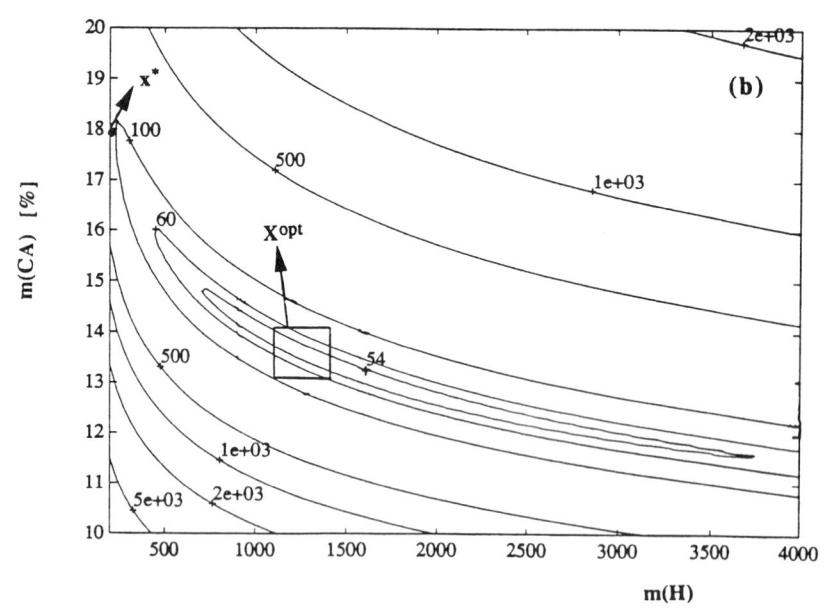

FIG. 5. Contour plots and optimal solutions for (a) $f(x)$ versus x and (b) $E(f(x)|X)$ versus $m(X)$.

with

$$f(\mathbf{X}) = E(y|\mathbf{X}) = \frac{\int_{i(X_1)}^{s(X_1)} \int_{i(X_2)}^{s(X_2)} p(\mathbf{x}) f(\mathbf{x}) \, dx_1 \, dx_2}{\int_{i(X_1)}^{s(X_1)} \int_{i(X_2)}^{s(X_2)} p(\mathbf{x}) \, dx_1 \, dx_2},$$

and $p(\mathbf{x})$ representing the (x_1, x_2) joint probability density function (for this particular study we assumed that both decision variables are independent and uniformly distributed).

In Fig. 5b we present the contour plots for $E(y|\mathbf{X})$ as a function of $\mathbf{m}(\mathbf{X})$. The corresponding optimal rectangle, $\mathbf{X}^*$, is

$$\mathbf{X}^* = \{x_1 \in [1114; 1414] \wedge x_2 \in [13.1; 14.1]\}. \tag{18}$$

This solution lies in a zone of the decision space that is quite distant from $\mathbf{x}^* = (200; 17.9)$ and targeting the operation around $\mathbf{x}^*$ results in clearly suboptimal performance.

This discrepancy illustrates some of the dangers associated with looking exclusively for pointwise solutions and performances, while neglecting the decision variables' variability. Although the specific problem-solving strategies developed in this chapter are aimed at the analysis of systems for which no good quantitative $f(\mathbf{x})$ and $p(\mathbf{x})$ exist a priori, the example presented shows the more general nature of the benefits that may derive from adopting the suggested alternative problem statements even when such models are available and used to find the final optimal hyperrectangle, $\mathbf{X}^*$.

Finally, it should be added that the conventional problem statement and pointwise solution format can be interpreted as a particular degenerate case of our more general formulations. As the minimum acceptable size for zones in the decision space decreases, the different performance criteria converge to each other and $\mathbf{X}^*$ gets closer and closer to $\mathbf{x}^*$. Both approaches become exactly identical in the extreme limiting case where $\Delta x_m = 0$, $m = 1, \ldots, M$, which is the particular degenerate case adopted in traditional formulations.

C. Taguchi Loss Functions as Continuous Quality Cost Models

The development of most of the optimization and operations research techniques was motivated and focused on the minimization of operating costs, which are usually expressed on a quantitative basis. However, when

y represents a quality related measure, performance has been traditionally evaluated through a categorical variable, whose values depend on whether the product is inside or outside the range of desired specifications. But it is recognized today that just being within any type of specification limits is not good enough, and the idea that any product is equally good inside a given range of values and equally bad outside it must be revised (Deming, 1986; Roy, 1990). This points to the need for assuming continuous performance metrics even when y is quality related. One of the most powerful contributions of Taguchi (1986) to quality management was the proposition of loss functions as ways of quantifying and penalizing on a continuous basis any deviations from a desired nominal target (Phadke, 1989; Clausing, 1993). Given a quality functional characteristic z, with a nominal target z^*, any deviation from z^* has some quality cost associated with it, and this cost increases gradually as we move away from the target (Fig. 6). To operationalize and quantify this quality degradation process, Taguchi loss functions express quality costs on a monetary basis, commensurate with operating costs, and define the particular quality cost associated with a generic z value, $L(z)$, as

$$L(z) = k(z - z^*)^2, \tag{19}$$

where k is a constant known as quality loss coefficient. The value of k is

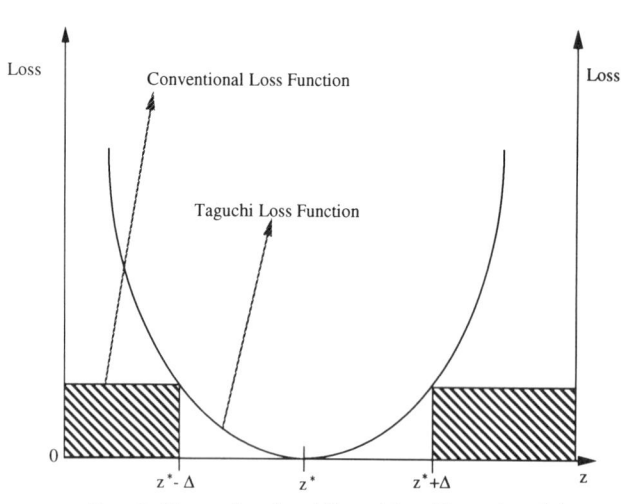

FIG. 6. Conventional and Taguchi quality cost models.

usually found by assigning a loss value to the specification limits, established at $z^* \pm \Delta$:

$$k = L(z^* \pm \Delta)/\Delta^2, \qquad (20)$$

$$L(z) = k(z - z^*)^2 = L(z^* \pm \Delta) \cdot (z - z^*)^2/\Delta^2. \qquad (21)$$

It is also important to realize that Taguchi loss functions not only bring into consideration both issues of location and dispersion of z but also provide a consistent format for combining them. By taking expectations on both sides of Eq. (19), and after a few algebraic rearrangements, we can show that the expected loss, $E[L(z)]$ is

$$E[L(z)] = k\left[\sigma_z^2 + (\mu_z - z^*)^2\right], \qquad (22)$$

where σ_z^2 is the z variance and μ_z is the z expectation.

For a zone **X** in the decision space to lead to a small conditional expected loss, $E[L(z)|\mathbf{X}]$, it must achieve both precise (reduced σ_z) and accurate ($\mu_z \cong z^*$) performance with respect to z. Finding such robust zones and operating on them results in inoculating the process against the transmission of variation from disturbances and the decision space to the performance space (Taylor, 1991).

If, besides the quality-related measure, z, one also wishes to include operating costs, ζ, in the analysis, because quality loss functions express quality costs on a monetary basis, commensurate with operating costs, the final global performance metric, y, which reflects total manufacturing cost, is simply the sum of both quality and operating costs (Clausing, 1993),

$$y = L(z) + \zeta. \qquad (23)$$

Consequently, the goal of our learning methodology is the identification of hyperrectangles in the decision space, **X**, that minimize expected total manufacturing cost, $E(y|\mathbf{X})$, a performance measure that combines in a consistent form and a quantitative basis both operating and quality costs.

D. LEARNING METHODOLOGY AND SEARCH PROCEDURE, S

In the previous paragraphs we defined the solution format ξ, performance criterion ψ, mapping procedure f, and performance metric y that characterize our learning methodology for systems with a quantitative metric y. Here we will assemble all these pieces together and briefly discuss the search procedure, S (further details can be found in Saraiva and Stephanopoulos, 1992c), that is employed to identify a set of final solutions, $\mathbf{X}^*$, which achieve low $\psi^{\text{est}}(\mathbf{X})$ scores. The two main stages of the learning procedure are examined in the following paragraphs.

1. Problem Formulation

The total loss function, y, given by Eq. (23), is not directly measured and has to be computed from information that is available and collected from the process, consisting of $(\mathbf{x}, z)$ pairs. After defining an adequate loss function, $L(z)$, and considering operating costs, ζ, one can identify the $(\mathbf{x}, y)$ pairs that correspond to each of the initial $(\mathbf{x}, z)$ data records.

Before starting the search for solutions, it is necessary to select among the M decision variables a subset of H variables, x_h, $h = 1, \ldots, H$, which influence significantly the system performance, and thus will be used by S and included in the definition of the final set of hyperrectangles, $\mathbf{X}^*$. For this preliminary choice of critical decision variables, other than his or her own specific process knowledge, the decisionmaker can count on a number of auxiliary techniques enumerated in Saraiva and Stephanopoulos (1992c).

This first stage of the learning procedure is concluded with the definition of a set of constraints related to

(a) Minimum acceptable width of operating windows $[w(X_h) \geq \Delta x_h, h = 1, \ldots, H]$
(b) Minimum number of data records, $N_{\min}$, that must be covered by any final solution, $\mathbf{X}^*[n(\mathbf{X}) \geq N_{\min}$ for $\mathbf{X}$ to be considered a feasible solution]. This constraint is identical to specifying a given maximum acceptable width for the $E(y|\mathbf{X})$ confidence interval that is obtained at a significance level, α, as defined by Eq. (15).

2. Search Procedure

Rather than finding the exact location of the single feasible hyperrectangle that optimizes $\psi^{\text{est}}(\mathbf{X})$, our primary goal is to conduct an exploratory analysis of the decision space, leading to the definition of a set of particularly promising solutions, $\mathbf{X}^*$, to be presented to the decisionmaker.

To identify this set of final feasible solutions, $\mathbf{X}^* \in \mathbf{I}^H$, with low $\psi^{\text{est}}(\mathbf{X})$ scores, we developed a greedy search procedure, S (Saraiva and Stephanopoulos, 1992c), that has resulted, within an acceptable computation time, in almost-optimal solutions for all the cases studied so far, while avoiding the combinatorial explosion with the number of $(\mathbf{x}, y)$ pairs of an exhaustive enumeration/evaluation of all feasible alternatives. The algorithm starts by partitioning the decision space into a number of isovolumetric and contiguous hyperrectangular seed cells, where for each cell the width associated with variable x_h is smaller than the corresponding Δx_h. Then, we gradually enlarge these seed cells, until they satisfy all imposed constraints, and further growth is found to degrade their estimated performance. Each initial cell is thus converted into a feasible solution candidate, **X**, and the corresponding $\psi^{est}(\mathbf{X})$ score is evaluated. Those feasible solution candidates receiving the lowest $\psi^{est}(\mathbf{X})$ are included in the set of final solutions, **X***, that is presented to the decisionmaker. It is the user's responsibility to analyze this set, make a selection among its elements, and thus choose the ultimate target zone to conduct the operation of the process.

E. CASE STUDY: PULP DIGESTER

In order to verify how close to a known true optimum the final solutions found by our learning methodology happen to be, we will describe here its application to a pulp digester, for which a perfect empirical model $f(\mathbf{x})$ is assumed to be available. Other applications are discussed in Saraiva and Stephanopoulos (1992c).

The original data format consists of (x_1, x_2, z, ω) records, where

x_1 stands for the *H-factor*.
x_2 is the *alkali charge*.
z is the pulp *kappa index*, with nominal target set at 30.0.
ω is the cooking yield, an indirect measure of operating cost that one wishes to maximize.

After defining the z loss function as

$$L(z) = 10(z - 30)^2 \quad \text{\$/ton of pulp} \qquad (24)$$

(where $ = U.S. dollars), and combining it with a commensurate measure of operating cost, expressed as a very simple function of ω

$$\zeta = 100 - \omega \quad \text{\$/ton of pulp}, \qquad (25)$$

one finally arrives at the identification of our total manufacturing cost performance metric,

$$y = 10(z - 30)^2 + 100 - \omega \quad \text{\$/ton of pulp}, \qquad (26)$$

which leads to the conversion of the original data records into the usual (**x**, y) format.

Let's consider that under the current operating conditions the values of **x** fall within a rectangle $\mathbf{X}_{current} = \{x_1 \in [200; 4000] \wedge x_2 \in [10; 20]\}$. Furthermore, we will assume that the two decision variables (x_1 and x_2) are independent and have uniform probability distributions. Using the available model, $f(\mathbf{x})$, we computed the current average total manufacturing cost, $E(y|\mathbf{X}_{current}) = 743.5$, a reference value that can be used to estimate the savings achieved with the implementation of any uncovered final solutions, $\mathbf{X}^*$.

To support the application of the learning methodology, $f(\mathbf{x})$ was used to generate 500 ($\mathbf{x}, z, \omega$) records of simulated operational data, transformed by Eq. (26) into an equivalent number of ($\mathbf{x}, y$) pairs. Finally, the following constraints were imposed to the search procedure, S:

(a) $w(X_1) \geq \Delta x_1 = 300$
    $w(X_2) \geq \Delta x_2 = 1.0$
(b) $N_{min} = 15$

Given the preceding problem definition, and after going through S, the final solution, $\mathbf{X}^*$, chosen for implementation is (Fig. 7):

$$\mathbf{X}^* = \{x_1 \in [910.2; 1566.3] \wedge x_2 \in [12.8; 14.6]\} \qquad E(y|\mathbf{X}^*) = 69.9. \tag{27}$$

Thus, $\mathbf{X}^*$ indeed leads to a quite significant average total cost reduction, because $E(y|\mathbf{X}^*) \ll E(y|\mathbf{X}_{current})$. Both $\mathbf{X}^*$ and $E(y|\mathbf{X}^*)$ are also close approximations (Fig. 7) to the true optimal solution given by Eq. (18), i.e., $\mathbf{X}_{opt}$ and $E(y|\mathbf{X}_{opt})$ are

$$\mathbf{X}_{opt} = \{x_1 \in [1114; 1414] \wedge x_2 \in [13.1; 14.1]\} \qquad E(y|\mathbf{X}_{opt}) = 52.2. \tag{28}$$

To benchmark our learning methodology with alternative conventional approaches, we used the same 500 ($\mathbf{x}, y$) data records and followed the usual regression analysis steps (including stepwise variable selection, examination of residuals, and variable transformations) to find an approximate empirical model, $f^{est}(\mathbf{x})$, with a coefficient of determination $R^2 = 0.79$. This model is given by

$$y \approx f^{est}(\mathbf{x}) = a \ln(x_1) + b \ln(x_2) + cx_2 + dx_1^2 + ex_2^2 + gx_1 \cdot x_2, \tag{29}$$

whose parameters were fitted by ordinary least squares.

By employing $f^{est}(\mathbf{x})$ in Eq. (17), we used this approximate model to find a final solution, $\mathbf{X}_{est}$ (Fig. 7), that satisfies the (a) constraints and

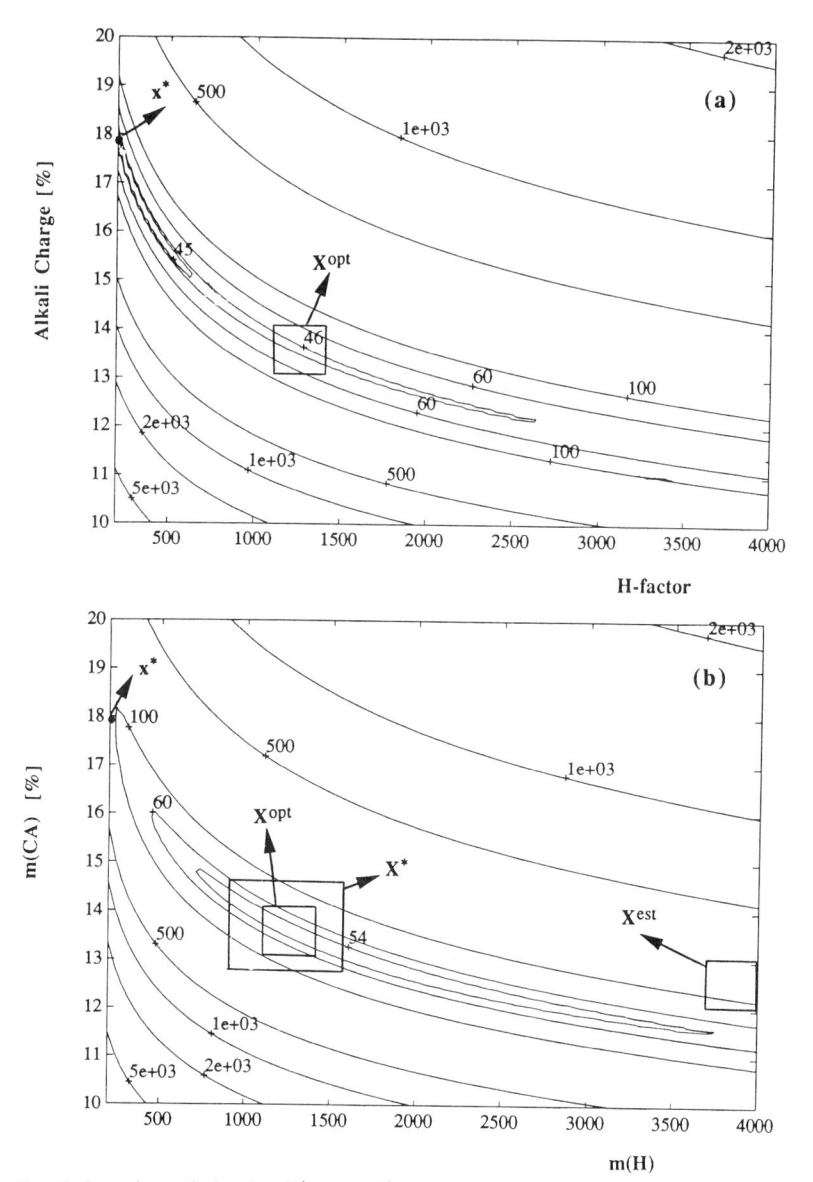

FIG. 7. Locations of $\mathbf{x}^*$, $\mathbf{X}^*$, $\mathbf{X}^{opt}$, and $\mathbf{X}^{est}$ in the decision space, and contour plots of (a) $y = f(\mathbf{x})$ versus $\mathbf{X}$, (b) $E(y|\mathbf{X})$ versus $\mathbf{m}(\mathbf{X})$.

minimizes $E[f^{\text{est}}(\mathbf{x})|\mathbf{X}]$:

$$\mathbf{X}_{\text{est}} = \{x_1 \in [3698; 3998] \land x_2 \in [12.1; 13.1]\} \qquad E(y|\mathbf{X}_{\text{est}}) = 145.6. \tag{30}$$

This solution leads to a considerably worse performance, $E(y|\mathbf{X})$, than $\mathbf{X}^*$, and it is also much more distant from the zone of the decision space where the true optimum, $\mathbf{X}_{\text{opt}}$, is located.

Results obtained with other $f^{\text{est}}(\mathbf{x})$ functional forms, for this and other similar problems, seem to indicate that even when only moderate amounts of data are available our direct sampling estimation procedure, f, and search algorithm, S, provide better final solutions than classical regression analysis followed by the use of Eq. (17) unless one is able to build from the data almost perfect $f^{\text{est}}(\mathbf{x})$ empirical models with the appropriate functional forms.

VI. Systems with Multiple Operational Objectives

Except for the combination of quality loss and operating costs given by Eq. (23), in the previous sections we assumed the system performance to be determined by a single objective. However, in the analysis of pieces of equipment or plant segments of nontrivial size/complexity, a multitude of objectives has usually to be taken into account in order to evaluate the system's global performance, and find ways to improve it.

In this section we describe extensions of the basic learning methodologies introduced in Sections IV and V that, while preserving the same premises and paradigms, enlarge considerably their scope by adding the capability to consider simultaneously multiple objectives. As before, and without loss of generality, we will focus our attention on the coexistence of several quality-related objectives.

Both situations with categorical and continuous, real-valued performance metrics will be considered and analyzed. Since Taguchi loss functions provide quality cost models that allow the different objectives to be expressed on a commensurate basis, for continuous performance variables only minor modifications in the problem definition of the approach presented in Section V are needed. On the other hand, if categorical variables are chosen to characterize the system's multiple performance metrics, important modifications and additional components have to be incorporated into the basic learning methodology described in Section IV.

A case study on the operational improvement of a plasma etching unit in microelectronics fabrication ends the section. This case study illustrates that if similar preference structures are used in both types of formulation, identical final solutions are found when either categorical or continuous performance evaluation modes are employed.

A. Continuous Performance Variables

Instead of a single quality-related performance variable, z, as in Section V, let's suppose that one has to consider a total of P distinct objectives and the corresponding continuous performance variables, z_i, $i = 1, \ldots, P$, which are components of a performance vector $\mathbf{z} = [z_1, \ldots, z_P]^T$. In such case, one has to identify the corresponding Taguchi loss functions, $L(z_i)$, $i = 1, \ldots, P$, for each of the performance variables:

$$L(z_i) = k_i(z_i - z_i^*)^2. \tag{31}$$

Since these loss functions express quality costs on a common and commensurate basis, extending the learning methodology of Section V to a situation with P objectives is straightforward. All one has to do is replace the original definition of the y performance metric [Eq. (23)] by the following more general version:

$$y = \sum_{i=1}^{P} k_i(z_i - z_i^*)^2 + \zeta. \tag{32}$$

Except for this modification, all the procedures and steps discussed in Section V carry over to the solution of multiobjective problems.

B. Categorical Performance Variables

Rather than a single objective, y, as in Section IV, we now have a total of P distinct categorical performance variables, y_i, $i = 1, \ldots, P$, associated with an equivalent number of objectives. Consequently, each data record is now composed of a $(\mathbf{x}, \mathbf{y})$ pair, where $\mathbf{y}$ is a performance vector defined by

$$\mathbf{y} = [y_1, \ldots, y_i, \ldots, y_P]^T. \tag{33}$$

The most important changes and adaptations that were introduced in order to handle such multiobjective problems are summarized in Table III. The solution space remains the same as for the single objective case.

TABLE III
SINGLE AND MULTIPLE CATEGORICAL PERFORMANCE VARIABLES

	Single objective	Multiple objectives
ξ	$\mathbf{X} \in \mathbf{I}^M$	$\mathbf{X} \in \mathbf{I}^M$
ψ	$y(\mathbf{X})$ or $p(y=j\|\mathbf{X})$	$y(\mathbf{X})$ or $p(y_i=j\|\mathbf{X})$
f	$\dfrac{n_j(\mathbf{X})}{n(\mathbf{X})}$	$\dfrac{n_{i,j}(\mathbf{X})}{n(\mathbf{X})}$
S	Induction of decision trees	Multiple agents and lexicographic search

However, since now we have P different objectives, the performance criteria, ψ, must include all of them. They may assume one of the following formats:

1. $\mathbf{y}(\mathbf{X}) = [y_1(\mathbf{X}), \ldots, y_i(\mathbf{X}), \ldots, y_P(\mathbf{X})]^T$,
   where $y_i(\mathbf{X})$ stands for the most likely y_i value within the zone of the decision space defined by $\mathbf{X}$.
2. $p(y_i = j|\mathbf{X})$, $i = 1, \ldots, P$ and $j = 1, \ldots, K_i$,
   the set of conditional probabilities for getting any given value for each of the y_i variables, with K_i representing the total number of different possible values that y_i can assume.

The mapping procedure, f, is identical to the one adopted for a single objective:

$$p^{\text{est}}(y_i = j | \mathbf{X}) = \frac{n_{i,j}(\mathbf{X})}{n(\mathbf{X})}, \qquad j = 1, \ldots, K_i, \tag{34}$$

where $n_{i,j}(\mathbf{X})$ is the total number of $(\mathbf{x}, \mathbf{y})$ pairs for which $\mathbf{x} \in \mathbf{X}$, and $y_i = j$.

The search procedure, S, requires major modifications in order to account for multiple objectives. As we will see, S becomes now a highly interactive process, with progressive articulation/elicitation of the user's preference structure, successive relaxation of aspiration levels and lexicographic construction of final solutions, $\mathbf{X}^*$, that lead to satisfactory joint performances according to all the P objectives. It includes P replicates of the basic learning methodology introduced in Section IV for a single objective, designated as *agents*, and a *coordination mechanism* that combines the results found by individual agents, in an attempt to identify conjunctions of zones uncovered by the agents that lead to the formation of the desired final solutions, $\mathbf{X}^*$. In the following paragraphs we will summarize the main steps of this search procedure (for details, see Saraiva and Stephanopoulos, 1992b).

1. Problem Definition

Besides the identification of the decision variables, x_m, $m = 1, \ldots, M$, and the performance variables, y_i, the user is asked to

(a) Rank the P objectives in order of decreasing relative importance.
(b) Provide constraints on minimum acceptable widths $[w(X_m) \geq \Delta x_m,$ $m = 1, \ldots, M]$, and coverage, $N_{\min}$.
(c) Identify minimum acceptability criteria or constraints in the performance space (e.g., no more than 10% of "very low" y_3 values will be tolerated) that must be satisfied irrespectively of how good or bad the corresponding performances for the other objectives may be.

2. Identification and Refinement of Initial Aspiration Levels

From past experience and an examination of the results provided by each of the P agents, a first tentative group of aspiration levels, $\mathbf{y}^*$, that one should aim to reach, is defined by the user ($y_1^* =$ "excellent$_1$," $y_2^* =$ "good-or-excellent$_2$," etc.):

$$\mathbf{y}^* = [y_1^*, \ldots, y_i^*, \ldots, y_P^*]^T. \tag{35}$$

These aspiration levels may be expressed either on an absolute ($y_1^* =$ "excellent$_1$") or on a probabilistic basis $[y_1^* \equiv p(y_1 =$ "excellent$_1$") ≥ 0.90].

Before beginning the search for feasible zones of the decision space where the preceding tentative aspiration levels can be achieved, a preliminary check for the possibility of existence of such a zone is conducted. If the perceived ideal, $\mathbf{y}^*$, does not pass this preliminary check, i.e., there is no commensurable solution to the multiobjective problem, the decisionmaker is asked to relax $\mathbf{y}^*$, in order to transform the problem into one with commensurable solutions. For instance, if the initial tentative perceived ideal were $\mathbf{y}^* = ($"excellent$_1$," "excellent$_2$," "excellent$_3$"), but there aren't any available $(\mathbf{x}, \mathbf{y})$ pairs with $\mathbf{y} = \mathbf{y}^*$, the user might adopt as a revised combination of aspiration levels $\mathbf{y}^* = ($"excellent$_1$," "excellent$_2$," "good-or-excellent$_3$"). This relaxation process, guided by the decisionmaker, continues until a final perceived ideal, $\mathbf{y}^*$, for which there are at least $N_{\min}(\mathbf{x}, \mathbf{y})$ pairs with $\mathbf{y} = \mathbf{y}^*$, can be identified. This is the set of aspiration levels that will be used to initiate the interactive search procedure for final solutions, $\mathbf{X}^*$.

3. Search for Solutions

The aspiration levels inherited from the previous step, $\mathbf{y}^*$, are used to guide the search process. Each agent i employs the corresponding aspiration level, y_i^*, and through the application of the learning methodology presented in Section IV tries to identify feasible hyperrectangles, $\mathbf{X}_i^*$, that lead to performance consistent with y_i^*.

Then, we try to combine the partial solutions uncovered by the several agents, $\mathbf{X}_i^*$, in order to find feasible final solutions, $\mathbf{X}^*$, that lead to joint satisfactory behavior in terms of all the objectives, and thus consistent with the current $\mathbf{y}^*$. This is achieved by building conjunctions of multiple $\mathbf{X}_i^*$, uncovered by different agents, according to a breadth-first type of search (Winston, 1984), that takes into account the relative importance assigned to the objectives. Following this lexicographic approach, final solutions $\mathbf{X}^*$ are gradually constructed, and partial paths expanded to accomplish less important goals only when the aspiration levels for the most important ones have already been satisfied. The construction process (whose details are given in Saraiva, 1993; Saraiva and Stephanopoulos, 1992b) relies heavily on the interaction with the decisionmaker to overcome dead-ends, examine arising conflicts, establish tradeoffs, and guide the procedure.

4. Validation of Results

After the search has been concluded, all the uncovered feasible final solutions, $\mathbf{X}^*$, leading to satisfactory joint performances, consistent with $\mathbf{y}^*$, are presented to the decisionmaker for close examination and for the selection of a particular hyperrectangle within this group for eventual implementation.

However, conflicts between the fulfillment of different objectives and aspiration levels may prevent any feasible zone of the decision space from leading to satisfactory joint performances. If the search procedure fails to uncover at least one feasible final solution, $\mathbf{X}^*$, consistent with $\mathbf{y}^*$, a number of options are available to the decisionmaker to try to overcome this impasse. Namely, the decisionmaker can revise the initial problem definition, by either

(a) Redefining any of the constraints originally imposed.
(b) Introducing further relaxations of aspiration levels, which result in new and less demanding perceived ideals, $\mathbf{y}^*$.
(c) Excluding from the search space of agents the particular decision variable, x_m, which creates the conflict among objectives.

Such revisions to the problem statement in order to overcome unsuccessful applications of the search procedure may have to be repeated a number of times before the problem possesses commensurable solutions, and one can find at least one final feasible solution, $\mathbf{X}^*$, that satisfies all the imposed constraints and achieves performances consistent with the current set of aspiration levels, $\mathbf{y}^*$.

C. Case Study: Operational Analysis of a Plasma Etching Unit

To conclude this section on systems with multiple objectives, we will consider a specific plasma etching unit case study. This unit will be analyzed considering both categorical and continuous performance measurement variables. Provided that similar preference structures are expressed in both instances, we will see that the two approaches lead to similar final answers. Additional applications of the learning methodologies to multiobjective systems can be found in Saraiva and Stephanopoulos (1992b, c).

1. System Characterization

This case study is based on real industrial data collected from a plasma etching plant, as presented and discussed in Reece *et al.* (1989). The task of the unit is to remove the top layer from wafers, while preserving the bottom one. Four different objectives and performance variables are considered:

(a) Maximize a measure of etching selectivity, z_1, expressed as the ratio of etching rates for the top and bottom layers.
(b) Minimize dispersion, z_2, of etching rate values across the wafer bottom-layer surface.
(c) Minimize dispersion, z_3, of etching rate values across the wafer, but now for the top layer.
(d) Maximize the average etching rate for the top layer, z_4.

A quantization of the z_i variables resulted in the definition of the following categorical performance variables, y_i:

Etching selectivity:

$$y_1 = \begin{cases} \text{``bad}_1\text{''} & \text{if} \quad z_1 \leq 3.4, \\ \text{``good}_1\text{''} & \text{if} \quad 3.4 < z_1 \leq 4.0, \\ \text{``excellent}_1\text{''} & \text{if} \quad 4.0 < z_1. \end{cases}$$

Bottom-layer etching dispersion:

$$y_2 = \begin{cases} \text{"bad}_2\text{"} & \text{if} \quad 7.7 < z_2, \\ \text{"good}_2\text{"} & \text{if} \quad 7.0 < z_2 \leq 7.7, \\ \text{"excellent}_2\text{"} & \text{if} \quad z_2 \leq 7.0. \end{cases}$$

Top-layer etching dispersion:

$$y_3 = \begin{cases} \text{"bad}_3\text{"} & \text{if} \quad 4.6 < z_3, \\ \text{"good}_3\text{"} & \text{if} \quad 3.0 < z_3 \leq 4.6, \\ \text{"excellent}_3\text{"} & \text{if} \quad z_3 \leq 3.0. \end{cases}$$

Average top-layer etching rate:

$$y_4 = \begin{cases} \text{"bad}_4\text{"} & \text{if} \quad z_4 \leq 1870, \\ \text{"good}_4\text{"} & \text{if} \quad 1870 < z_4 \leq 2000, \\ \text{"excellent}_4\text{"} & \text{if} \quad 2000 < z_4. \end{cases}$$

Similarly, for the case where a continuous performance metric, y, is employed, the following loss functions were defined (Saraiva and Stephanopoulos, 1992c):

(a) $L(z_1) = 1.924(4.321 - z_1)^2$;
(b) $L(z_2) = 0.033(z_2 - 4.50)^2$;
(c) $L(z_3) = 0.022(z_3 - 0.31)^2$;
(d) $L(z_4) = 5.1387 \cdot 10^{-7}(2595.0 - z_4)^2$;

and $y = \sum_{i=1}^{4} L(z_i)$.

The three decision variables, and the corresponding ranges of values in $\mathbf{X}_{\text{current}}$, are

x_1: power at which the unit is operated, with values ranging between 75 and 150 W (watts)

x_2: pressure in the apparatus, ranging from 200 to 255 mtorr (millitorr)

x_3: flow of etchant gas, varying between 20 and 40 sccm (cubic centimeters per minute of gas flow at standard temperature and pressure conditions)

Since only 20 data records were collected from the system during the execution of the designed experiments conducted by Reece et al. (1989), we used their response surface models, deliberately contaminated with small Gaussian noise terms, to generate a total of 500 $(\mathbf{x}, \mathbf{z})$ pairs (assuming that the three variables, x_1, x_2, x_3, have independent and uniform probability distributions). Finally, the following constraints were considered:

(a) As minimum acceptable window sizes, values close to 10% of the $\mathbf{X}_{current}$ ranges are used, leading to $\Delta x_1 = 7.5$, $\Delta x_2 = 5.5$, and $\Delta x_3 = 2.05$.

(b) As minimum coverage, we set $N_{min} = 5$.

2. Categorical Performance Variables

The initial tentative perceived ideal, $\mathbf{y}^*$, was set at

$$\mathbf{y}^* = [\text{"excellent}_i\text{"}]^T, \quad i = 1, \ldots, 4.$$

After successive interactive relaxations, all of them leading to an insufficient number of $(\mathbf{x}, \mathbf{y})$ pairs that jointly satisfy the aspiration levels, we finally came down to the following revision of the perceived ideal:

$$\mathbf{y}^* = [\text{"good-or-excellent}_i\text{"}]^T, \quad i = 1, \ldots, 4.$$

After going through the complete search procedure, given the perceived ideal shown above, the following final solution, $\mathbf{X}_1^*$, was selected:

$$\mathbf{X}_1^* = \{x_1 \in [134.6, 149.1] \wedge x_2 \in [235.1, 243.2] \wedge x_3 \in [20.9, 25.4]\}. \tag{36}$$

A projection of this solution into the (x_1-x_2) plane is shown in Fig. 8, together with the available $(\mathbf{x}, \mathbf{y})$ pairs that verify the condition imposed over x_3 values. One can qualitatively confirm the validity of condition (36): $\mathbf{X}_1^*$ is indeed a good approximation of the zones in the x_1-x_2 plane that by visual inspection one would associate with leading simultaneously to "good-or-excellent$_i$" performances for all the four objectives.

3. Continuous Performance Variables

The solution found when the plasma etching was analyzed in terms of continuous performance metrics is also presented in Fig. 8, and is given by

$$\mathbf{X}_2^* = \{x_1 \in [129.79; 140.39] \wedge x_2 \in [230.17; 243.22]$$

$$\wedge x_3 \in [20.22; 23.18]\}. \tag{37}$$

This solution is similar to the one found [see hyperrectangle defined by Eq. (36)] previously, when categorical performance evaluation variables were employed. Since the preference structures expressed under both

——— Solution Uncovered with Categorical y_i
- - - - Solution Uncovered with Continuous y

Objective 1: ● Bad$_1$; ☐ Good-or-Excellent$_1$

Objective 2: ● Good-or-Excellent$_2$; ☐ Bad$_2$

Objective 3: ● Good-or-Excellent$_3$; ☐ Bad$_3$

Objective 4: ● Bad$_4$; ☐ Good-or-Excellent$_4$

FIG. 8. Data records and final solutions.

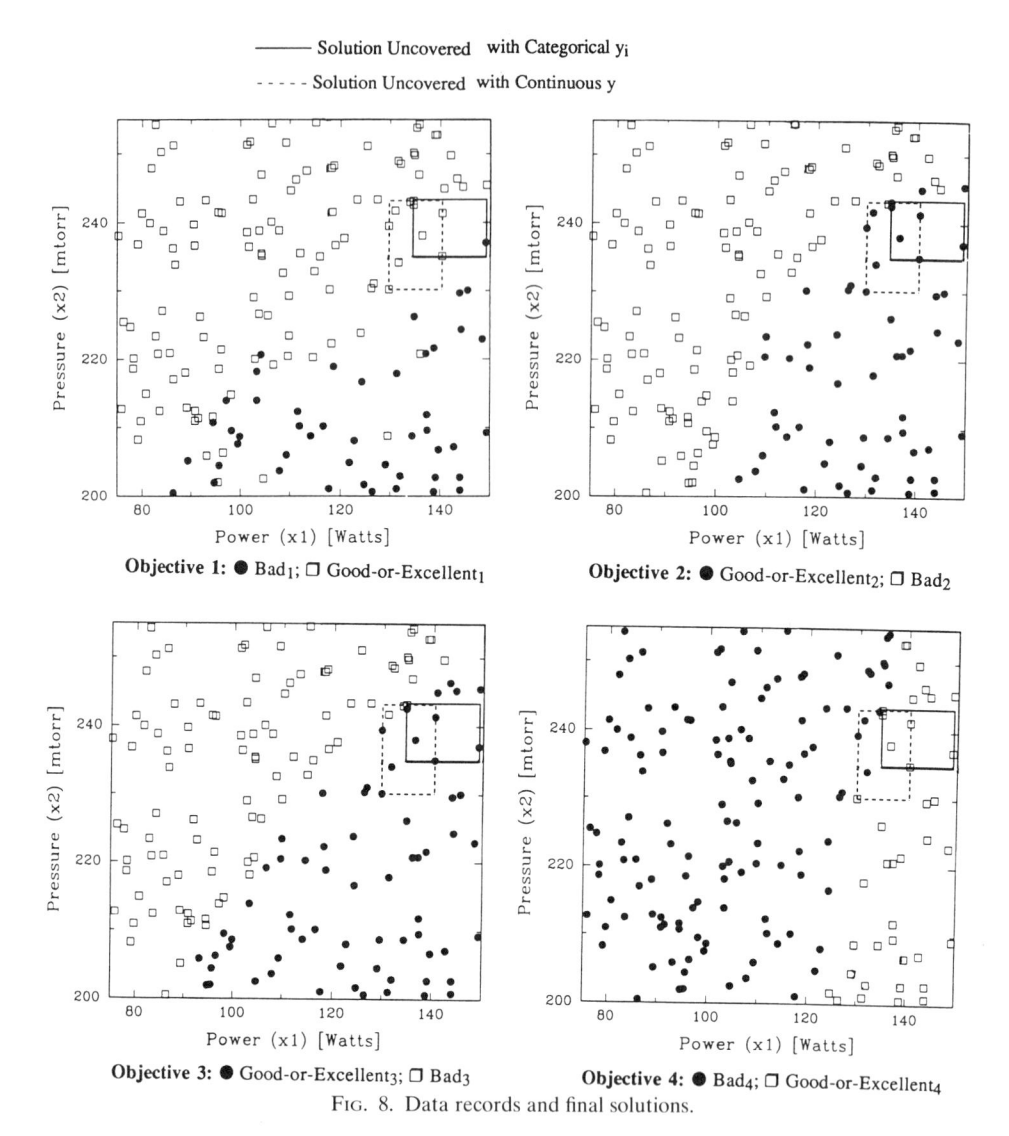

formats were chosen to be consistent with each other, this is what one should expect and desire to happen.

Additional studies documented in Saraiva and Stephanopoulos (1992b,c) also illustrate how the introduction of changes in preference structures is translated into displacements of the final uncovered solutions in the decision space.

VII. Complex Systems with Internal Structure

In previous sections we covered a variety of possible applications, including single or multiple objectives, continuous or categorical performance variables. However, so far it has always been assumed that the systems studied are simple systems without any type of internal structure, isolated from the remaining world, and self-sufficient for decisionmaking purposes. In this section we discuss additional extensions of the basic learning methodologies, in order to address complex systems composed of a number of interconnected subsystems. Although situations with categorical performance variables can be treated in similar ways, requiring only minor changes and adaptations, we will consider here only continuous performance metrics.

First, we discuss the problem statements and key features of the learning architecture that are specific to complex systems. This is followed by a brief presentation of the search procedures that are used to build a final solution. The section ends with a summary of the application of the learning architecture to the analysis of a Kraft pulp mill.

A more detailed description of this section's contents can be found in Saraiva (1993) or Saraiva and Stephanopoulos (1992d).

A. Problem Statement and Key Features

Complex manufacturing systems, such as an unbleached Kraft pulp plant (Fig. 9), are almost always characterized by some type of internal structure, composed of a number of interconnected subsystems with their own data collection and decisionmaking responsibilities. This raises a number of additional issues, not addressed in previous sections. For instance, if the learning methodology described in Section VI is applied to the digester module of a pulp plant (Fig. 9), it is possible for the final selected solution, $X_{digester}$, to include ranges of desired values of sulfidity or other composition properties of the white liquor that enters the digester. This being the case, and since an adjustment of the liquor composition has to be achieved elsewhere in the plant, the request over white liquor sulfidity values has to be propagated backward, to the causticizing area, and eventually from this module to the one preceding it, before a final solution can be found.

In the development of a learning architecture able to extend the methodologies introduced in other sections to complex systems, we looked

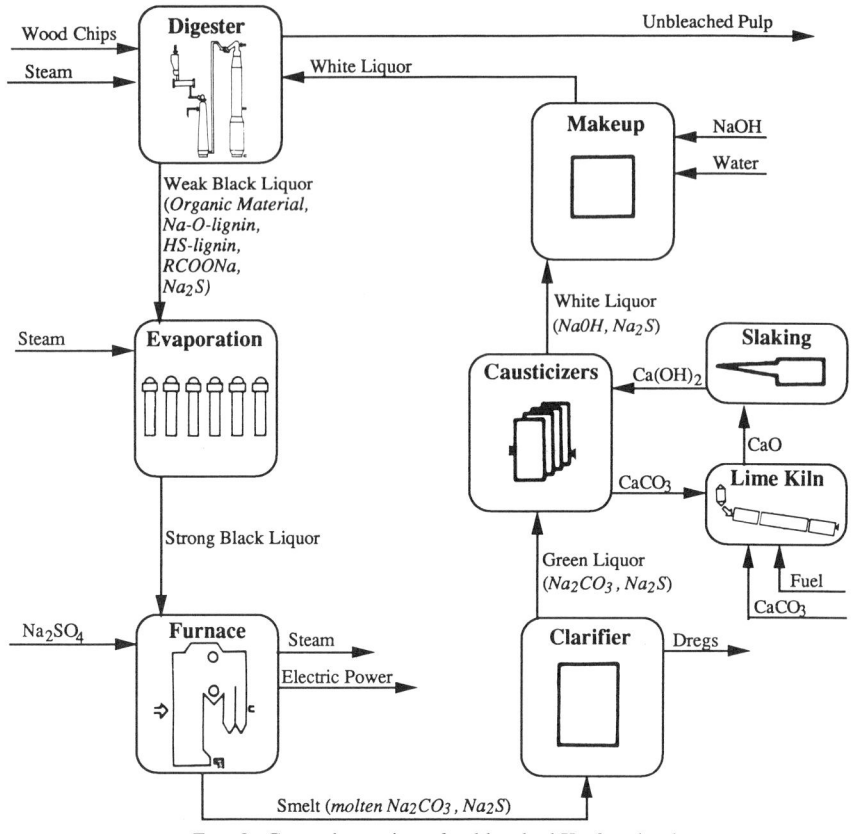

FIG. 9. General overview of unbleached Kraft pulp plant.

explicitly (for reasons presented and justified in Saraiva and Stephanopoulos, 1992c) for approaches that

(a) Are modular and decentralized, allowing each subsystem to take the initiative or make its own decisions, and assigning coordination roles to the upper hierarchical level.
(b) Support and reflect the existing organizational decisionmaking structures, responsibilities, data collection, and analysis activities.

Most of the existing tools to improve process operations fail to provide a systematic and formal process of handling complex systems, or do so in ways that do not fulfill the preceding set of requests. In this paragraph we provide a more formal characterization of a complex system and its several

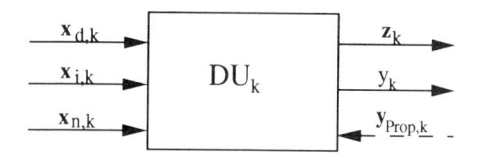

FIG. 10. Schematic representation of a decision unit.

subsystems, which will then be used to introduce our problem statement, and compare it with analogous conventional formulations.

1. Complex Systems as Networks of Interconnected Subsystems

The basic building block in the definition of a complex system, as well as the key element in our learning architecture, is what we will designate as an *infimal decision unit* or subsystem (Mesarović *et al.*, 1970; Findeisen *et al.*, 1980), DU_k (Fig. 10). These decision units will in general correspond to a particular piece of equipment or section of the plant. The overall system is represented by a single *supremal decision unit* (Mesarović *et al.*, 1970; Findeisen *et al.*, 1980), DU_0, and contains a total of K interconnected infimal decision units (Fig. 11), DU_k, $k = 1, \ldots, K$.

For each of the infimal decision units, DU_k, one has to consider several groups of input and output variables (Fig. 10). Among the inputs are

(a) A vector of decision variables, $x_{d,k}$, which are variables that fall under the scope of authority and can be directly manipulated by DU_k.
(b) A vector of connection variables, $x_{i,k}$, containing those variables that link consecutive infimal decision units, because they are simultaneously inputs to DU_k (although not under its control) and outputs from the preceding decision unit, DU_{k+1}.

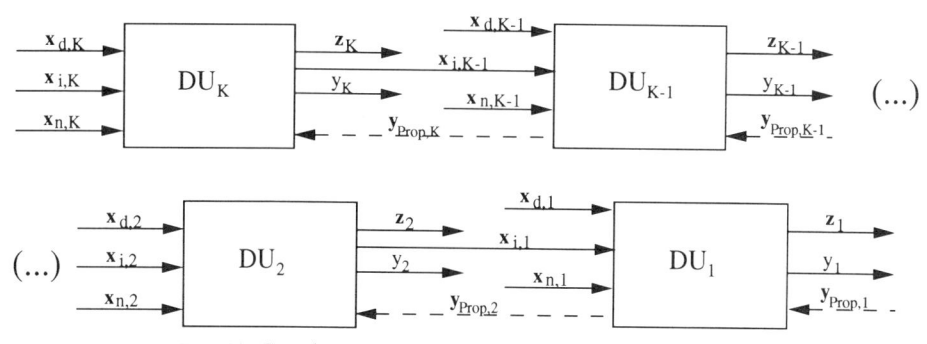

FIG. 11. Complex system as a sequence of infimal decision units.

(c) A vector of disturbance factors, $x_{n,k}$, including variables that reside within the boundaries of DU_k, but nonetheless over which neither DU_k nor any of the other decision units have any control.

As for possible outputs, one has

(a) A vector of performance variables, $z_k = [z_{1,k}, \ldots, z_{p(k),k}]$ [*note:* $p(k)$ stands for the total number of subsystem k objectives], that reflect operating costs, quality related measures and/or the violation of existing constraints on either the input or output spaces.
(b) A scalar global measure of performance for DU_k, y_k, which depends on z_k and results from the combination on a common basis of operating, quality, and constraint violation costs. This measure is global in the sense that it aggregates together all the operational objectives of the decision unit, but local in the sense that it is limited in scope to DU_k goals.
(c) A vector of temporary propagation loss functions, $y_{\text{Prop},k}$, used to transmit the requests over the values of connection variables from one decision unit to the ones that precede it, during the backpropagation process that takes place in the construction of a final solution (for details about the procedure adopted to define propagation loss functions, see Saraiva, 1993; Saraiva and Stephanopoulos, 1992d). Once that solution has been found, these propagation loss functions cease to exist.

2. Conventional and Alternative Problem Definitions

In order to define and compare in a concise way the different problem formulations, let's designate as

n the total number of decision variables, distributed among the K subsystems, $x_{d,j}$, $j = 1, \ldots, n$.

$\mathbf{x}_{\text{DP}}$ a generic pointwise decision policy, i.e., a vector whose components are values of decision variables,

$$\mathbf{x}_{\text{DP}} = [x_{d,1}, \ldots, x_{d,j}, \ldots, x_{d,n}].$$

$\mathbf{X}_{\text{DP}}$ a generic interval vector decision policy, whose components are ranges of decision variables

$$\mathbf{X}_{\text{DP}} = [X_{d,1}, \ldots, X_{d,n}].$$

TABLE IV

CONVENTIONAL APPROACHES AND SUGGESTED ALTERNATIVE

	Conventional approaches	Suggested alternative
ξ	$\mathbf{x}_{DP} \in \mathscr{R}^M$	$\mathbf{X}_{DP} \in \mathbf{I}^M$
ψ	$y_0(\mathbf{x}_{DP})$ or $E(y_0\|\mathbf{x}_{DP})$	$E(\mathbf{y}_k\|\mathbf{X}_{DP})$ and $\sigma(\mathbf{y}_k\|\mathbf{X}_{DP})$
f	Technique-dependent	Sample averages and standard deviations
S	Optimization	Top–down and bottom–up

$\mathbf{y}_k$ a performance vector whose components are all the unit k performance variables,

$$\mathbf{y}_k = [\mathbf{z}_k, y_k], \quad k = 0, 1, \ldots, K.$$

The main differences between conventional approaches and our learning architecture are summarized in Table IV, and are discussed below:

1. As final solution formats, interval vector decision policies, $\mathbf{X}_{DP}$, replace their pointwise counterparts, $\mathbf{x}_{DP}$. Thus, a decision policy, $\mathbf{X}_{DP}$, in the context of this section is an interval vector whose components are intervals of decision variables associated with one or more of the infimal decision units. No connection variables or disturbance factors are involved in their definition;

2. As performance criteria, ψ, the traditional centralized optimization approach considers an overall objective function, y_0, as its single evaluation measure, and the role assigned to the search procedure is the identification of a pointwise decision policy that minimizes $y_0(\mathbf{x}_{DP})$ or $E(y_0|\mathbf{x}_{DP})$. However, any single aggregate measure alone, such as y_0, can not provide a complete and meaningful evaluation of performance for a complex plant that includes several subsystems, a large variety of objectives and local performance measurement variables. Although such aggregate measures may be used to facilitate the search for a promising solution, they should not be interpreted as leading to the very best possible answer, neither should the solutions found be implemented without a detailed analysis of how they will affect all subsystems and the achievement of their own local objectives. Thus, in our approach each subsystem is assumed to have its own goals, and these are taken into account explicitly in the definition of the performance criteria, ψ, which include the conditional expectations and standard deviations of the several goal achievement measures associated with the multiple subsystems, as

well as the system as a whole:

$$E(\mathbf{y}_k|\mathbf{X}_{DP}) = \{E(z_{1,k}|\mathbf{X}_{DP}),\ldots,E(z_{p(k),k}|\mathbf{X}_{DP}),E(y_k|\mathbf{X}_{DP})\}^T,$$
$$k = 0,1,\ldots,K, \quad (38)$$

$$\sigma(\mathbf{y}_k|\mathbf{X}_{DP}) = \{\sigma(z_{1,k}|\mathbf{X}_{DP}),\ldots,\sigma(z_{p(k),k}|\mathbf{X}_{DP}),\sigma(y_k|\mathbf{X}_{DP})\}^T,$$
$$k = 0,1,\ldots,K. \quad (39)$$

3. Traditional techniques use $f(\mathbf{x}_{DP})$ quantitative models to perform the mapping from the solution to the performance space. These models essentially reduce the problem to a simple system, having the decision variables $x_{d,j}$ as inputs and y_0 as output, because they allow one to express the $y_0(\mathbf{x}_{DP})$ or $E(y_0|\mathbf{x}_{DP})$ estimates as a function of the decision variables, $\psi^{est} = f(\mathbf{x}_{DP})$. On the other hand, our mapping procedures, f, are based on the construction of direct sampling estimates. Given a decision policy, $\mathbf{X}_{DP}$, both $E(\mathbf{y}_k|\mathbf{X}_{DP})$ and $\sigma(\mathbf{y}_k|\mathbf{X}_{DP})$ are estimated by identifying the available data records for which $\mathbf{x} \in \mathbf{X}_{DP}$, and computing the corresponding sample averages and standard deviations for all the $z_{i,k}$ and y_k variables.

4. In conventional centralized approaches the goal of the search procedures is to find a final feasible decision policy, $\mathbf{x}_{DP}^*$, that optimizes $f(\mathbf{x}_{DP})$. This solution is then imposed to the several subsystems for implementation, although these were not directly involved in its construction. Other similar methodologies include problem decomposition strategies (Lasdon, 1970; Biegler, 1992) that explore particular properties of the complex system structure and dynamic programming (Bellman and Dreyfus, 1962; Roberts, 1964; Nemhauser, 1966). They share the same basic assumptions, problem statement and solution formats as centralized approaches, although, for the sake of computational efficiency gains, a multistage and sequential identification of the final result, $\mathbf{x}_{DP}^*$, is adopted as the problem solving strategy.

Our search procedures represent a departure from the above type of paradigm. Rather than simply accepting and implementing a decision policy found by DU_0, that optimizes an overall measure of performance, the infimal subsystems and corresponding plant personnel play an active role in the construction and validation of solutions. One tries to build a consensus decision policy, $\mathbf{X}_{DP}$, validated by all subsystems, DU_k, $k = 1,\ldots,K$, as well as by the whole plant, DU_0, and only when that consensus has been reached does one move toward implementation. Within this context, the upper-level decision unit, DU_0, assumes a coordination role, and does not have the power to impose solutions to the several subsystems. Two different search procedures, used to find such consensus decision policies, $\mathbf{X}_{DP}$, and designated respectively as *bottom–up* and *top–down*, are described in the following paragraphs.

3. Final Problem Statement

The ultimate goal of our learning architecture is to uncover at least one decision policy, $\mathbf{X}_{DP}$, that

(a) Is feasible, i.e., satisfies constraints imposed over the minimum acceptable size of operating windows $[w(X_{d,j}) \geq \Delta x_{d,j}, j = 1, \ldots, n]$ and coverage $[n(\mathbf{X}_{DP}) \geq N_{min}]$.
(b) Leads to a significant improvement over the current levels of performance for one or more of the decision units.
(c) Is accepted and validated by all decision units involved with or affected by the use of $\mathbf{X}_{DP}$ as the zone to conduct the operation.

A decision policy that satisfies all these requirements is designated as an active decision policy, $\mathbf{X}_{DP}^*$.

To declare a decision policy, $\mathbf{X}_{DP}$, as either unacceptable, acceptable or leading to a significant improvement, each decision unit compares its current levels of performance with the ones that are expected within $\mathbf{X}_{DP}$. The current levels of performance for unit k are provided by the $E(\mathbf{y}_k|\mathbf{X}_{current})$ and $\sigma(\mathbf{y}_k|\mathbf{X}_{current})$ estimates obtained from a sample of data records, just as in the case of $E(\mathbf{y}_k|\mathbf{X}_{DP})$ and $\sigma(\mathbf{y}_k|\mathbf{X}_{DP})$. A comparison of these reference performance values for $\mathbf{X}_{current}$ with the ones achieved by $\mathbf{X}_{DP}$ is made by each decision unit. As a result of this comparison, a final evaluation of $\mathbf{X}_{DP}$ by decision unit k leads to one of the three possible outcomes:

1. $f_k(\mathbf{X}_{DP}) \gg f_k(\mathbf{X}_{current})$, meaning that significant improvement is expected.
2. $f_k(\mathbf{X}_{DP}) \approx f_k(\mathbf{X}_{current})$, in case $\mathbf{X}_{DP}$ is accepted and validated by decision unit k, although no significant improvements are expected.
3. $f_k(\mathbf{X}_{DP}) \ll f_k(\mathbf{X}_{current})$, meaning that $\mathbf{X}_{DP}$ can not be accepted by unit k, because it would result in performance deterioration down to levels that fall below what the unit can tolerate for at least one of its performance variables.

Thus, an active decision policy, $\mathbf{X}_{DP}^*$, is a feasible decision policy such that

$$\exists_{k \in \{0,1,\ldots,K\}} \mid f_k(\mathbf{X}_{DP}) \gg f_k(\mathbf{X}_{current});$$

$$\forall_{k \in \{0,1,\ldots,K\}} \{f_k(\mathbf{X}_{DP}) \approx f_k(\mathbf{X}_{current}) \lor f_k(\mathbf{X}_{DP}) \gg f_k(\mathbf{X}_{current})\}.$$

B. Search Procedures

Two different search procedures (bottom–up and top–down) can be followed to build active decision policies, $\mathbf{X}_{DP}^*$.

In the bottom–up approach the initiative to start the learning process is taken by one of the infimal decision units. Since solutions found at this unit may include connection variables, the request for given values of these variables is propagated backward, to unit $k + 1$, through temporary loss functions. After successive backpropagation steps, the participation of several other DU_k and the operators associated with them, a final decision policy, accepted and validated by all infimal decision units, is eventually found. Then, this policy is brought to the attention of the supremal decision unit, DU_0, who is responsible for detecting whether it leads to an improved performance of the system as a whole. If so, the uncovered policy is an active decision policy, and one can proceed with its implementation.

On the other hand, the top–down approach starts the learning process at the supremal decision unit, DU_0, and only on a second stage does it move down to the infimal decision units for approval and validation.

The next paragraphs provide a brief description of both the bottom–up and top–down search procedures (for further details, see Saraiva, 1993; Saraiva and Stephanopoulos, 1992d).

1. Bottom–Up Approach

The bottom–up approach contains two distinct stages. First, by successive backpropagation steps one builds a decision policy. Then, this uncovered policy is evaluated and refined, and its expected benefits confirmed before any implementation actually takes place. This two-stage process is conceptually similar to dynamic programming solution strategies, where first a decision policy is constructed by backward induction, and then one finds a realization of the process for the given policy, in order to check its expected performance (Bradley *et al.*, 1977).

a. Decision Policy Construction. Learning activities may be initiated by any of the infimal decision units, DU_k, to which one applies the ***basic*** learning methodology introduced in Sections V and VI, leading to the identification of a particular final solution, $\mathbf{X}_k$.

If $\mathbf{X}_k$ involves only ranges of decision variables attached to unit k, $x_{d,k}$, it defines a decision policy, and thus one can move directly to the validation and refinement phase.

However, $\mathbf{X}_k$, besides decision variables, may also include ranges of connection variables, $x_{i,k}$, that link units $k + 1$ and k. That being the case, the learning process has to be propagated backward, toward decision unit DU_{k+1}. To induce this propagation, one has first to identify temporary loss functions (Saraiva, 1993; Saraiva and Stephanopoulos, 1992d) for all connection variables present in the definition of $\mathbf{X}_k$. These temporary propagation loss functions are combined with other DU_{k+1} goals, and *basic* is now applied to decision unit DU_{k+1}. Further propagations, from unit $k + 1$ to unit $k + 2, k + 2$ to $k + 3$, etc., are identical to the one that occurred from DU_k to DU_{k+1}. This backpropagation through different decision units continues until one eventually reaches an infimal decision unit, DU_{k+j}, where a solution, $\mathbf{X}_{k+j}$, involving only ranges of decision variables, is found. When that happens, a final decision policy, $\mathbf{X}_{DP}$, can be immediately constructed by simply assembling all of its pieces together: it is the conjunction of the decision variable intervals, distributed among several decision units, that were uncovered during the upstream propagation process.

b. Validation and Refinement. Because the construction of $\mathbf{X}_{DP}$ resulted from the contributions of multiple infimal decision units, taking into consideration their specific goals and imposed constraints, it may be already an active decision policy. However, and even if that is the case, before proceeding to any implementation, it is necessary to evaluate the benefits that would derive from operating within $\mathbf{X}_{DP}$.

The process of $\mathbf{X}_{DP}$ validation and refinement starts with a final detailed analysis of its realization, through the computation of $E(\mathbf{y}_k|\mathbf{X}_{DP})$ and $\sigma(\mathbf{y}_k|\mathbf{X}_{DP})$, $k = 0, 1, \ldots, K$, estimates. If for some reason one or more infimal decision units judge the implementation of $\mathbf{X}_{DP}$ to be unacceptable, an additional attempt is made to refine the decision policy and find a revised version of it, $\mathbf{X}_{DP}^{final}$, that is accepted and validated by all infimal decision units, through additional applications of *basic* to the decision units in conflict with the initial $\mathbf{X}_{DP}$ version (see Saraiva, 1993; Saraiva and Stephanopoulos, 1992d, for details).

Once agreement and consensus have been reached by all infimal decision units, for $\mathbf{X}_{DP}^{final}$ to be declared an active decision policy, it is necessary to bring it to the attention of DU_0. The effects of a possible implementation of $\mathbf{X}_{DP}^{final}$ on the system as a whole are examined, to check that it also translates into improved performance from a global perspective. If this final validation test is passed, $\mathbf{X}_{DP}^{final}$ represents an active decision policy, and one can proceed to its implementation.

2. Top–Down Approach

In the top–down approach the supremal decision unit, DU_0, starts the learning process by itself, and identifies a decision policy, $\mathbf{X}_{DP}$. Then, in a second stage, one moves down to the infimal subsystems, seeking their support and validation for $\mathbf{X}_{DP}$.

Let's assume that the inputs to the supremal decision unit are a subset of all the decision variables attached to infimal decision units, consisting of those $x_{d,k}$ variables that are believed to be particularly influential with respect to the operation of the overall system. Then, an application of *basic* to DU_0 results directly in the identification of a decision policy, $\mathbf{X}_{DP}$. This decision policy is then passed down to the lower level in the hierarchy, where it is submitted to a process of validation and refinement by all infimal decision units that is identical to the one that takes place in the bottom–up approach.

C. Case Study: Operational Analysis of a Pulp Plant

The overall system that we will analyze comprises the unbleached Kraft pulp line, chemicals and energy recovery zones of a specific paper mill (Melville and Williams, 1977). We will employ a somewhat simplified but still realistic representation of the plant, originally developed in a series of research projects at Purdue University (Adler and Goodson, 1972; Foster *et al.*, 1973; Melville and Williams, 1977). The records of simulated operation data, used to support the application of our learning architecture, were generated by a reimplementation, with only minor changes, of steady-state models (for each individual module and the system as a

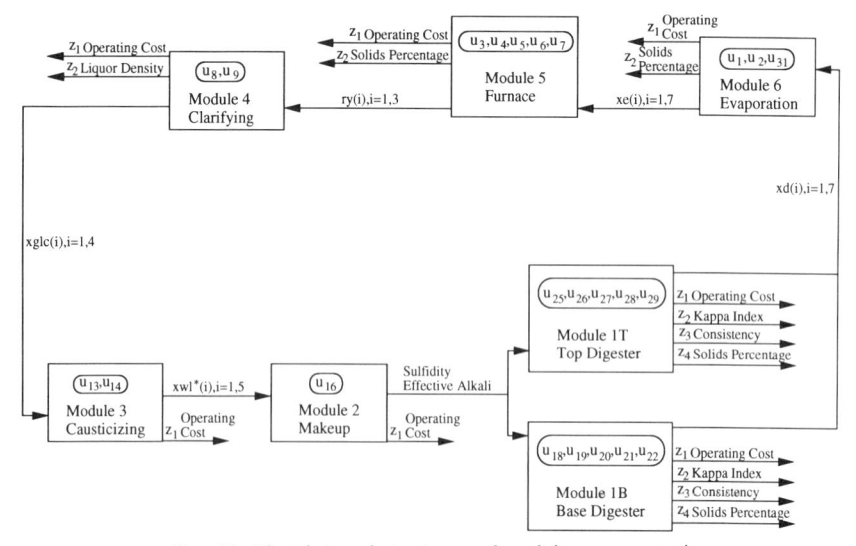

Fig. 12. Plant internal structure and modular representation.

whole) presented in the references cited above. A more detailed description of this case study can be found in Saraiva (1993), or Saraiva and Stephanopoulos (1992d).

The global structure of a Kraft pulp plant was illustrated in Fig. 9. The corresponding modular representation adopted in this study is shown schematically in Fig. 12. It includes 7 infimal decision units, with a total of 23 different decision variables distributed among them [to keep the notation consistent with Melville and Williams (1977), decision variables are designated as u_d]. A complete list of these variables, including their physical meaning, measurement units and ranges of values in $\mathbf{X}_{current}$, is provided in Table V. There are also 28 connection variables, linking successive infimal subsystems, and 16 local performance variables, $z_{i,k}$. All the performance measures (operating costs, quality losses, penalty functions) are expressed on a common basis of U.S. dollars per ton of air-dried pulp produced ($/TADP).

TABLE V

LIST OF ALL 23 DECISION VARIABLES AND CORRESPONDING WINDOWS UNDER CURRENT OPERATING CONDITIONS

u_d	u_d description	Units	Range
u_1	Oxidation tower efficiency	Adimensional	[0.0; 0.9]
u_2	Steam flow to evaporators	lb/h	$[0.9 \cdot 10^5; 1.25 \cdot 10^5]$
u_3	Sodium sulfate addition	lb/h	[2000; 4000]
u_4	Black liquor temperature at nozzles	°F	[245; 250]
u_5	Primary airflow to furnace	lb/h	$[3 \cdot 10^5; 4 \cdot 10^5]$
u_6	Secondary airflow to furnace	lb/h	$[1.75 \cdot 10^5; 2.75 \cdot 10^5]$
u_7	Furnace flue gas temperature	°F	[550; 750]
u_8	Steam/condensate added to smelt tank	lb/TADP	[4000; 5000]
u_9	Washing water flow to dregs filter	lb/TADP	[0.0; 75.0]
u_{13}	Water fraction in lime mud slurry	Adimensional	[0.3; 0.4]
u_{14}	Water spray flow to white liquor filter	lb/TADP	[1000; 2000]
u_{16}	White liquor effective alkali	lb/gal	[0.8; 0.9]
u_{18}	Lower heater temperature	°F	[295; 310]
u_{19}	Digester blow flow	gpm[a]	[1100; 1250]
u_{20}	Washing circulation temperature	°F	[225; 280]
u_{21}	Black liquor extraction flow	gpm	[1150; 1250]
u_{22}	White liquor added to digester	gpm	[400; 500]
u_{25}	Lower heater temperature	°F	[295; 308]
u_{26}	Digester blow flow	gpm	[800; 950]
u_{27}	Washing circulation temperature	°F	[225; 280]
u_{28}	Black liquor extraction flow	gpm	[950; 1150]
u_{29}	White liquor added to digester	gpm	[400; 500]
u_{31}	Steam flow to flash system	lb/h	[0; 10000]

[a] Gallons per minute.

1. Top–Down Approach

When examined at the supremal level, the system's primary goal is the production of pulp with the desired kappa indices. Consequently, as DU_0 performance variables we will consider the following:

(a) $z_{1,0}$, total operating cost, which is the sum of all the 7 subsystems operating costs, $z_{1,k}$, $k = 1,\ldots,7$
(b) $z_{2,0}$, kappa index of the pulp produced at the base digester
(c) $z_{3,0}$, kappa index of the pulp produced at the top digester

Furthermore, we will consider the 23 different decision variables, u_d, as the DU_0 inputs.

Both operating costs and quality losses were combined together, leading to the following overall performance measure, y_0, for the supremal decision unit:

$$y_0 = z_{1,0} + (z_{2,0} - 100)^2 + 5(z_{3,0} - 60)^2. \quad (40)$$

A preliminary analysis of the available DU_0 data records showed that y_0 is currently dominated by the behavior of $z_{3,0}$, and its deviations from the target. After going through the search procedure described in Section VII, the following final active decision policy, $\mathbf{X}_{DP}^{final}$, was identified:

$$\mathbf{X}_{DP}^{final} = \{u_2 \in [112{,}000; 122{,}000] \wedge u_3 \in [2060; 2700]$$
$$\wedge u_4 \in [245.3; 247.3] \wedge u_{25} \in [304.2; 307.8]\}. \quad (41)$$

In Table VI we compare the performance achieved through the implementation of the above strategy with that defined by the current values. An implementation of $\mathbf{X}_{DP}^{final}$ results in a significant decrease of the y_0 average, primarily as a consequence of a reduction in the average cost associated with the operation of the top digester, y_{1T}. On its own hand, the y_{1T} average decrease derives from the fact that $\mathbf{X}_{DP}^{final}$ centers the average kappa index of the top-digester pulp much closer to its target of 60, while also reducing its standard deviation. All these results are consistent with the observation made earlier that pointed to $z_{3,0}$ as the key performance variable that conditions the current levels of overall system performance, $E(y_0|\mathbf{X}_{current})$.

2. Bottom–Up Approach

The learning process was initiated at the top-digester infimal decision unit, leading to a solution, $\mathbf{X}_{1T}$, that involves local decision variables and a range of white liquor sulfidity (fraction of active reactants in the white

TABLE VI
COMPARISON OF PERFORMANCE MEASURES BEFORE AND AFTER IMPLEMENTATION OF $\mathbf{X}_{DP}^{final}$

	Average ($\mathbf{X}_{current}$)	Average ($\mathbf{X}_{DP}^{final}$)
y_0	1786.5	895.6
y_{1T}	1500.6	695.3
y_{1B}	281.5	187.8
y_2	8.8	9.2
y_3	7.4	7.8
y_4	185.6	168.8
y_5	120.0	50.1
y_6	5.8	4.8
Kappa index (top)	67.4	59.9
Kappa index (base)	102.9	102.0

liquor that are present as HS^- rather than as OH^-) values. Successive upstream propagations of this request had to be performed before a solution involving only decision variables, $\mathbf{X}_5$, was found at the furnace module, thus leading to the identification of a decision policy, $\mathbf{X}_{DP}$, which consists of the conjunction of all the decision variable ranges identified up to that point.

It is worth noticing that the path that was followed in the construction of $\mathbf{X}_{DP}$ is coherent and logical with respect to a physical understanding of the plant. It was found at the top digester that one of the most important variables, conditioning the location and dispersion of the pulp kappa index, is the amount of sulfur present in the white liquor added to the digester. But sodium sulfate enters the plant to compensate for sulfur losses at the recovery furnace, and thus it is basically within this infimal decision unit that the levels of sulfidity can be adjusted. Accordingly, the backpropagation process involved requests over the values of sulfur flows in several intermediate streams, and it stopped only at the furnace module, where the decision policy construction was concluded, leading to a $\mathbf{X}_{DP}$ that involves the amount of sodium sulfate added to the furnace.

After submitting $\mathbf{X}_{DP}$ through the validation and refinement stage, the final uncovered active decision policy, $\mathbf{X}_{DP}^{final}$, was given by

$$\mathbf{X}_{DP}^{final} = \{u_3 \in [3200; 3635] \wedge u_4 \in [247.5; 248.7] \wedge u_8 \in [4800; 5000]$$
$$\wedge u_{18} \in [306, 309] \wedge u_{25} \in [304.5; 308.0] \wedge u_{29} \in [460.0; 490.0]\} \quad (42)$$

Table VII summarizes the levels of performance that are achieved within $\mathbf{X}_{DP}^{final}$, and compares them with the performance corresponding to the current values.

TABLE VII
Comparison of Performance Measures Before and After Implementation of $\mathbf{X}_{\text{DP}}^{\text{final}}$

	Average ($\mathbf{X}_{\text{current}}$)	Average ($\mathbf{X}_{\text{DP}}^{\text{final}}$)
y_0	1786.5	489.1
y_{1T}	1500.6	387.4
y_{1B}	281.5	89.2
y_2	8.8	11.8
y_3	7.4	8.6
y_4	185.6	14.9
y_5	120.0	91.7
y_6	5.8	4.9
Kappa index (top)	67.4	60.7
Kappa index (base)	102.9	99.5

3. Brief Comparison of Final Decision Policies

In the previous paragraphs it was shown how by following the top–down and bottom–up approaches one arrived at the construction of two distinct and promising decision policies, Eqs. (41) and (42). Both of these decision policies include intervals of values for certain critical decision variables (e.g., u_3, u_4, u_{25}). Since there is some consistency between the infimal and supremal decision unit goals, this communality of variables should be expected. However, different infimal decision unit local goals were taken into account in the construction of these policies, and choices among several possible solutions were made by the user during that construction. Thus, it does not come as a surprise that the two final decision policies also involve ranges of different decision variables. Similarly, the performances achieved are not entirely identical.

A final comparison of the results obtained with each policy, as well as under the current operating conditions, is given in Fig. 13. Both the top–down and the bottom–up approaches uncovered promising decision policies, which lead to considerable improvements over the current plant performance levels. When one compares in a more detailed way the differences in infimal decision unit performances for the two policies, one can see that they reflect the participation of the corresponding subsystems in their construction. Finally, it should be added that, besides performance related issues, the upstream propagation associated with the bottom–up approach provided some important insight into understanding the main causes of dispersion and location for the top-digester pulp kappa index. Specifically, it was uncovered that the kappa index is highly dependent on

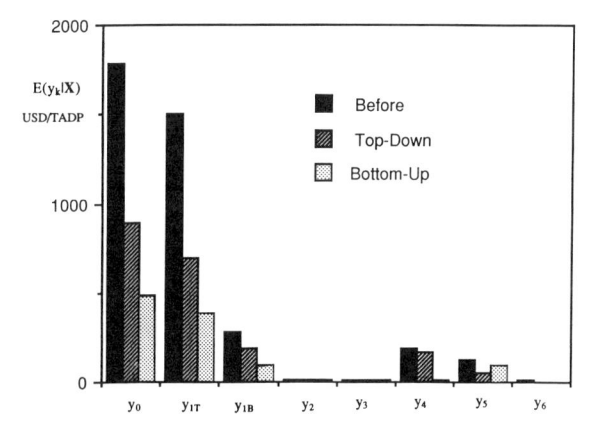

FIG. 13. Final comparison of performances for both decision policies.

the white liquor sulfidity, a relationship translated later on into the definition of intervals for decision variables associated with the recovery boiler, which is the module where one has the ability to manipulate the amounts of sulfur present in all the plant liquor streams. This type of knowledge would have been hard or impossible to acquire by merely examining the final decision policies or through an analysis of the top–down approach alone.

VIII. Summary and Conclusions

In this chapter we revisited an old problem, namely, exploring the information provided by a set of (x, y) operation data records and learn from it how to improve the behavior of the performance variable, y. Although some of the ideas and methodologies presented can be applied to other types of situations, we defined as our primary target an analysis at the supervisory control level of (x, y) data, generated by systems that cannot be described effectively through first-principles models, and whose performance depends to a large extent on quality-related issues and measurements.

We have introduced modified statements and solution formats for the preceding problem, with hyperrectangles in the decision space replacing the conventional pointwise results. The advantages and implications of adopting this alternative language to express final solutions were discussed, and it was also shown that traditional formulations can be interpreted as a particular degenerate case of the suggested more general problem definitions, where one searches for ranges of decision variables rather than single values.

To address the modified problem statements and uncover final solutions with the desired alternative formats, data-driven nonparametric learning methodologies, based on direct sampling approaches, were described. They require far fewer assumptions and a priori decisions on the part of the user than most conventional techniques. These practical frameworks for extracting knowledge from operating data present the final uncovered solutions to the decisionmaker in formats that are both easy to understand and implement.

We presented extensions and variations of the basic learning methodologies aimed at enlarging their flexibility and cover a number of different situations, including systems where performance is evaluated by categorical or continuous variables, with single or multiple objectives, simple or complex plants containing some type of internal structure and composed of a number of interconnected subsystems.

The potential practical capabilities of the described learning methodologies, and their attractive implementational features from an industrial point of view, were illustrated through the presentation of a series of case studies with both real-world industrial and simulated operating data.

Acknowledgments

The author would like to acknowledge financial an other types of support received from the Leaders for Manufacturing program at MIT, Fulbright Program, Rotary Foundation, Comissão Permanente da INVOTAN, Fundação Luso-Americana para o Desenvolvimento, and Comissão Cultural Luso-Americana. Special thanks also to Professor George Stephanopoulos, who always provided the right amount of support and guidance, while at the same time allowing me to have all the freedom that I needed to pursue my own research dreams and try to convert them into reality.

References

Adler, L., and Goodson, R., "An Economic Optimization of a Kraft Pulping Process," Laboratory for Applied Industrial Control, Report 48. Purdue University, West Lafayette, IN, 1972.

Alefeld, G., and Herzberger, J., "Introduction to Interval Computations." Academic Press, New York, 1983.

Bellman, R., and Dreyfus, S., "Applied Dynamic Programming." Princeton University Press, Princeton, NJ, 1962.

Biegler, L., Optimization strategies for complex process models. *Adv. Chem. Eng.* **18**, 197 (1992).
Bradley, S., *et al.*, "Applied Mathematical Programming." Addison-Wesley, Reading, MA, 1977.
Breiman, L., *et al.*, "Classification and Regression Trees." Wadsworth, Belmont, CA, 1984.
Clausing, D., "Total Quality Development." ASME Press, New York, 1993.
Daniel, C., and Wood, F., "Fitting Equations to Data." 2nd ed. Wiley, New York, 1980.
Deming, W., "Out of the Crisis." Massachusetts Institute of Technology, Center for Advanced Engineering Study, Cambridge, MA, 1986.
Duda, R., and Hart, P., "Pattern Classification and Scene Analysis." Wiley, New York, 1973.
Ellingsen, W., Implementation of advanced control systems. *AIChE Symp. Ser.* **159**, 150 (1976).
Findeisen, W., *et al.*, "Control and Coordination in Hierarchical Systems." Wiley, New York, 1980.
Foster, R., *et al.*, "Optimization of the Chemical Recovery Cycle of the Kraft Pulping Process," Laboratory for Applied Industrial Control, Report 54. Purdue University, West Lafayette, IN, 1973.
Fu, K., "Sequential Methods in Pattern Recognition and Machine Learning." Academic Press, New York, 1968.
Gaines, B., The trade-off between knowledge and data in knowledge acquisition. *In* "Knowledge Discovery in Databases" (G. Shapiro and W. Frawley, eds.), p. 491. MIT Press, Cambridge, MA, 1991.
Garcia, C., and Prett, D., Advances in industrial model-predictive control. *In* "Chemical Process Control, CPC-III." (Morari, M. and McAvoy, T. J., eds.). CACHE-Elsevier, New York, 1986.
Goodman, R., and Smyth, P., Decision tree design using information theory. *Knowl. Acquis.* **2**, 1 (1990).
Hayes, R., *et al.*, "Dynamic Manufacturing: Creating the Learning Organization." Free Press, New York, 1988.
Hunt, E., "Concept Learning: An Information Processing Problem." Wiley, New York, 1962.
James, M., "Classification Algorithms." Wiley, New York, 1985.
Juran, J., "Managerial Breakthrough." McGraw-Hill, New York, 1964.
Klein, J., "Revitalizing Manufacturing." R.D. Irwin, Homewood, IL, 1990.
Kodratoff, Y., and Michalski, R., eds., "Machine Learning: An Artificial Intelligence Approach." Vol. 3. Morgan Kaufmann, San Mateo, CA, 1990.
Lasdon, L., "Optimization Theory for Large Systems." Macmillan, New York, 1970.
Lasdon, L., and Baker, T., The integration of planning, scheduling and process control. *In* "Chemical Process Control, CPC-III." (Morari, M. and McAvoy, T. J., eds.). CACHE-Elsevier, New York, 1986.
Latour, P., Comments on assessment and needs. *AIChE Symp. Ser.* **159**, 161 (1976).
Latour, P., Use of steady-state optimization for computer control in the process industries. *In* "On-line Optimization Techniques in Industrial Control" (Kompass, E. J. and Williams, T. J., eds.). Technical Publishing Company, 1979.
Launks, U., *et al.*, On-line optimization of an ethylene plant. *Comput. Chem. Eng.* **16**, S213 (1992).
Melville, S., and Williams, T., "Application of Economic Optimization to the Chemical Recovery System of a Kraft Pulping Process," Laboratory for Applied Industrial Control, Report 107. Purdue University, West Lafayette, IN, 1977.
Mesarović, M., *et al.*, "Theory of Hierarchical, Multilevel Systems." Academic Press, New York, 1970.

Moore, R., "Methods and Applications of Interval Analysis." SIAM, Philadelphia, 1979.

Moore, R., What and who is in control. *In* "The Second Shell Process Control Workshop" (D.M. Prett, C.E. Garcia, and B.L. Ramaker, eds.). Butterworth, Stoneham, MA, 1990.

Moret, B., Decision trees and diagrams. *ACM Comput. Surv.* **14**(4), 593 (1982).

National Research Council, "The Competitive Edge. "National Academy Press, Washington, DC, 1991.

Nemhauser, G., "Introduction to Dynamic Programming." Wiley, New York, 1966.

Phadke, M., "Quality Engineering Using Robust Design." Prentice Hall, Englewood Cliffs, NJ, 1989.

Quinlan, J., Induction of decision trees. *Mach. Learn.* **1**, 81 (1986).

Quinlan, J., Simplifying decision trees. *Int. J. Man-Mach. Stud.* **27**, 221 (1987).

Quinlan, J., "C4.5: Programs for Machine Learning." Morgan Kaufmann, San Mateo, CA, 1993.

Reece, J., Daniel, D. and Bloom R., Identifying a plasma etch process window. *In* "Understanding Industrial Designed Experiments" (Schmidt S. and Launsby R., eds.), 2nd ed. AIR Academy Press, Colorado Springs, CO, 1989.

Roberts, S., "Dynamic Programming in Chemical Engineering." Academic Press, New York, 1964.

Roy, R., "A Primer on the Taguchi Method." Van Nostrand-Reinhold, Princeton, NJ, 1990.

Saraiva, P., Data-driven learning frameworks for continuous process analysis and improvement. Ph.D. Thesis, Massachusetts Institute of Technology, Dept. Chem. Eng., Cambridge, MA, 1993.

Saraiva, P., and Stephanopoulos, G., Continuous process improvement through inductive and analogical learning. *AIChE J.* **38**(2), 161 (1992a).

Saraiva, P., and Stephanopoulos, G., "Learning to Improve Processes with Multiple Pattern Recognition Objectives," Working paper. Massachusetts Institute of Technology, Dept. Chem. Eng., Cambridge, MA, 1992b.

Saraiva, P., and Stephanopoulos, G., "An Exploratory Data Analysis Robust Optimization Approach to Continuous Process Improvement," Working paper. Massachusetts Institute of Technology, Dept. Chem. Eng., Cambridge, MA, 1992c.

Saraiva, P., and Stephanopoulos, G., "Data-Driven Learning Architectures for Process Improvement in Complex Systems with Internal Structure," Working paper. Massachusetts Institute of Technology, Dept. Chem. Eng., Cambridge, MA, 1992d.

Sargent, R., The future of digital computer based industrial control systems. *In* "Industrial Computing Control After 25 Years" (Kompass, E. J. and Williams, T. J., eds.), p. 63. Technical Publishing Company, 1984.

Senge, P., "The Fifth Discipline." Currency, 1990.

Shannon, C., and Weaver, W., "The Mathematical Theory of Communication," 11th printing. University of Illinois Press, Urbana, 1964.

Shapiro, G., and Frawley, W., eds., "Knowledge Discovery in Databases." MIT Press, Cambridge, MA, 1991.

Shavlik, J., and Dietterich, T., eds., "Readings in Machine Learning." Morgan Kaufmann, San Mateo, CA, 1990.

Sheridan, T., "45 Years of Man-Machine Systems." Massachusetts Institute of Technology, Dept. Mech. Eng., Cambridge, MA, 1985.

Shiba, S., *et al.*, "The Four Revolutions of Management Thinking: Planning and Implementation of TQM for Executives." Productivity Press, 1993.

Sinnar, R., Impact of model uncertainties and nonlinearities on modern controller design. *In* "Chemical Process Control, CPC-III." (Morari, M. and McAvoy, T. J., eds.), p. 53. CACHE-Elsevier, 1986.

Sonquist, J., *et al.*, "Searching for Structure." University of Michigan, Ann Arbor, Michigan, 1971.
Taguchi, G., "Introduction to Quality Engineering." Asian Productivity Association, 1986.
Taylor, W., "What Every Engineer Should Know About Artificial Intelligence." MIT Press, Cambridge, MA, 1989.
Taylor, W., "Optimization and Variation Reduction in Quality." McGraw-Hill, New York, 1991.
Tukey, J., "Exploratory Data Analysis." Addison-Wesley, Reading, MA, 1977.
Turban, E., "Decision Support and Expert Systems." Macmillan, New York, 1988.
Utgoff, P., Perception trees: A case study in hybrid concept representations. *In* "Proceedings of AAAI88," Vol. 2, p. 601. Morgan Kaufmann, San Mateo, CA, 1988.
Winston, P., "Artificial Intelligence," 2nd ed. Addison-Wesley, Reading, MA, 1984.

EMPIRICAL LEARNING THROUGH NEURAL NETWORKS: THE WAVE-NET SOLUTION

Alexandros Koulouris, Bhavik R. Bakshi,[1] and George Stephanopoulos

**Laboratory for Intelligent Systems in Process Engineering
Department of Chemical Engineering
Massachusetts Institute of Technology
Cambridge, Massachusetts 02139**

I. Introduction	438
II. Formulation of the Functional Estimation Problem	441
A. Mathematical Description	444
B. Neural Network Solution to the Functional Estimation Problem	449
III. Solution to the Functional Estimation Problem	451
A. Formulation of the Learning Problem	451
B. Learning Algorithm	465
IV. Applications of the Learning Algorithm	471
A. Example 1	471
B. Example 2	474
C. Example 3	477
V. Conclusions	479
VI. Appendices	480
A. Appendix 1	480
B. Appendix 2	481
C. Appendix 3	482
References	483

Empirical learning is an ever-lasting and ever-improving procedure. Although neural networks (NN) captured the imagination of many researchers as an outgrowth of activities in artificial intelligence (AI), most of the progress was accomplished when empirical learning through NNs was cast within the rigorous analytical framework of the *functional estimation problem*, or *regression*, or *model realization*. Independently of the

---

[1] Present address: Ohio State University, Department of Chemical Engineering, Columbus, OH 43210.

name, it has been long recognized that, due to the inductive nature of the learning problem, to achieve the desired accuracy and generalization (with respect to the available data) in a dynamic sense (as more data become available) one needs to seek the unknown approximating function(s) in functional spaces of varying structure. Consequently, a recursive construction of the approximating functions at multiple resolutions emerges as a central requirement and leads to the utilization of wavelets as the basis functions for the recursively expanding functional spaces. This chapter fuses the most attractive features of a NN: representational simplicity, ability for universal approximation, and ease in dynamic adaptation, with the theoretical soundness of a recursive functional estimation problem, using wavelets as basis functions. The result is the *Wave-Net* (wavelets, network of), a multiresolution hierarchical NN with localized learning. Within the framework of a *Wave-Net* where adaptation of the approximating function is allowed, we have explored the use of the L^∞ error measure as the design criterion. Within the framework of a *Wave-Net* one may cast any form of data-driven empirical learning to address a variety of modeling situations encountered in engineering problems such as design of process controllers, diagnosis of process faults, and planning and scheduling of process operations. This chapter will discuss the properties of a *Wave-Net* and will illustrate its use on a series of examples.

I. Introduction

Estimating an unknown function from its examples is a central problem in statistics, where it is known as the problem of *regression*. Under a different name but following the same spirit, the problem of *learning* has attracted interest in the AI research since the discovery that many intelligent tasks can be represented through functional relationships and, therefore, learning those tasks involves the solution of a regression problem (Poggio and Girosi, 1989). The significance, however, of the problem is by no means limited to those two fields. *Modeling*, an essential part of science and engineering, is the process of deriving mathematical correlations from empirical observations and as such, obeys the same principles as the regression or learning problem. Many methods developed for the solution of these problems are extensively used as tools to advance understanding and allow prediction of physical behavior.

In recent years the merging between the statistical and AI points of view on the same problem has benefited both approaches. Statistical regression techniques have been enriched by the addition of new methods and learning techniques have found in statistics the framework under which their properties can be studied and proved. This is especially true for neural networks (NNs), which are the main product of AI research in the field and the most promising solution to the problem of functional estimation. Despite their origination as biologically motivated models of the brain, NNs have been established as nonlinear regression techniques and studied as such. Their main features are their capability to represent nonlinear mappings, their ability to learn from data and finally their inherent parallelism which allows fast implementation. Every application where an unknown nonlinear function has to be reconstructed from examples is amenable to a NN solution. This explains the wide spread of NNs outside the AI community and the explosion of their applications in numerous disciplines and for a variety of tasks. In chemical engineering and especially in process systems engineering, NNs have found ground for applications in process control as models of nonlinear systems behavior (Narendra and Parthasarathy, 1990; Ungar et al., 1990; Ydstie, 1990; Hernandez and Arkun, 1992; Bhat and McAvoy, 1990; Psichogios and Ungar, 1991; Lee and Park, 1992), fault diagnosis (Hoskins and Himmelblau, 1988; Leonard and Kramer, 1991), operation trend analysis (Rengaswamy and Venkatasubramaniam, 1991), and many other areas. In all these applications, the exclusive task, NNs are required to perform, is *prediction*. From the interpretation point of view, NNs are blackbox models. They fail to provide explicitly any physical insight simply because their representation does not coincide and is not motivated by the underlying phenomenon they model. The only expectation from their use is accuracy of their predictions, and this is the criterion on which the merits of their use can be judged.

The mathematical point of view on the analysis of NNs did not only aim at depriving them from their mystery related to their biological origin, but was also supposed to provide guarantees, if any, on their performance and guide the user with their implementation. Many theorems have been recruited to support the approximation capabilities of NNs. The universal approximation property has been proved for many different activation functions used in NNs such as sigmoids (Hornik et al., 1989), or radial basis functions (RBFs) (Hartman et al., 1990). At the same time, it was shown that such a property is quite general and not difficult to prove for many other sets of basis functions not used in NNs. Rates of convergence to the real function have also been derived (Barron, 1994), but these are limited to special conditions on the space of functions where the unknown function belongs. The central theme of all these theorems is that given a large enough network and enough data points any unknown function can be approximated arbitrarily well. Despite these theorems, the potential user is still puzzled by issues such as what type of network to implement, how many nodes or hidden layers to use, and how to interconnect them. The issues to these questions are still derived on empirical grounds or through a trial-and-error procedure. The theorems have, however, clearly shown that it is exactly these choices related to the above questions that determine the approximating capabilities and accuracy of a given network.

Under the new light that the mathematical analysis and practical considerations have brought, new directions in NN research have been revealed. The list of basis functions used in NNs expands steadily with new additions, and methods for overcoming the empiricism in determining the network architecture are explored (Bhat and McAvoy, 1992; Mavrovouniotis and Chang, 1992). Following this spirit, a novel NN architecture, the wavelet network or *Wave-Net* has been recently proposed by Bakshi and Stephanopoulos (1993). *Wave-Nets* use wavelets as their activation functions and exploit their remarkable properties. They apply a hierarchical localized procedure to evolve in their structure, guided by the local approximation error. *Wave-Nets* are data-centered, and all important decisions on their architecture are decided constructively by the data and that is their main attractive property.

In this study the problem of estimating an unknown function from its examples is revisited. Its mathematical description is attempted to map as closely as possible the practical problem that the potential NN user has to face. The objective of the chapter is twofold: (1) to draw the framework in which NN solutions to the problem can be developed and studied, and (2) to show how careful considerations on the fundamental issues naturally lead to the *Wave-Net* solution. The analysis will not only attempt to justify the development of the *Wave-Net*, but will also refine its operational characteristics. The motivation for studying the functional estimation problem is the derivation of a modeling framework suitable for process control. The applicability of the derived solution, however, is not limited to control implementations.

The remainder of this chapter is structured as follows. In Section II the problem of deriving an estimate of an unknown function from empirical data is posed and studied in a theoretical level. Then, following Vapnik's original work (Vapnik, 1982), the problem is formulated in mathematical terms and the sources of the error related to any proposed solution to the estimation problem are identified. Considerations on how to reduce these errors show the inadequacy of the NN solutions and lead in Section III to the formulation of the basic algorithm whose new element is the pointwise presentation of the data and the dynamic evolution of the solution itself. The algorithm is subsequently refined by incorporating the novel idea of structural adaptation guided by the use of the L^∞ error measure. The need for a multiresolution framework in representing the unknown function is then recognized and the wavelet transform is proposed as the essential vehicle to satisfy this requirement. With this addition, the complete algorithm is presented and identified as the modified *Wave-Net* model. Modeling examples (Section IV) demonstrate the properties of the derived solution and the chapter concludes with some final thoughts (Section V).

II. Formulation of the Functional Estimation Problem

In its more abstract form, the problem is to estimate an unknown function $f(\mathbf{x})$ from a number of, possibly noisy, observations $(\mathbf{x}_i, y_i)$, where $\mathbf{x}$ and y correspond to the vector of *input* or *independent variables* and the *output variable* of that function, respectively. To avoid confusion later on, the terms *regression*, *functional estimation* or *modeling* will all equivalently refer to the process of obtaining a *solution* or *approximating function* or *model* from the set of available data. The function $f(\mathbf{x})$ will be referred as the *real* or *target* function. The objective is to use the derived estimate to make predictions on the behavior of the real function in new, unseen situations. In almost all cases, the function $f(\mathbf{x})$ provides a mathematical description of a physically sound, but otherwise unknown, relationship between $\mathbf{x}$ and y. However, the form of the approximating function sought is in no way motivated by the underlying physical phenomenon and, consequently, from the point of view of interpretability, the proposed solutions are "blackbox" models. The only basic requirements for $f(\mathbf{x})$ related to its physical origination are that (1) it exists and (2) it is continuous with respect to all its arguments and that (3) the dimensionality of the input space is known.

Before attempting to formalize and solve the problem in mathematical terms, it is instructive to recognize the characteristics that are inherent to the functional estimation task and not attributable to the particular tool used to solve it. This will also set the ground on which different solutions to the problem can be analyzed and compared.

First, it can be easily recognized that finite data are always available within a bounded region of the input space. *Extrapolation* of any approximating function beyond that region is meaningful only when the model is derived by physical considerations applicable in the entire input space. In any other case, extrapolation is equivalent to postulating hypotheses that cannot be supported by evidence and, therefore, are both meaningless and dangerous. For that reason, we will confine the search for the unknown function in the fraction of the input space where data are available.

A very important characteristic of the problem is its *inductiveness*. The task is to obtain globally valid models by generalizing partial, localized and, many times, incorrect information. Because of its inductive nature, the problem inherits some disturbing properties. No solution can ever be guaranteed and the validity of any model derived by any means is not amenable to mathematical proofs. The model validation can only be perceived as a *dynamic* procedure triggered by any available piece of data and leading to an endless cycle between model postulating and model testing. This cycle can potentially prove the inefficiencies of any proposed model, but can never establish unambiguously its correctness.

Additional complications result from the fact that the functional estimation problem is *ill-posed* (Poggio and Girosi, 1989). Our only way to check whether a given function is a potential solution, is by measuring the *accuracy* of the fit for the available data. It is, however, clear that, for every given set of data points, there exists an infinite number of, arbitrarily different from the real, functions that can approximate the data arbitrarily well. Indeed, the implementation of every available tool for data regression, including NNs, will result in an equal number of different approximating functions. All these functions are equally plausible solutions to the functional estimation problem. The question that naturally arises is whether there exists some criterion based on which potential solutions can be screened out. What is certain is that the requirement for accuracy (independently of how it is mathematically measured) is not adequate in defining a unique solution to the problem and definitely not an appropriate basis for comparing the performance of various regression methods.

Where mathematical logic fails, intuition takes over. Intuitively, not all approximating functions for a given data set are equally plausible solutions to the functional estimation problem. The attribute that distinguishes the intuitively plausible from the unfavorable solutions is the *smoothness* of the approximating curve (in this context, smoothness does not refer to the mathematical property of differentiability). It is natural to seek the "best" solution in the face of the simplest, smooth-looking function that approximates well the available data. This conforms with a celebrated philosophical principle known as *Occam's razor*, which has found wide applications in the AI area (Kearns and Vazirani, 1994) and which, in simple terms, favors the shortest hypotheses that can explain the observations. In our context, shortness is equivalent to smoothness where the latter has so far only an intuitive connotation. For example, the approximating curve in Fig. 1a is a perfectly reasonable model of the data points, despite the fact that it differs considerably from the real function that produced those points. It is also remarkable to notice that, for that example and the given data, the

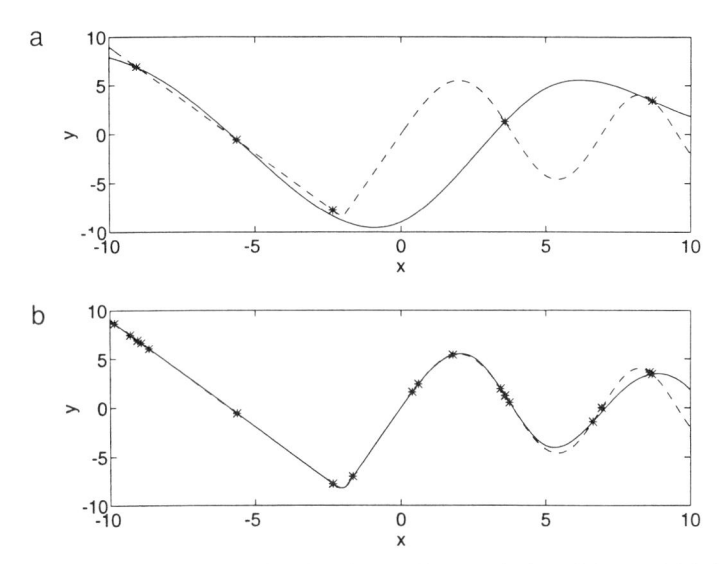

FIG. 1. Example of a functional estimation problem: evolution of the model (solid line) and comparison with the real function (dashed line) as more data (asterisks) become available.

real function is intuitively inferior (i.e., unnecessarily complicated) as an approximating function compared to its incorrect model. The real function becomes both the most accurate and smoothest (therefore, most favorable) solution when more reveal in data become available as it is the case in Fig. 1b. The fact that the model in Fig. 1a failed for additional data does not disqualify it as the "best" (in a weak, empirical sense) solution when only the initial points were available. The encouraging observation is that, by favoring intuitively simple approximating functions, the real function gradually emerges as it is evident by comparing the models in Figs. 1a and 1b.

The simple example in Fig. 1 indicates very clearly that the way the functional estimation problem should be attacked is by dynamically searching for the smoothest function that fits accurately the data. What we need is tools that will automatically produce approximating curves like the ones shown in Fig. 1. Of course, that will require us to first express the problem and its solution in mathematical terms and give precise mathematical meaning to the ambiguous attributes "accurate" and "smooth." The degree to which each proposed algorithm will prove able to mathematically represent those terms and appropriately satisfy the requirements for accuracy and smoothness will be the basis for judging its success.

A. MATHEMATICAL DESCRIPTION

Mathematically, the problem of deriving an estimate of a function from empirical data can be stated as follows (Vapnik, 1982):

1. Functional Estimation Problem

Given a set of data points $(\mathbf{x}_i, y_i) \in R^{k \times 1}$ $i = 1, 2, \ldots, l$ drawn with some unknown probability distribution, $P(\mathbf{x}, y)$, find a function $g^*(\mathbf{x})\colon X \to Y$ belonging to some class of functions G that minimizes the *expected risk functional*

$$I(g) = \int_{X \times Y} \mu[y, g(\mathbf{x})] P(\mathbf{x}, y)\, d\mathbf{x}\, dy, \qquad (1)$$

where $\mu(\cdot, \cdot)$ is a metric in the space Y. We will restrict this definition to the case where y is given as a nonlinear function of $\mathbf{x}$ corrupted by some additive noise independent of $\mathbf{x}$:

$$y = f(\mathbf{x}) + d, \qquad (2)$$

where d follows some unknown probability function $\mathbf{P}(d)$ and is also bounded: $|d| < \delta$. In that case, $P(\mathbf{x}, y) = P(\mathbf{x}) P(y|\mathbf{x})$ and $P(y|\mathbf{x}) = \mathbf{P}[y - f(\mathbf{x})]$. The real function, $f(\mathbf{x})$, belongs to the space $C(X)$ of continuous functions defined in a closed hypercube $X = [a, b]^k$, where a and b are, respectively, the lower and upper bounds of the hypercube and k is the dimensionality of the input space. Therefore, $f(\mathbf{x})$ is bounded in both its L^2 and L^∞-norm:

$$\|f\|_2 = \int_X f^2(\mathbf{x})\, d\mathbf{x} < \infty,$$

$$\|f\|_\infty = \sup_x |f(\mathbf{x})| < \infty.$$

No other a priori assumptions about the form or the structure of the function will be made. For a given choice of g, $I(g)$ in Eq. (1) provides a measure of the real approximation error with respect to the data in the entire input space X. Its minimization will produce the function $g^*(\mathbf{x})$ that is closest to G to the real function, $f(\mathbf{x})$ with respect to the, weighted by the probability $P(\mathbf{x}, y)$ metric μ. The usual choice for μ is the Euclidean distance. Then $I(g)$ becomes the L^2-metric:

$$I(g) = \int_{X \times Y} [y - g(\mathbf{x})]^2 P(\mathbf{x}, y)\, d\mathbf{x}\, dy \qquad (3)$$

In an analogous way, $I(g)$ can be defined in terms of all L^n norms with

$1 \leq n < \infty$. With a stretch of the notation, we can make Eq. (1) correspond to the L^∞-metric:

$$I(g) = \sup_x |y - g(\mathbf{x})|, \quad (4)$$

which is a meaningful definition $[I(g) < \infty]$ only because we have assumed the noise to be bounded and, therefore, $y - g(\mathbf{x})$ is finite.

The minimization of the expected risk given by Eq. (1) cannot be explicitly performed, because $P(\mathbf{x}, y)$ is unknown and data are not available in the entire input space. In practice, an estimate of $I(g)$ based on the empirical observations is used instead with the hope that the function that minimizes the *empirical risk* $I_{emp}(g)$ (or *objective function*, as it is most commonly referred) will be close to the one that minimizes the real risk $I(g)$.

Let $(\mathbf{x}_i, y_i) i = 1, \ldots, l$ be the available observations. The empirical equivalents of Eqs. (3) and (4) are respectively

$$I_{emp}(g) = \sum_{i=1}^{l} [y_i - g(\mathbf{x}_i)]^2, \quad (5)$$

$$I_{emp}(g) = \max_{i=1,\ldots,l} |y_i - g(\mathbf{x}_i)|. \quad (6)$$

The magnitude of $I_{emp}(g)$ will be referred as the *empirical error*. All regression algorithms, by minimizing the empirical risk $I_{emp}(g)$, produce an estimate, $\hat{g}(\mathbf{x})$, which is the solution to the functional estimation problem.

a. Decisions Involved in the Functional Estimation Problem. As it was theoretically explained earlier, due to the inductive nature of the problem, the only requirements we can impose on the model, $\hat{g}(\mathbf{x})$ are accuracy and smoothness with respect to the data. Equations (5) and (6) provide two of the possible mathematical descriptions of the accuracy requirement. The smoothness requirement is more difficult to describe mathematically and, in fact, does not appear anywhere in the problem definition. We will postpone the mathematical description of smoothness for later. The important question now is what tools are available for controlling the accuracy and smoothness of the approximating function, $\hat{g}(\mathbf{x})$.

The mathematical description of the problem involves the following elements:

(a) The data, $(\mathbf{x}_i, y_i) i = 1, \ldots, l$
(b) The function space, G, where the solution is sought
(c) The empirical risk, $I_{emp}(g)$
(d) The algorithm for the minimization of $I_{emp}(g)$

The problem definition itself does not pose any restrictions on the decisions described above. For all practical purposes all these choices are arbitrary. That, in fact, demonstrates the ill-posedness of the problem, which results in a variety of solutions depending on the particular tool used and the specific choices that are forced.

The data are usually given a priori. Even when experimentation is tolerated, there exist very few cases where it is known how to construct good experiments to produce useful knowledge suitable for particular model forms. Such a case is the identification of linear systems and the related issue on data quality is known under the term *persistency of excitation* (Ljung, 1987).

The critical decisions in the modeling problem are related to the other three elements. The space G is most often defined as the linear span of a finite number, m, of basis functions, $\theta(\mathbf{x})$, each parametrized by a set of unknown *coefficients* $\mathbf{w}$ according to the formula

$$G = \left\{ g(\mathbf{x}) | g(\mathbf{x}) = \sum_{k=1}^{m} c_k \theta_k(\mathbf{w}, \mathbf{x}), \mathbf{c} \in R^k, \mathbf{w} \in R^n \right\}, \qquad (7)$$

where $\mathbf{c}$ is the vector of independent *weights*. All models we consider will be of the form given by Eq. (7). The number of basis functions in G will be referred as the *size of G*. The choice of the basis functions $\theta(\mathbf{x})$ is a problem of *representation*. Any a priori knowledge on the unknown function, $f(\mathbf{x})$, could effectively be exploited to reduce the space G where the model is sought. In most practical situations, however, it is unlikely that such a condition is available. Indeed, in the problem definition we have only assumed $f(\mathbf{x})$ to be continuous and bounded. That sets G equal to the L^{∞} space, which, however, is not helpful at all, since the L^{∞} space is already too large.

The important question is what restrictions the accuracy and smoothness requirements impose on the function space G, i.e., if smoothness and accuracy (a) can be achieved with any choice of G. (b) Can "optimally" be satisfied in *all* cases for a specific choice of G.

First, with respect to the type of basis functions used in G, smoothness is by no means restrictive. As it is intuitively clear and proved in practice, weird nonsmooth basis functions have to be excluded from consideration but beyond that, all "normal" bases are able to create smooth approximations of the available data. Accuracy is not a constraint either. Given enough basis functions, arbitrary accuracy for the prediction on the data is possible.

Both smoothness and accuracy are possible for given selection of the basis functions. It is, however, clear that a tradeoff between the two appears with respect to the size of G. Each set of basis functions resolves optimally this tradeoff for different sizes of G. This gives rise to an interesting interplay between the size of G and the number of data points. This is appropriately exemplified by the so-called *bias vs. variance* tradeoff in statistics (Barron, 1994), or *generalization vs. generality* tradeoff in the AI literature (Barto, 1991).

If G is "small," good generalization can be achieved (in the sense that with few degrees of freedom, the behavior of the approximating function in between the data points is restricted) but that introduces bias in the model selection. The solution might look smooth, but it is likely to exhibit poor accuracy of approximation even for the data used to derive it. This is, for example, usually the case when linear approximations are sought to nonlinear functions. Increased accuracy can be achieved by making G large enough. If that is the case, generality can also be achieved, since a large set of functions can potentially approximate well different sets of data. The problem is, however, that in a large G more than one and probably quite different functions in G are able to approximate well or even interpolate perfectly the data set. The variance of the potential solutions is large and the problem can be computationally ill-posed. With such a variety of choices, the risk of deriving a nonsmooth model is large. This is the problem most often encountered with the space of polynomials as basis functions and is also related to the problem of *overfitting*. Data are overfitted when smoothness is sacrificed for the sake of increased accuracy. Because overfitting usually occurs when a large number of basis functions are used for fitting the data, the two have become almost synonymous. It should be noted, however, that this might not always be the case. There can be sets of basis functions (like the piecewise linear Haar basis) that are so poor as approximation schemes, that many of them are required to fit decently (but not overfit) a given set of data. On the other hand, the newly developed fractal-like functions (see, for example, the wavelets created by Daubechies, 1992) provide examples of structures that even in small numbers give rise to approximating functions that make the data look definitely overfitted.

The conclusion is that for every particular set of basis functions and given data, there exists an "appropriate" size of G that can approximate both accurately and smoothly this data set. A decisive advantage would be if there existed a set of basis functions, which could probably represent *any* data set or function with *minimal complexity* (as measured by the number of basis functions for given accuracy). It is, however, straightforward to construct different examples that acquire minimal representations with respect to different types of basis functions. Each basis function for itself is the most obvious positive example. A Gaussian (or discrete points from a Gaussian) can minimally be represented by the Gaussian itself as the basis function and so on. It is, consequently, hopeless to search for the unique *universal approximator*. The question boils down to defining for each selected basis and data the appropriate size of G that yields both an accurate and smooth approximation of the data.

2. Sources of the Generalization Error

The model, $\hat{g}(\mathbf{x})$, that any regression algorithm produces will, in general, be different from the target function, $f(\mathbf{x})$. This difference is expressed by the *generalization error* $\|f(\mathbf{x}) - \hat{g}(\mathbf{x})\|$, where $\|\cdot\|$ is a selected functional norm. The generalization error is a measure of the prediction accuracy of the model, $\hat{g}(\mathbf{x})$, for unseen inputs. Although this error cannot in practice by measured (due to the inductive character of the problem), the identification of its sources is important in formulating the solution to the problem.

There are two sources to the generalization error (Girosi and Anzellotti, 1993):

1. The *approximation error* that stems from the "finiteness" of the function space, G
2. The *estimation error* due to the finiteness of the data.

The unknown function, $f(\mathbf{x})$ is an infinite-dimensional object requiring an infinite number of basis functions for an exact representation. In Fourier analysis, any square-integrable function is exactly represented by an infinite number of trigonometric functions. Similarly, a polynomial of infinite degree is needed for an exact reconstruction of any generic function. On the other hand, the approximating space G, for computations to be feasible, can be spanned only by a finite set of basis functions and characterized only by a finite number of adjustable parameters. Independently of the basis used, the size o G has to be finite. The selection of any function from G [including the "best" approximating function $g^*(\mathbf{x})$] will inevitably introduce some approximation error compared to the target function. Even worse, due to the finiteness of the available data that does not allow the minimization of $I(g)$, a suboptimal solution, $\hat{g}(\mathbf{x})$, rather than $g^*(\mathbf{x})$, will be derived as the model. The difference between $\hat{g}(\mathbf{x})$ and $g^*(\mathbf{x})$ corresponds to the estimation error which further increases the generalization error. To eliminate the approximation error, G has to be infinite-dimensional. Similarly, infinite number of data are required to avoid the estimation error. Both requirements are impractical and that forces us to an important conclusion that the expectation to perfectly reconstruct the real function has somehow to be relaxed.

B. Neural Network Solution to the Functional Estimation Problem

For an introduction to NNs and their functionality, the reader is referred to the rich literature on the subject (e.g., Rumelhart *et al.*, 1986; Barron and Barron, 1988). For our purposes it suffices to say that NNs represent nonlinear mappings formulated inductively from the data. In doing so, they offer potential solutions to the functional estimation problem and will be studied as such.

Mathematically, NNs are weighted combinations of activation functions (units), $\theta(\mathbf{x})$, parametrized by a set of unknown coefficients. In that respect, they fall exactly under the general model form given by Eq. (7). The use of all NNs, for deriving an approximating function $g(\mathbf{x})$ given a set of data, follows a common pattern summarized by the following general NN algorithm:

Step 1. Select a family of basis functions, $\theta(\mathbf{x})$.

Step 2. Use some procedure to define G (the number of basis functions, m, and possibly the coefficients, $\mathbf{w}$).

Step 3. Learning Step: Solve the objective function minimization problem with respect to the weights $\mathbf{c}$ (and the coefficients, $\mathbf{w}$, if not defined in the previous step).

Although the minimization of the objective function might run to convergence problems for different NN structures (such as backpropagation for multilayer perceptrons), here we will assume that step 3 of the NN algorithm unambiguously produces the best, unique model, $\hat{g}(\mathbf{x})$. The question we would like to address is what properties this model inherits from the NN algorithm and the specific choices that are forced.

1. Properties of the Approximations

In recent years some theoretical results have seemed to defeat the basic principle of induction that no mathematical proofs on the validity of the model can be derived. More specifically, the *universal approximation* property has been proved for different sets of basis functions (Hornik *et al.*, 1989, for sigmoids; Hartman *et al.*, 1990, for Gaussians) in order to justify the bias of NN developers to these types of basis functions. This property basically establishes that, for every function, there exists a NN model that exhibits arbitrarily small generalization error. This property, however, should not be erroneously interpreted as a guarantee for small generalization error. Even though there might exist a NN that could approximate well any given function, once the NN structure is fixed (step 2), the universality of the approximation is lost. Any given NN can approximate well only a limited set of functions, and it is only sheer luck that determines whether the unknown function belongs in that set. On the other hand, the universal approximation property is quite general and can be proved for different sets of basis functions, such as polynomials or sinusoids, even if they are not used in NNs. Basically, every nonlinearity can be used as a universal approximator (Kreinovich, 1991). This conforms with our earlier conclusion that all "normal" basis functions are eligible as approximation schemes, and no one can be singled out as an outstanding tool. As the study of the NN history reveals, the use of most bases in NNs has been motivated by other than mathematical reasons.

2. Error Bounds

Furthermore, there exist some theoretical results (Barron, 1994; Girosi, 1993) which provide a priori order of magnitude bounds on the generalization error as a function of the size of the network and the number of the data points. For all these results, however, an assumption on the function space, where the target function belongs, is invariably made. Since it is unlikely that the user will have a priori knowledge on issues like the magnitude of the first Fourier moment of the real function (Barron, 1994) or if the same function is of a Sobolev type (Girosi, 1993), the convergence proofs and error bounds derived under those assumptions have limited practical value. It can be safely concluded that NNs offer no more guarantees with respect to the magnitude of the generalization error than any other regression technique.

The next question is whether the construction of NNs at least guarantees accurate and smooth approximating functions with respect to the available data. The critical issue here is, given a specific selection of the basis functions, how their number is determined (step 2). The established practice reveals that the selection of the NN size is based on either empirical grounds or statistical techniques. These techniques use different aspects of the data, such as the size of the sample or its distribution in the input space, to guide the determination of the network structure. Some techniques, such as the k-means clustering algorithm (Moody and Darken, 1989), are essentially reflections of the empirical rule that the basis functions should be placed in proportion to the data density in the input space. That is an intuitive argument, but also easy to defy. Large density in the input space does not necessarily entail the existence of a complex feature of the underlying function requiring many basis functions to be represented.

It is interesting to notice that the decision on the network size is taken in the absence of any consideration on both the accuracy and the smoothness of the resulting model. On one hand, this might lead to unnecessary complexity and nonsmoothness of the model if the size is larger than needed. This is the case with the regularization approach to radial basis function networks (RBFNs) (Poggio and Girosi, 1989), which assigns a basis function to every data point available. On the other hand, because of the arbitrariness in structuring the network, the accuracy can be poor. The discovery of the network inefficiency, though some validation technique using additional empirical (testing) data, results to a repeated cycle of trial-and-error tests between steps 2 and 3 of the NN algorithm. This is a strong indication that the NN algorithm does not provide in the first place any guarantees of accuracy and smoothness for the approximating function.

Finally, it would be interesting to see if, with an increasing number of data and allowing the NN to find the optimal coefficient values, the real function can ever be closely approximated. Because of the arbitrariness of the NN structuring, the approximation error for the selected network can be arbitrarily large. As long as adaptation of the structure is not allowed, the gap between the model and the real function can never close. The adaptation of the coefficient values (reduction of the estimation error) is certainly not the remedy, if the initial selection of G is inappropriate. It should be noted that there exists some work (e.g., Bhat and McAvoy, 1992) on the problem of optimally structuring the network. However, the problem is dealt at the initial stage (step 1) and the proposed solutions do not include any considerations for structural adaptation at the learning stage (step 3). In most of the cases, the definition of the network architecture involves few initial data points and the solution of a considerable optimization problem that is difficult to duplicate in an on-line fashion when more data are available. As long as the network structuring becomes a static (with respect to the data) decision, the preceding considerations still hold independently of the method used for determining the NN structure.

III. Solution to the Functional Estimation Problem

A. FORMULATION OF THE LEARNING PROBLEM

The analysis of NNs has shown that, in order to assure both accuracy and smoothness for the approximating function, the solution algorithm will have to allow the model (essentially its size) to evolve *dynamically* with the data. This calls not only for a new algorithm but also for a new problem definition. The new element is the stepwise presentation of the data to the modeling algorithm and the dynamic evolution of the solution itself. All elements in the problem have to be indexed by a time-like integer l, which will correspond to the amount of data available and that, by abuse of notation, will be referred as the *instant*. The new problem is called the *learning problem* and is defined as follows:

Learning Problem. At every instant l let $Z_l = \{(\mathbf{x}_i, y_i) \in R^{k \times 1} i = 1, 2, \ldots, l\}$ be a set of data drawn with some unknown probability distribution, $P(\mathbf{x}, y)$ and G_l, a space of functions. Find a function $\hat{g}_l(\mathbf{x}): X \to Y$ belonging to G_l that minimizes the empirical risk $I_{emp}(g)$.

There is a "time" continuity of all the variables in the problem, which is expressed by the following relations:

(a) $Z_l = Z_{l-1} \cup \{(\mathbf{x}_l, y_l)\}$
(b) $G_l \supseteq G_{l-1}$
(c) $G_l = A(G_{l-1})$, where A is the algorithm for the adaptation of the space, G_l
(d) $\hat{g}_l(\mathbf{x}) = M(Z_l, G_l)$, where M is the algorithm for the generation of the model, $\hat{g}_l(\mathbf{x})$

The first relation represents the data buildup, while the next two reassure that the space G_l evolves over its predecessor, G_{l-1} according to algorithm A. At every instant, the model results by applying algorithm M on the available data, Z_l, and the chosen input space, G_l.

With such an approach to the functional estimation problem, we escape from the cycle of deriving models and testing them. With every amount of data available, the simplest and smoothest model is sought. Every new data point builds constructively on the previous solution and does not defy it, as it is the case with statistical techniques such as cross-validation. The expectation is that eventually a function close to the real one will be acquired. The procedure is ever-continuing and is hoped to be ever-improving. Not unjustifibly, we call such a procedure *learning* and the computational scheme that realizes it, a *learning scheme*. The learning scheme consists of four elements:

1. A dynamic memory of previous data, Z_l
2. An approximating function or model, $\hat{g}_l(\mathbf{x})$
3. A built-in algorithm A for the adaptation of the space, G_l
4. A built-in algorithm M for the generation of the model, $\hat{g}(\mathbf{x})$

Independently of the amount of data and the way they are acquired, the definition of the learning problem simulates a continuous data flow. Such a framework is of particular importance for the cases where the data are indeed available sequentially in time. Process modeling for adaptive control is one important application, which fits perfectly into this framework. The very nature of feedback control relies on and presupposes the explicit or implicit measurement of the process inputs and outputs; therefore, a continuous *teacher* to the process model is available. Accurate process modeling is essential for control. However, process identification is a hard problem. The derivation of a suitable mechanism which can fully exploit the continuous flow of useful information has been our main motivation for developing the learning framework itself and its solution.

1. Derivation of Basic Algorithm

The solution of the learning problem requires the construction of an algorithm, A, for adaptation of the structure and that in turn presupposes the selection of a proper criterion, which will indicate the need for adaptation. Unlike NNs, such a criterion will have to be based on the empirical error to avoid the dangers of overdetermining the size of the network, which might cause overfitting, or underdetermining it, which will result in large approximation error. The framework calls for an appropriately defined *threshold* on the empirical error, as the criterion which will serve as a guide for the structural adaptation.
We are now ready to propose the basic algorithm for the solution of the learning problem.

Basic Algorithm

Step 1. Select a family of basis functions, $\theta(\mathbf{x})$.
Step 2. Select a form for the empirical risk, $I_{emp}(g)$, and establish a threshold, ε, on the empirical error.
Step 3. Select a minimal space, G_0.
Step 4. Learning step: For every available data point, apply algorithm M to calculate the model and estimate the empirical error on all available data. If it exceeds the defined threshold, use algorithm A to update the structure of G until threshold is satisfied.

In the remaining sections, specific answers will be given for the selection of the

(a) Functional space, G.
(b) Basis functions $\theta(\mathbf{x})$.
(c) The empirical risk, $I_{emp}(g)$, and the corresponding error threshold, ε.

With every specification of the above parameters, the basic algorithm will be refined and its properties will be studied, until the complete algorithm is revealed. However, each of the presented algorithms can be considered as a point of departure, where different solutions from the finally proposed can be obtained.

2. Selection of the Function Space

The solution to the learning problem should provide the flexibility to search for the model in increasingly larger spaces, as the inadequacy of the smaller spaces to approximate well the given data are proved. This immediately calls for a hierarchy in the space of functions. Vapnik (1982) has introduced the notion of *structure* as an infinite ladder of finite-dimensional nested subspaces:

$$S_0 \subset S_1 \subset S_2 \subset \cdots S_n \subset \cdots. \tag{8}$$

For reasons that will become clear later on, we will refer to the index j of each subspace as the *scale*. For continuity to be satisfied, the basis functions that span each subspace S_j have to be a superset of the set of basis functions spanning the space S_{j-1}. Vapnik solves the minimization problem in each subspace and uses statistical arguments to choose one as the final solution to the approximation problem for a given set of data. He calls his method *structural minimization*. Although we have adopted his idea of the structure given by Eq. (8), we have implemented it in a fairly different spirit. At every instant, the space G_l, where the model is sought, is set equal to one of the predefined spaces S_j. With more data available and depending on the prediction accuracy, the model is sought in increasingly larger spaces. In this way, we establish a strictly forward move into the structure and allow the incoming data dictate the pace of the move.

One example of a structure (8) is the space of polynomials, where the ladder of subspaces corresponds to polynomials of increasing degree. As the index j of S_j increases, the subspaces become increasingly more complex where complexity is referred to the number of basis functions spanning each subspace. Since we seek the solution at the lowest index space, we express our bias toward simpler solutions. This is not, however, enough in guaranteeing smoothness for the approximating function. Additional restrictions will have to be imposed on the structure to accommodate better the notion of smoothness and that, in turn, depends on our ability to relate this intuitive requirement to mathematical descriptions.

Although other descriptions are possible, the mathematical concept that matches more closely the intuitive notion of smoothness is the *frequency* content of the function. Smooth functions are sluggish and coarse and characterized by very gradual changes on the value of the output as we scan the input space. This, in a Fourier analysis of the function, corresponds to high content of low frequencies. Furthermore, we expect the frequency content of the approximating function to vary with the position in the input space. Many functions contain high-frequency features dispersed in the input space that are very important to capture. The tool used to describe the function will have to support *local* features of *multiple resolutions* (variable frequencies) within the input space.

Smoothness of the approximating function with respect to the data forces analogous considerations. First, we expect the approximating function to be smoother where data are sparse. Complex behavior is not justified in a region of the input space where data are absent. On the other hand, smoothness has to be satisfied in a dynamic way with every new data point. The model adaptation, triggered by any point, should not extend beyond the neighborhood of "dominance" of the new data point and, at the same time, should not be limited to a smaller area. It is not justifiable to change the entire model every time one point is not predicted accurately by the model, nor is it meaningful to induce so local changes that only a single point is satisfied. Again the need for *localization both in space and frequency* shows up. This is not surprising, since the very nature of the problem is a game between the localization of the data and the globality of the sought model. The problem itself forces us to think in multiple resolutions.

Here is how we can achieve a multiresolution decomposition of the input space. Let $I_0 = [a,b]^k$ be the entire bounded input space according to the problem definition. In each dimension, the interval $[a,b]$ is divided into two parts that are combined into 2^k subregions, $I_{1,\mathbf{p}}$, where $\mathbf{p}$ is k-dimensional vector that signifies the position of the subregion and whose elements take the values 1 or 2. Let I_1 signify the set of all these subregions: $I_1 = \{I_{1,\mathbf{p}} | \mathbf{p} = (q,r), q, r = 1 \text{ or } 2)$, which will be referred as the scale 1 *decomposition* of the input space. The union of all these subregions forms the entire input space: $\cup I_{1,\mathbf{p}} = I_0$. the subdivision of each subregion in I_1 follows the same pattern, resulting in a *hierarchy* of decompositions $I_0, I_1, I_2, \ldots$ each corresponding to a division of the input space in a different resolution. An example of the resulting ladder of subregions is given in Fig. 2 for the two-dimensional case.

There exists a one-to-one correspondence between the structure of subspaces and the hierarchy of input space decompositions. This already

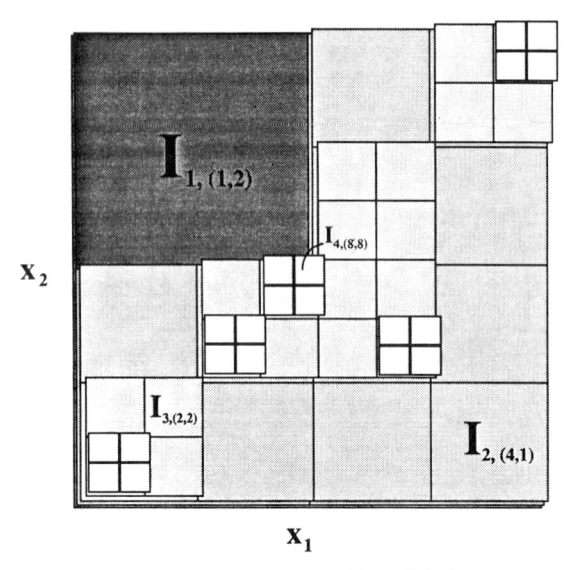

FIG. 2. Multiresolution decomposition of the input space.

poses a restriction on the basis functions that span the subspaces. They have to be local and have bounded support. Such a requirement excludes popular in regression functional bases such as polynomials and sigmoids that are global. Each subregion in I_j corresponds to the support of a basis function in S_j. In this way, the problem is decomposed into smaller subproblems where it is solved almost independently depending on the degree of overlap of the functions spanning those regions. The extreme case is when there is no overlap such as when piecewise approximation in each subregion is used. In this analysis, we have adopted for simplicity a predefined dyadic subdivision of the input space. However, this is not a unique choice. Different formulas can be invented to break the dyadic regularity and even formulate the grid of subregions not in a predefined way but in parallel with the data acquisition. The additional requirement for the basis functions used is the construction of the basis to support this localized multiresolution structure.

Given a multiresolution decomposition of the input space, the algorithm A for the adaptation of the model structure, identifies the region $I_{j+1,\mathbf{p}}$ in the next scale where the incoming data point belongs and adds the corresponding basis function that spans this region. With this addition, the basic algorithm can now be modified as follows.

Algorithm 1

Step 1. Select a family of basis functions, $\theta(\mathbf{x})$, supporting a multiresolution decomposition of the input space.

Step 2. Select a form for the empirical risk, $I_{emp}(g)$, and establish a threshold, ε, on the empirical error.

Step 3. Select a minimal space, G_0, and input space decomposition, I_0.

Step 4. Learning step: For every available data point, apply algorithm M to calculate the model and estimate the empirical error on all available data. If it exceeds the defined threshold, use algorithm A to update the structure of G, by inserting the lowest scale basis function that includes the new point. Repeat until threshold is satisfied.

3. Selection of the Error Threshold

Algorithm 1 requires the a priori selection of a threshold, ε, on the empirical risk, $I_{emp}(g)$, which will indicate whether the model needs adaptation to retain its accuracy, with respect to the data, at a minimum acceptable level. At the same time, this threshold will serve as a termination criterion for the adaptation of the approximating function. When (and if) a model is reached so that the generalization error is smaller than ε, learning will have concluded. For that reason, and since, as shown earlier, some error is unavoidable, the selection of the threshold should reflect our preference on how "close" and in what "sense" we would like the model to be with respect to the real function.

Given a space G, let $g^*(\mathbf{x})$ be the "closest" model in G to the real function, $f(\mathbf{x})$. As it is shown in Appendix 1, if $f \in G$ and the L^∞ error measure [Eq. (4)] is used, the real function is also the best function in G, $g^* = f$, independently of the statistics of the noise and as long as the noise is symmetrically bounded. In contrast, for the L^2 measure [Eq. (3)], the real function is not the best model in G if the noise is not zero-mean. This is a very important observation considering the fact that in many applications (e.g., process control), the data are corrupted by non-zero-mean (load) disturbances, in which cases, the L^2 error measure will fail to retrieve the real function even with infinite data. On the other hand, as it is also explained in Appendix 1, if $f \notin G$ (which is the most probable case), closeness of the real and "best" functions, $f(\mathbf{x})$ and $g^*(\mathbf{x})$, respectively, is guaranteed only in the metric that is used in the definition of $I(g)$. That is, if $I(g)$ is given by Eq. (3), $g^*(\mathbf{x})$ can be close to $f(\mathbf{x})$ only in the L^2-sense and similarly for the L^∞ definition of $I(g)$. As is clear, L^2

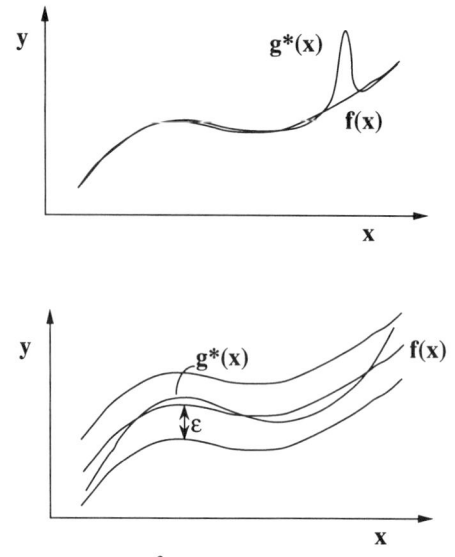

FIG. 3. L^2 vs. L^∞ closeness of functions.

closeness does not entail L^∞ closeness and vice versa. The question now is what type of closeness is preferable.

Figure 3 gives two examples of L^2 and L^∞ closeness of two functions. The L^2 closeness leaves open the possibility that in a small region of the input space (with, therefore, small contribution to the overall error) the two functions can be considerably different. This is not the case for L^∞ closeness, which guarantees some minimal proximity of the two functions. Such a proximity is important when, as in this case, one of the functions is used to predict the behavior of the other, and the accuracy of the prediction has to be established on a pointwise basis. In these cases, the L^∞ error criterion (4) and its equivalent [Eq. (6)] are superior. In fact, L^∞ closeness is a much stricter requirement than L^2 closeness. It should be noted that whereas the minimization of Eq. (3) is a quadratic problem and is guaranteed to have a unique solution, by minimizing the L^∞ expected risk [Eq. (4)], one may yield many solutions with the same minimum error. With respect to their predictive accuracy, however, all these solutions are equivalent and, in addition, we have already retreated from the requirement to find the one and only real function. Therefore, the multiplicity of the "best" solutions is not a problem.

There is one additional reason why the L^∞ empirical risk is a better objective function to use. With the empirical risk given by Eq. (6), which is by definition a pointwise measure, it is clear how to define in practice the numerical value of the threshold that will be applied to every data point to assess the model accuracy. That would not be the case if the L^2 error were used whose absolute magnitude is difficult to correlate with the goodness of the approximation.

In light of the previous discussion and contrary to the established practice, we propose the use of the maximum absolute error (6) as the empirical risk to be minimized because it offers the following advantages:

1. As long as the noise is symmetrically bounded, the L^∞ error is minimized by the true function, $f(\mathbf{x})$, independently of the statistics of the noise.
2. Minimization of the L^∞ error ensures L^∞ closeness of the approximating function to the real function.
3. It is straightforward to define a numerical value for the threshold on the L^∞ error.

The first two points raise the question of convergence of the model to the real function, or its best model, $g^*(\mathbf{x})$. In the following analysis we will address the convergence question, first assuming that the model with the desired accuracy lies in a known space G and then when this space is attempted to be reached with the hierarchical adaptation procedure of Algorithm 1.

Given a choice for G, the question is, if and under which conditions the model, $\hat{g}(\mathbf{x})$, converges to $g^*(\mathbf{x})$ as the number of points increases to infinity, or, in other words, if it is possible to completely eliminate the estimation error. The answer will emerge after addressing the following questions.

4. As the Sample Size Increases, Does $I_{emp}(g)$ Converge to $I(g)$?

Under very general conditions, it is true that the empirical mean given by Eq. (5), converges to the mathematical expectation (3), as $l \to \infty$ (Vapnik, 1982). On the other hand, it is clear that

$$\max_{i=1,\dots,l} |y_i - g(\mathbf{x}_i)| \to \sup_x |y - g(\mathbf{x})| \quad \text{as} \quad l \to \infty$$

and convergence of $I_{emp}(g)$, given by Eq. (4) to $I(g)$, given by Eq. (6), is also guaranteed.

5. Does Convergence of $I_{emp}(g)$ to $I(g)$ Entail Convergence of $\hat{g}$ to g^*?

If the answer were positive, that would ensure that, as we minimize the $I_{emp}(g)$ and more data become available, we approach the best approximating function, $g^*(\mathbf{x})$. Unfortunately, in general, this is not true and that means that as more data become available the estimation error does not necessarily decrease. Convergence can, however, be proved in a weaker sense for the L^∞ case, if we retreat from the requirement to acquire $g^*(\mathbf{x})$ at infinity and be satisfied with *any* solution that satisfies an error bound in its L^∞-norm distance from the real function $f(\mathbf{x})$.

Theorem 1. *Let $\varepsilon > 0$ and $G_\varepsilon = \{g \in G | I(g) < \varepsilon\}$ be a nonempty subset of G and ε, a small positive number. The minimization of the empirical risk [Eq. (6)] will converge in G_ε as $l \to \infty$.*

The proof is given in Appendix 2. The theorem basically states that if there exists a function in G that satisfies $I(g) < \varepsilon$, then by minimizing the L^∞ error measure, this function will be found. The generalization error for this function is guaranteed to be less than $\varepsilon - \delta$. The theorem, however, presupposes the existence of the function with the desired accuracy. This, in general, cannot be guaranteed a priori. In the following, we will prove that the adaptation methodology developed for Algorithm 1 will converge to a subspace where the existence of such a function is guaranteed. For the proof, we have to make the following assumptions for the ladder of subspaces defined by Eq. (8):

Assumption 1. At every subspace S_{j2} there exists a better approximating function than any function in S_{j1} with $j_2 > j_1$:

$$\forall j_2 > j_1, \exists g' \in S_{j2} \text{ such that } I(g') < I(g) \; \forall g \in S_{j1}.$$

Assumption 2. There exists a subspace S_{j^*} with $j^* < \infty$ where a function with the desired approximation accuracy exists:

$$\exists j^* \text{ and } g^* \in S_{j^*} \text{ such that } I(g^*) < \varepsilon,$$

or, in other words, G_ε in S_{j^*} is nonempty.

On the basis of these assumptions we can prove the following theorem:

Theorem 2. *Algorithm 1 will converge to S_{j^*}, and no further structural adaptation will be needed for any additional data points.*

The proof is given in Appendix 3. Theorem 2 indicates that by forcing $I_{\text{emp}}(g)$ to be less than ε at all times and by following Algorithm 1, we are guaranteed to reach a subspace S_{j^*} where a solution with $I(g) < \varepsilon$ exists. According to Theorem 1, the algorithm will converge into the set G_ε and all subsequent solutions to the minimization of $I_{\text{emp}}(g)$ will obey $I(g) < \varepsilon$ which is the criterion we want to be satisfied. The generalization error will be $\varepsilon - \delta$. Therefore, by defining an error threshold $\varepsilon + \delta$, we will eventually reach a solution with a generalization error less than ε. The significance of Algorithm 1 according to Theorem 2, is that it provides a formal way to screen out functional spaces where solutions with the desired accuracy are not present, and that it eventually concludes on the smaller subspace where such a solution exists. This result is possible by the use of the L^∞ error measure and the structural adaptation procedure. It should be reminded that if the space G were to be determined a priori and statically (as it is the case for NNs), such a convergence would not be possible unless, luckily, that space already contained a function with the desired accuracy.

It should be noted, however, that the convergence result does not entail a monotonic decrease of the generalization error. The algorithm moves to spaces where potentially better solutions can be obtained, but the solution that it chooses is not necessarily better than the previous one. This is too much to ask and indeed no approximation scheme, whether it applies structural adaptation or not, can guarantee strict improvement as the number of data points increases to infinity. On the other hand, no reassurance on the number of data needed for convergence in S_{j*} and G_ε are provided. This issue is related to the data quality, i.e., the degree to which the incoming data reveal the inadequacies of the existing model. If the data are not "good enough," convergence can be delayed beyond the point where data are available. In this case, however, the solution is still satisfactory as long as, by construction, it serves properly the accuracy—smoothness tradeoff, which, as stated earlier, is the only criterion in assessing the goodness of the approximation.

We can now revise Algorithm 1 with the definition of the error threshold.

Algorithm 2

Step 1. Select a family of basis functions, $\theta(\mathbf{x})$, supporting a mutliresolution decomposition of the input space.

Step 2. Select a threshold ε, with $\varepsilon > 0$ and $\varepsilon > \delta$, on the empirical error defined by Eq. (6).

Step 3. Select a minimal space, G_0, and input space decomposition, I_0.

Step 4. Learning step: For every available data point, apply algorithm M to calculate the model by solving the minimization problem, $\hat{g} = \min_g I_{\text{emp}}(g) = \min_g \max_{i=1,\ldots,l} |y_i - g(x_i)|$, $g_j \in S_j$, and estimate the empirical error on all available data, $I_{\text{emp}}(g)$. If $I_{\text{emp}}(g) > \varepsilon$, use algorithm A to update the structure of G, by inserting the lowest scale basis function that includes that point. Repeat until threshold is satisfied.

6. Representation of the Functional Spaces

The only element of the learning algorithm that remains undetermined is the basis functions that span the subspaces according to Eq. (8). The basic requirement for the basis functions are to be local, smooth and to have adjustable support to fit the variable size subregions of the input space. The Gaussian function, for example, with variable mean and variance has all the required properties. The convergence theorems stated in the previous section, however, show that the basis used is also required to exhibit well-defined properties as an approximation scheme. This stems from the requirements (implied by Assumptions 1 and 2) that, as we move on to higher index subspaces, we increase our ability to fit better not only the data (which is the easy part) but the unknown function as well. This is a major difference of this approach with other algorithms, such as the resource-allocating network (RAN) algorithm (Platt, 1991) where on-line adaptation of the structure is performed but not in a well-defined framework that will guarantee improved approximation capabilities of the augmented network. The additional approximation properties are sought in the framework of *multiresolution analysis* and the *wavelet representation*. For details on both these tools, the reader is referred to the work of Mallat (1989), Daubechies (1992), Strang (1989), and also Bakshi and Stephanopoulos (1993) for a NN-motivated treatment of wavelets. Here, we will state briefly the basic facts that prove that the wavelet transform indeed fits the developed framework and will emphasize more its properties as an approximation scheme.

a. Multiresolution Analysis. Let $f(x) \in L^2(R)$ be a one-dimensional function of finite energy. Following Mallat (1989), a multiresolution analysis of $f(x)$ is defined as a sequence of successive approximations of $f(x)$ resulting from the projection of $f(x)$ on finer and finer subspaces of functions, S_j. Le $f(x)$ be the approximation of $f(x)$ at resolution 2^j or scale j. As the scale decreases, $f^j(x)$ looks coarser and smoother and devoid of any detail of high frequency. Inversely, with the scale increasing, the approximation looks finer and finer until the real function is recovered at infinite scale:

$$\lim_{j \to \infty} f^j(x) = f(x). \tag{9}$$

Multiresolution analysis conforms with the definition of structure (8). More importantly, it guarantees that by moving to higher subspaces (scales), better approximations of the unknown functions can potentially be obtained, which is the additional property sought.

Regarding the representation of the subspaces, S_j, there exists a unique function $\phi(x)$, called the *scaling function*, whose translations and dilations span those subspaces:

$$S_j = \left\{ \sum_n c_{jn} \phi_{jn}(x) \mid j, n \in Z, c_{jn} \in R \right\}, \quad \text{where}$$

$$\phi_{jn}(x) = \sqrt{2^j} \phi(2^j x - n) \quad (10)$$

The scaling function, $\phi(x)$, has either local support or decays very fast to zero. For all practical purposes, it is a local function. By translating and dilating that function we are able to cover the entire input space in multiple resolutions, as it is required.

The framework, however, as introduced so far is of little help for our purpose since the shift from any subspace to its immediate in hierarchy would require to change entirely the set of basis functions. Although $\phi_{jn}(x)$ are all created by the same function, they are different functions and, consequently, the approximation problem has to be solved from scratch with any change of subspace. The theory of wavelets and its relation to multiresolution analysis provides the ladder that allows the transition from one space to the other.

It can be proved that the approximation $f^{j+1}(x)$ of $f(x)$ in scale $j+1$ can be written as a combination of its approximation at the lower scale j with additional detail represented by the wavelet transform at the same scale:

$$S_{j+1} = \left\{ \sum_n c_{jn} \phi_{jn}(x) + \sum_n d_{jn} \psi_{jn}(x) \mid j, n \in Z, c_{jn}, d_{jn} \in R \right\}, \quad \text{where}$$

$$\psi_{jn}(x) = \sqrt{2^j} \psi(2^j x - n) \quad (11)$$

and $\psi(x)$ is the *wavelet function* determined in unique correspondence to the scaling function. With this addition, the shift to higher scales for improved approximation involves the addition of a new set of basis functions that are the wavelets at the same scale and whose purpose is to capture the high-frequency detail ignored in the previous scale. This analysis can be easily extended to dimensions higher than one.

There exist different pairs of wavelets and scaling functions. One such pair is shown in Fig. 4. This is the "Mexican hat" pair (Daubechies, 1992), which draws its name by the fact that the scaling function looks like the

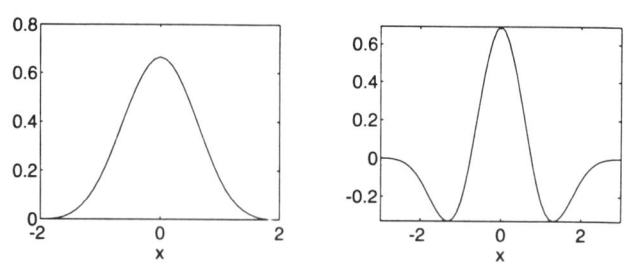

FIG. 4. "Mexican hat" scaling function and wavelet.

Gaussian and the wavelet like the Gaussian second derivative ("Mexican hat" function).

b. *Wavelet Properties*. All wavelets and scaling functions share the property of *localization in space and in frequency*, as required. Although such a property (in the weak sense that we are using it as a means to represent the smoothness requirement) can be satisfied with other functions as well, it is only with wavelets that it finds an exact mathematical description. The Fourier transform of the wavelets is a bandpass filter. The higher the scale, the more the passband extends over higher frequencies. That explains how the wavelets of high scales are able to capture the fine, high-frequency detail of any function they represent. The combination of space and frequency localization is a unique property for bases of $L^2(R)$ and it is what made wavelets a distinguished mathematical tool for applications where a multiresolution decomposition of a function or a signal is required.

As approximation schemes, wavelets trivially satisfy the Assumptions 1 and 2 of our framework. Both the L^2 and the L^∞ error of approximation is decreased as we move to higher index spaces. More specifically, recent work (Kon and Raphael, 1993) has proved that the wavelet transform converges uniformly according to the formula

$$\sup_x |f^j(x) - f(x)| \leq C 2^{-jp}, \qquad (12)$$

where j is the scale, C a constant, and p the order of the approximation. The latter corresponds to the number of vanishing moments of the wavelet (Strang, 1989):

$$\int x^m \psi(x)\, dx = 0 \quad \text{for } m = 0, \ldots, p - 1, \qquad (13)$$

and is an important indicator of the accuracy of each wavelet as an approximator. As can be easily verified by Eq. (12), the larger the number of vanishing moments, the faster the wavelet transform converges to the real function. The approximation error decreases with the scale and for any arbitrary small number ε there exists a scale $j^* = -(1/p)\log_2(\varepsilon/C)$, where the approximation error is less than ε. To use the NN terminology, wavelets are universal approximators since any function in $L^2(R^k)$ can be approximated arbitrarily well by its wavelet representation. Wavelets, however, offer more than any other universal approximator in that, they provide, by construction, a systematic mechanism (their hierarchical structure) to achieve the desired accuracy.

The different pairs of wavelet and scaling functions are not equivalent as approximators. Besides the number of vanishing moments, another distinguishable property is that of orthogonality. Orthogonality results in considerable simplifications in the structural adaptation procedure, since every new orthogonal wavelet inserted in the approximating function introduces independent information. From the approximation point of view, orthogonality is equivalent to more compact and, therefore, more economical representations of the unknown functions. There is, however, a price to pay. With the exception of the discontinuous Haar wavelet, there do not exist wavelets that are compactly supported, symmetrical and orthogonal. In addition, all orthogonal and compactly supported wavelets are highly nonsmooth and, therefore, are not suitable for approximation problems. For this reason, *biorthogonal* and *semiorthogonal* wavelets have been constructed (Cohen, 1992; Feauveau, 1992) that retain the advantages of orthogonality and, at the same time, satisfy the smoothness requirement.

The space–frequency localization of wavelets has lead other researchers as well (Pati, 1992; Zhang and Benveniste, 1992) in considering their use in a NN scheme. In their schemes, however, the determination of the network involves the solution of complicated optimization problem where not only the coefficients but also the wavelet scales and positions in the input space are unknown. Such an approach evidently defies the on-line character of the learning problem and renders any structural adaptation procedure impractical. In that case, those networks suffer from all the deficiencies of NNs for which the network structure is a static decision.

B. LEARNING ALGORITHM

With the selection of wavelets as the basis functions the learning algorithm can now be finalized.

Algorithm 3

Step 1. Select a family of scaling functions and wavelets.
Step 2. Select a threshold, ε, with $\varepsilon > 0$ and $\varepsilon > \delta$, on the empirical error defined by Eq. (6).
Step 3. Select an initial scale, j_0.
Step 4. Learning step: For every available data point, apply algorithm M to calculate the model by solving the minimization problem, $\hat{g} = \min_g I_{emp}(g) = \min_g \max_{i=1,\ldots,l} |y_i - g(\mathbf{x}_i)|$, $g_j \in S_j$, and estimate the empirical error on all available data, $I_{emp}(g)$. If $I_{emp}(g) > \varepsilon$, use algorithm A to update the structure of G, by inserting the lowest scale wavelet that includes that point. Repeat until threshold is satisfied.

The algorithm is schematically presented in Fig. 5. The implementation of the algorithm presupposes the a priori specification of

(a) The scaling function and wavelet pair.
(b) The initial scale j_0.
(c) The value of the error threshold, ε.

For the moment, there are no guidelines for the selection of the particular basis functions for any given application. The important issue here is that the properties of the wavelets will be inherited by the approximating

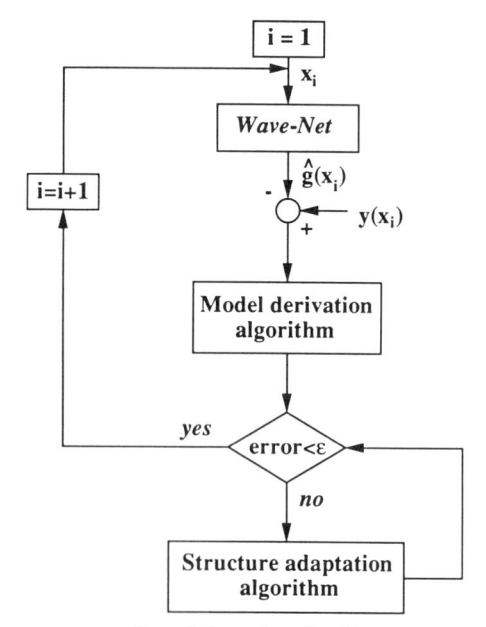

FIG. 5. Learning algorithm.

function and, therefore, if there exist prerequisites on the form of the approximating function (such as the number of existing derivatives), they should be taken into account before the selection of the wavelet. The initial scale, where the first approximation will be constructed, can be chosen as the scale in which few scaling functions (even one) fit within the input range, I_0, of the expected values of **x**. Finally, the selection of the error bound depends on the accuracy of the approximation sought for the particular application. In any case, it should be equal or greater than the bound on the expected noise, δ.

1. The Model and Its Derivation

The approximating function constructed by the previous algorithm is of the form

$$g(x) = \sum_{k=1}^{m} c_k \theta_k(\mathbf{x}), \tag{14}$$

where $\theta(\mathbf{x})$ stands for both the scaling functions at scale j_0 and wavelets at the same or higher scales. Equation (14) can be easily identified as the mathematical representation of the *Wave-Net* (Bakshi and Stephanopoulos, 1993). With the use of the L^∞ error measure, the coefficients are calculated by solving the minimization problem:

$$\mathbf{c} = \arg\min \max_{i=1,\ldots,l} \left| y_i - \sum_{k=1}^{m} c_k \theta_k(\mathbf{x}_i) \right|. \tag{15}$$

The problem [Eq. (15)] is a minimax optimization problem. For the case (as it is here) where the approximating function depends linearly on the coefficients, the optimization problem [Eq. (15)] has the form of the *Chebyshev approximation* problem and has a known solution (Murty, 1983). Indeed, it can be easily shown that with the introduction of the dummy variables z, z_i, z_i^* the minimax problem can be transformed to the following linear program (LP):

$$\min_{\mathbf{c}} z,$$

$$z_i = -y_i + \sum_{k=1}^{m} c_k \theta_k(\mathbf{x}_i) + z \geq 0, \tag{16}$$

$$z_i^* = y_i - \sum_{k=1}^{m} c_k \theta_k(\mathbf{x}_i) + z \geq 0.$$

The transformation to this LP program is graphically depicted in Fig. 6 for the case when **c** is a scalar. For each data pair $(\mathbf{x}_i, y_i)$ the term $|y_i - \Sigma c_k \theta_k(\mathbf{x}_i)|$ represents two $(m+1)$-dimensional hyperplanes, $z = y_i - \Sigma c_k \theta_k(\mathbf{x}_i)$ and $z = -y_i + \Sigma c_k \theta_k(\mathbf{x}_i)$. For the scalar case, these correspond

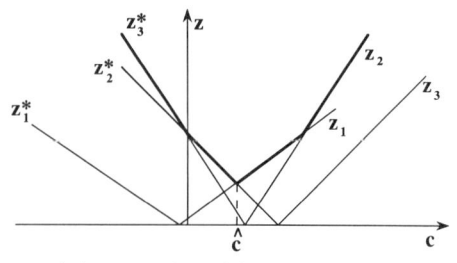

FIG. 6. Geometric interpretation of the Chebyshev approximation problem.

to two lines of opposite slope that meet on the c axis as shown in Fig. 6. For every value of $\mathbf{c}$, according to problem (15), we need to identify the data pair whose corresponding hyperplane has the maximum value, z. The locus of these points is the polyhedron that inscribes the common space above each hyperplane. If an LP formulation, the space above every hyperplane corresponds to an inequality constraint according to Eq. (16), and the intersection of all these constraints is the feasible region. The bottom of the convex polyhedron that inscribes the feasible region is the value of c that minimizes both Eqs. (15) and (16). Its distance from c axis is the minimum value of z, or the minimum possible error. The optimization problem is convex and is guaranteed to have a solution. It can be easily solved using standard LP techniques based on the dual simplex algorithm.

The selection to minimize absolute error [Eq. (6)] calls for optimization algorithms different from those of the standard least-squares problem. Both problems have simple and extensively documented solutions. A slight advantage of the LP solution is that it does not need to be solved for the points for which the approximation error is less than the selected error threshold. In contrast, the least squares problem has to be solved with every newly acquired piece of data. The LP problem can effectively be solved with the dual simplex algorithm, which allows the solution to proceed recursively with the gradual introduction of constraints corresponding to the new data points.

2. Variations on the Structural Adaptation Algorithm

The multiresolution framework allows us to reconsider more constructively some of the features of the structure adaptation algorithm. First, a strictly forward move in the ladder of subspaces [Eq. (8)] is not necessary. Due to localization, the structural correction can be sought in higher spaces before all functions in the previous space are exhausted. This results in an uneven distribution of variable scale features in the input space, which, in turn, means considerable savings in the number of basis functions used.

In a k-dimensional space, one way to construct the scaling function and the wavelets is by taking the k-term tensor products of the one-dimensional parts. For example, in two-dimensional (2D) space the scaling function is $\phi(x_1)\phi(x_2)$ and there are three wavelets: $\phi(x_1)\psi(x_2)$, $\psi(x_1)\phi(x_2)$, and $\psi(x_1)\psi(x_2)$. This result in $2^k - 1$ wavelets all sharing the same support, $I_j, \mathbf{p}$, in the input space. Every time adaptation of the structure is needed at a given position and scale, there are $2^k - 1$ choices and a selection problem arises. By solving locally a small optimization problem, we can identify the wavelet whose insertion in the model minimizes the empirical error and use this one to update the structure. The fact, however, that we are not seeking a complete wavelet representation allows us to consider alternatively empirical selection methods that are not computationally expensive. For example, we might choose to use only the one symmetrical wavelet $\psi(x_1)\psi(x_2)\ldots\psi(x_k)$ at every scale and ignore all the others. On the other hand, there exist methods for constructing high-dimensional wavelets that do not use the tensor products (*nonseparable* wavelets; Kovačević and Vetterli, 1992) and, therefore, automatically overcome this selection problem.

Another variation on the algorithm can be applied by allowing the freedom to select the initial space S_0 outside the multiresolution framework. In many cases, a model of the unknown function is already available by first-principles modeling, or there might exist some a priori knowledge (or bias) on the structure of the unknown function (linear, polynomial, etc.). In all these cases, we would like to combine the available knowledge with the NN model to reduce the blackbox character of the latter and increase its predictive power. There has been some recent work in this direction (Psichogios and Ungar, 1992; Kramer *et al.*, 1992). The multiresolution framework allows us to naturally unify the a priori model with the NN by allowing us to select the initial space S_0 in accordance to our preference. Then, the model in S_0 provides an initial global representation of the unknown function, and the learning algorithm is used to expand and correct locally this model. This procedure is in agreement with the notion of learning as a way to extend what is already known, and is also of considerable practical value.

3. Derivation of Error Bounds

As shown earlier, by imposing a threshold on the L^∞ empirical error and applying the learning algorithm, an approximating function with generalization error less that the threshold will eventually be reached. Therefore, the selected threshold can be regarded as a *global* measure of the approximating capability of the derived model. The multiresolution decomposition of the input space, however, allows us to impose *local* error thresholds and, therefore, seek for variable approximation accuracy in different areas of the input space according to the needs of the problem.

The error threshold can be allowed to vary not only on a spatial but also on a temporal basis as well. Sometimes, the bound on the magnitude of the noise, δ, is not easy to be known, in which cases, the value of the threshold (which must be greater than δ) cannot be determined. In some other situations, it might be risky to impose a tight threshold from the beginning when relatively few data points are available because this might temporarily lead to overfitting. An empirical but efficient way to alleviate those problems is to enforce the error threshold gradually. In other words, we can devise some sequence of error thresholds that satisfy the relationship

$$\varepsilon_1 > \varepsilon_2 > \cdots > \varepsilon_n = \varepsilon$$

and apply them sequentially in time. With this approach, conservative models with respect to the prediction accuracy are initially derived. But as the threshold tightens and data become more plentiful, the model is allowed to follow the data more closely. Since the final value for the threshold is still ε, the theoretical results derived in previous section still hold.

The value of the threshold provides a global upper bound on the expected error of approximation. As stated earlier, guaranteeing any error bound is impossible unless some a priori assumption is imposed on the real function. Such an assumption could effectively be an upper and lower bound on the magnitude of the first derivative of the function. These bounds naturally arise when the unknown function describes a physical phenomenon, in which case, no steep gradients should be expected. If such bounds are available, the value of function between two neighboring data points is restricted between the intersection of two conic regions whose boundaries correspond to the upper and lower values of the first derivative. This conic region can effectively be considered as a guaranteed bound on the error, which, however, might or might not satisfy the desired accuracy as expressed by the value of the selected error threshold.

A tighter and local *estimate* on the generalization error bound can be derived by observing locally the maximum *encountered* empirical error. Consider a given dyadic multiresolution decomposition of the input space and, for simplicity, let us assume piecewise constant functions as approximators. In a given subregion of the input space, $I_{j,\mathbf{p}}$, let $Z_{l,\mathbf{p}}$ be the set of data at instant l that are present into this region. The maximum encountered empirical error in this area is equal to

$$\eta_{l,\mathbf{p}} = \frac{\max(y_{l,\mathbf{p}}) - \min(y_{l,\mathbf{p}})}{2},$$

and also satisfies the relationship: $\eta_{l,\mathbf{p}} < \varepsilon$. This value constitutes a local error bound on the expected accuracy of the approximation unless otherwise proved by the incoming data. By observing the evolution of the local error bound, we can draw useful conclusions on the accuracy of the approximation in the different areas of the input space.

IV. Applications of the Learning Algorithm

In this section, the functionality and performance of the proposed algorithm are demonstrated through a set of examples. In all examples, the "Mexican hat" wavelet (Fig. 4) was used. An illustration of the effectiveness of the algorithm was given with the example of Fig. 1. The approximating curves were indeed derived by applying the algorithm using an error threshold equal to .5. The example was introduced in Zhang and Benveniste (1992) and its solution using *Wave-Nets* is more thoroughly explained in Bakshi *et al.* (1994).

A. EXAMPLE 1

The first example is the estimation of the 2D function (from Narendra and Parthasarathy, 1990)

$$f(x_1, x_2) = \frac{x_1}{(1 + x_1^2)} + x_2^3. \tag{17}$$

A graph of the real function is presented in Fig. 7. A set of examples was created by random selections of the pairs (x_1, x_2) in the range $I = [0, 1]^2$ and the error threshold, ε, was set equal to .1.

Figure 8 presents the evolution of the model during training. More specifically, the size of the network (number of basis functions) and the generalization error are plotted as a function of data presented to the learning algorithm. It can be observed that although the number of data is small, the generalization error is high and the network adaptation proceeds in high rates to compensate for the larger error. The generalization

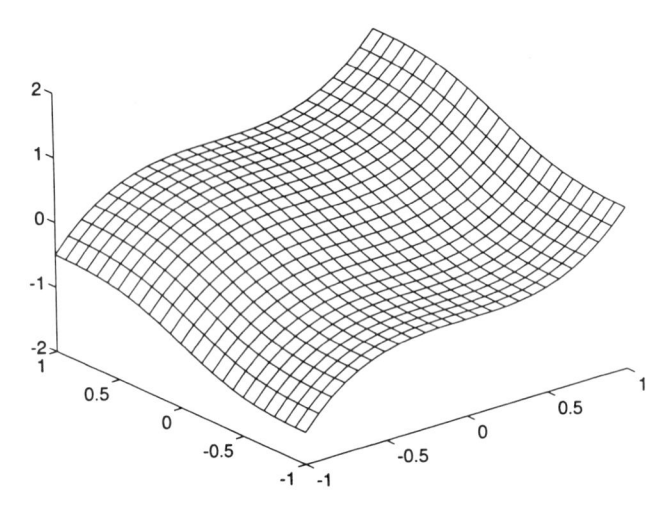

FIG. 7. Real function for Example 1.

error does not decrease monotonically, but it exhibits a clear descending trend. Large changes in the error coincide with data points that trigger structural adaptation (for example, after about 100 and 160 data points). That means that at those points the algorithm shifts to subspaces where better solutions are available and, most importantly, picks one of them resulting in drastic reductions of the error. After almost 160 points, no

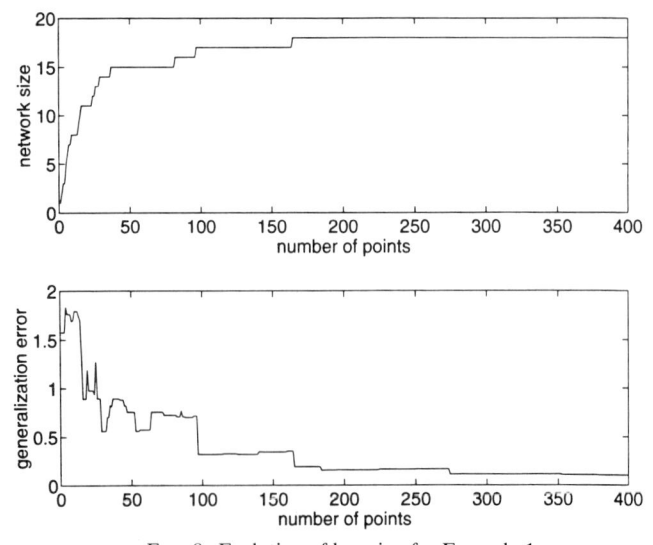

FIG. 8. Evolution of learning for Example 1.

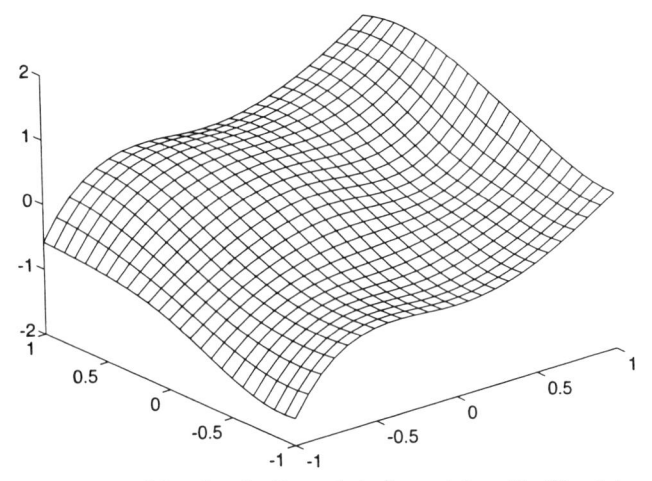

FIG. 9. Model surface for Example 1 after training with 400 points.

further structural adaptation is required and the network size reaches a steady state. The flat sections in the generalization error correspond to periods where no model adaptation is needed. After the 160th data point, coefficient adaptation is required only for three points until the 350th data when the generalization error drops for the first time below .1. At this point, learning has concluded, and a model with the specified accuracy has

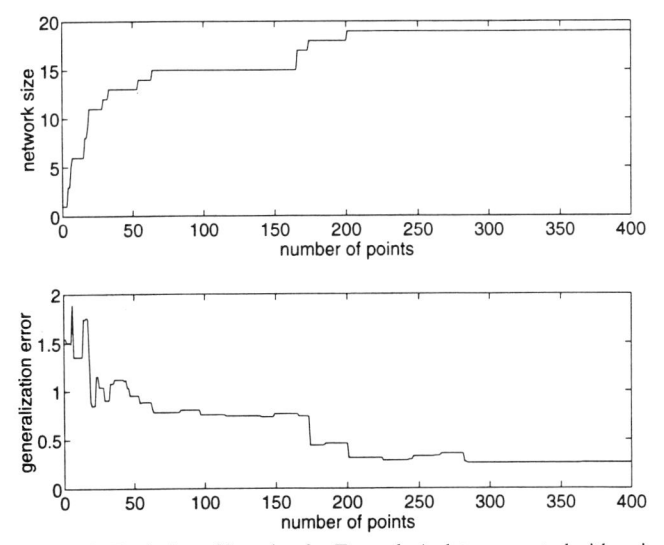

FIG. 10. Evolution of learning for Example 1: data corrupted with noise.

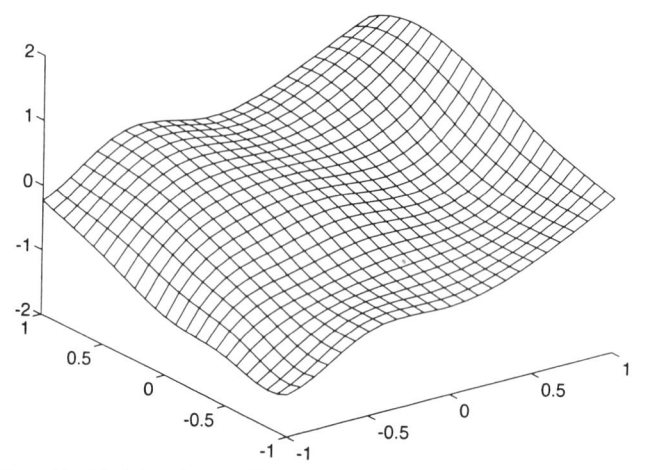

FIG. 11. Model surface for Example 1 after training with 400 noisy points.

been found. This model is shown in Fig. 9. It should be noticed, as is clear by comparing Fig. 7 and 9, that the error of .1 is the worst accuracy achieved. The approximation is much closer in the largest portion of the input space.

The same example was solved for the case when noise is present at the output, y. The maximum amplitude of the noise, δ, was .1 and the error threshold, ε, was set to .2 to reflect our desire to achieve a generalization error of .1($= \varepsilon - \delta$). The results (shown in Fig. 10) are similar with the no-noise case, except that the decrease of the generalization error and the derivation of the final model are slower. This is expected, since the presence of noise degrades the quality of information carried by the data as compared to equal amount of noise-free data. After 400 points the generalization error is .26 and, therefore, a model with the desired accuracy is yet to be achieved. The model (shown in Fig. 11), although not as accurate as before, provides a decent approximation of the real function. More importantly, the use of a larger threshold allows greater tolerance for the observed empirical error and, in this way, allows the model to avoid overfitting the noisy data.

B. Example 2

Consider a continuous-stirred-tank reactor (CSTR) with cooling jacket where a first order exothermic reaction takes place. It is required to derive a model relating the extent of the reaction with the flowrate of the heat transfer fluid. This is a nonlinear identification example very popular in chemical engineering literature. In this study we adopt the approach followed by Hernandez and Arkun (1992). The dimensionless differential equations that describe the material and energy balances in the reactor are

$$\frac{dx_1}{dt} = -x_1 + Da(1-x_1)\exp\left(\frac{x_2}{1+x_2/\gamma}\right),$$

$$\frac{dx_2}{dt} = -x_2 + BDa(1-x_1)\exp\left(\frac{x_2}{1+x_2/\gamma}\right) + b(u-x_1), \quad (18)$$

$$y = x_1$$

where x_1 (or y) is the measured extent of the reaction, and x_2 the dimensionless temperature of the reactor. The input, u, is the dimensionless flowrate of the heat transfer fluid through the cooling jacket. The input (shown in Fig. 12) was constructed as a concatenation of step changes and random signals created by adding a pseudo random binary signal (PRBS) signal between -1 and 1 and a random variable with uniform distribution between $-.5$ and $.5$. The set of differential equations [Eq. (18)] was solved numerically with the initial conditions corresponding to

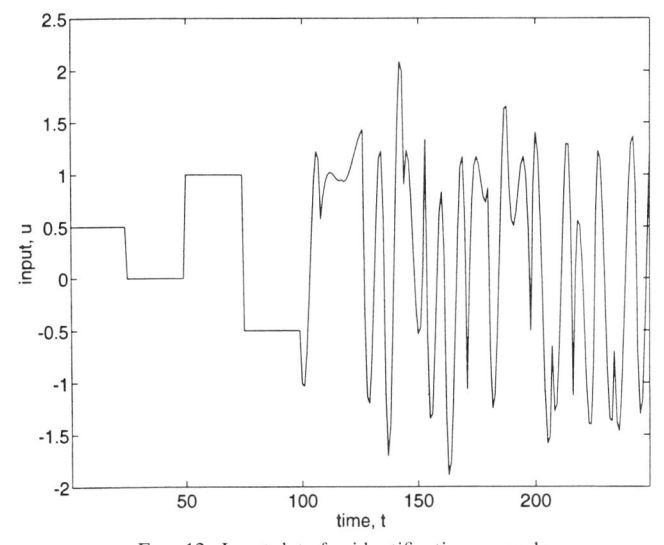

FIG. 12. Input data for identification example.

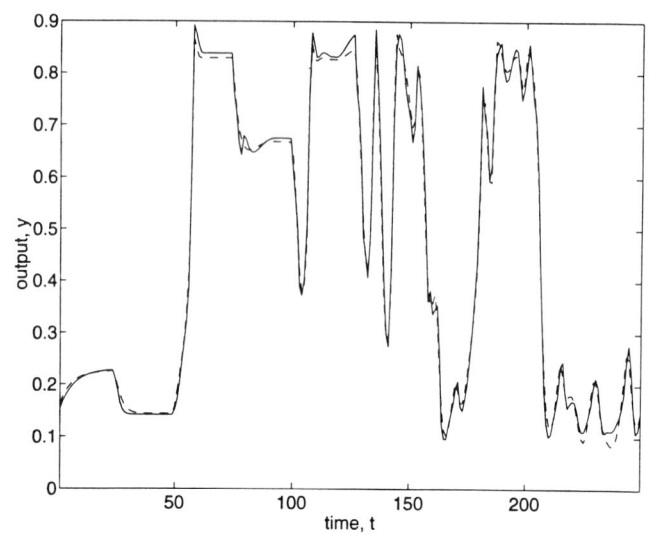

FIG. 13. Comparison of model predictions (solid line) with training data (dashed line) for identification example.

the equilibrium point $x_1 = .144$, $x_2 = .855$, and $u = .0$. The values of the output, created in this way, are represented by the dashed line in Fig. 13. The derived data were used to construct an input/output process model of the form

$$y(t+1) = f[(y(t), y(t-1), u(t))] \qquad (19)$$

by using the *Wave-Net* algorithm to learn the unknown function $f(\cdot, \cdot, \cdot)$. The objective was to use the derived model to predict the behavior of the system in the future for different input values.

In Fig. 13, the model predictions are compared with the real output after training the network with 250 data points. The error threshold was set to .03 and the resulting model contained 37 basis functions with equal number of unknown coefficients. This model was used without further training to predict the system output for the next 750 time instants. The results are shown in Fig. 14, where a fairly good accuracy can be observed. The maximum error associated with the predictions is .11. It is greater than the defined threshold but it is much better than the maximum error ($= .28$) that would result by predicting the value of $y(t+1)$ equal to $y(t)$. This shows that the model has indeed captured the underlying dynamics of the physical phenomenon and is not merely predicting the future as a zero-order extrapolation of the present.

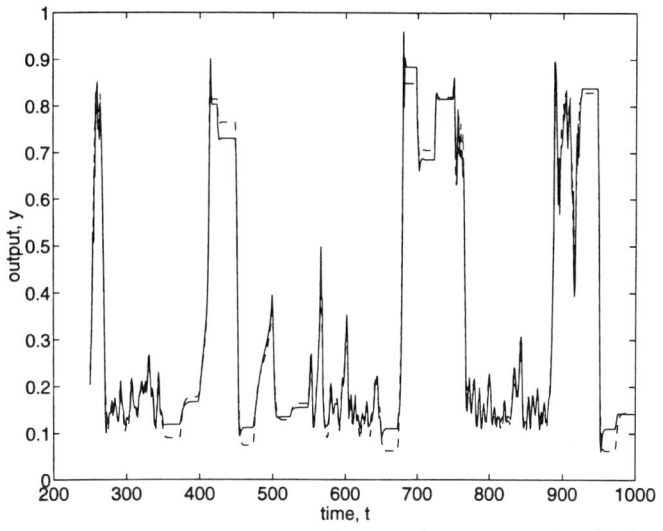

EMPIRICAL LEARNING THROUGH NEURAL NETWORKS 477

FIG. 14. Comparison of model predictions (solid line) with testing data (dashed line) for identification example.

C. EXAMPLE 3

The final example introduced in Zhang and Benveniste (1992) is a two-dimensional function with distinguished localized features. It is analytically represented by the formula

$$f(x_1, x_2) = (x_1^2 - x_2^2)\sin(5x_1) \tag{20}$$

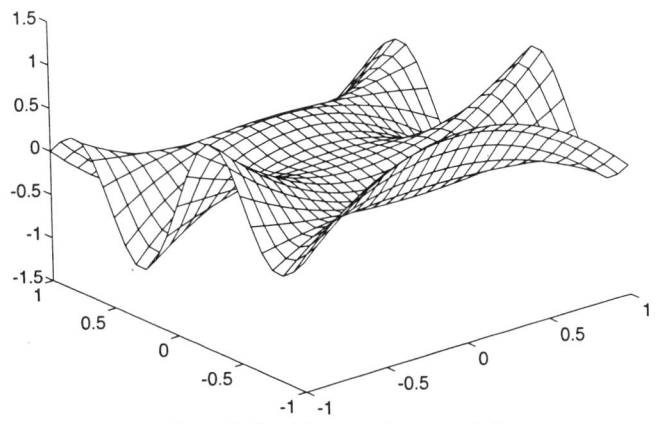

FIG. 15. Real function for example 3.

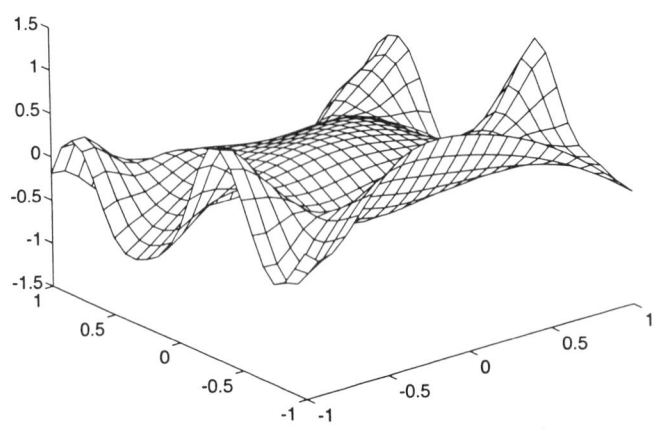

FIG. 16. Model surface for example 3 after training with 500 points.

and graphically represented in Fig. 15. After setting $\varepsilon = .2$ and using 500 data points for training, the resulting model exhibited a generalization error of .22. The model, shown in Fig. 16, has, within the predefined accuracy, captured the important features of the unknown function. This was possible by including during training more high-scale wavelets along the x_1 axis where the significant features lie. The coverage of the input space by the basis functions is shown in Fig. 17, where the circles

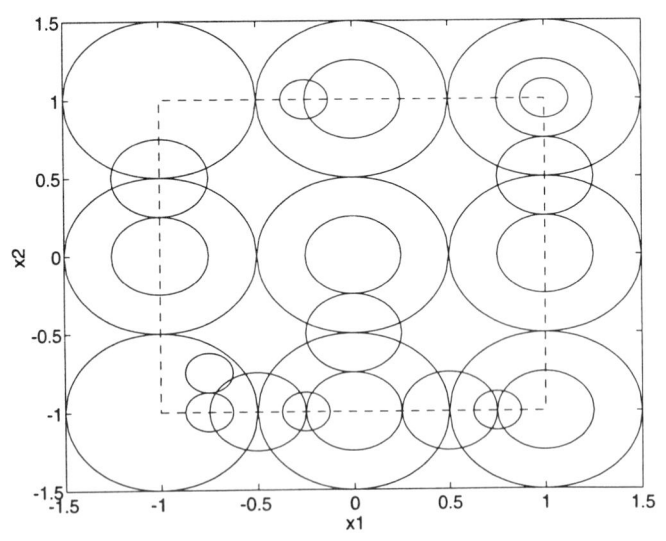

FIG. 17. Coverage of the input space (square) by the support (circles) of the basis functions for the model of Example 3.

correspond to the "dominant" support of each basis function and the square bounds the area where data were available. It is not surprising that most of the basis functions lie in the boundaries of the input space because this is where the biggest errors are most often encountered.

V. Conclusions

In this chapter the problem of estimating an unknown function from examples was studied. We emphasized the issues that stem from the practical point of view in posing the problem. In practice, one starts with a set of data and wants to construct an approximating function, for which, almost always, nothing is known a priori. The main point of the chapter is that the unknown function cannot be found but can be learned. Learning is an ever-lasting and ever-improving procedure. It requires the existence of a teacher that is the data and an evaluative measure. To ensure continuous improvement, the model has to be susceptible to new data and structural adaptation is an essential requirement to achieve this. NNs are neither data-centered nor flexible enough to enable one to learn from them since the important decisions for a given application are made in the absence of data. The derived approximating function has to be coarse and conservative where data are sparse and to follow closely the fluctuations of the data that might indicate the presence of a distinguished feature; localization and control of smoothness are important elements of the approximating scheme.

All the issues raised by these practical considerations have found rigorous answers within the *Wave-Net* framework. The chapter advocates the following points:

(a) The need for on-line model adaptation in parallel with the data acquisition
(b) The use of the local approximation error measured by the L^∞ error to guide the model adaptation
(c) The use of the multiresolution framework to formalize the intuitive preference to simple and smooth approximating functions
(d) The use of bases, like wavelets, with well-defined properties as approximation schemes

The derivation of process models for adaptive control falls exactly within the framework of the estimation problem studied in this chapter. Control-related implementation are natural extensions to the current work and are studied in another publication (Koulouris and Stephanopoulos, 1995). Some computational details and variations of the adaptation algorithm have also to be evaluated to improve the efficiency of the computations.

VI. Appendices

A. APPENDIX 1

1. Case 1: $f \in G$.

When $I(g)$ is given by Eq. (3), the solution to the minimization problem is the *regression function* (Vapnik, 1982):

$$g^* = \int yP(y|\mathbf{x}) \, dy. \tag{21}$$

Since $y = f(\mathbf{x}) + d$, $P(y|z\mathbf{x}) = P[y - f(\mathbf{x})] = \mathbf{P}(d)$, and Eq. (21) yields

$$g^* = \int [f(\mathbf{x}) + d] P[y - f(\mathbf{x})] \, dy = \int f(\mathbf{x}) \mathbf{P}[y - f(\mathbf{x})] \, dy + \int d\mathbf{P}(d) \, dd$$

$$= f(x) + \bar{d}.$$

That is, the real function $f(\mathbf{x})$ is the solution to the minimization of Eq. (3) only in the absence of noise ($d = 0$) or when the noise has zero mean ($\bar{d} = 0$). This is, in fact, true for all L^n norms with $2 < n < \infty$ under the condition that n moments of the noise are zero. When the L^∞ metric given by Eq. (4) is used, we will prove that $f(\mathbf{x})$ yields the minimal value of $I(g)$, independently of the statistics of the noise and as long as the noise is symmetrically bounded.

$$\text{If } g = f, I(f) = \sup_x |y - f(\mathbf{x})| = \sup|d| = \delta,$$

$$\text{If } g \neq f, I(g) = \sup_x |y - g(\mathbf{x})| = \sup_x |f(\mathbf{x}) - g(\mathbf{x}) + d|$$

$$= \sup_x |f(\mathbf{x}) - g(\mathbf{x})| + \delta > \delta. \tag{22}$$

The reason for the last inequality is that for every $\mathbf{x}$ the maximum value of $|f(\mathbf{x}) - g(\mathbf{x}) + d|$ is $|f(\mathbf{x}) - g(\mathbf{x})| + \delta$ and the supremum value for all $\mathbf{x}$ is $\sup_x |f(\mathbf{x}) - g(\mathbf{x})| + \delta$.

2. Case 2: $f \notin G$.

It can be easily proved that when $f(\mathbf{x})$ is the regression function (Vapnik, 1982)

$$\int [y - g(x)]^2 P(\mathbf{x}, y) \, d\mathbf{x} \, dy = \int [y - f(\mathbf{x})]^2 P(\mathbf{x}, y) \, d\mathbf{x}, dy$$

$$+ \int [f(\mathbf{x}) - g(\mathbf{x})]^2 P(\mathbf{x}) \, d\mathbf{x},$$

$$I(g) = I(f) + \mu_2(g, f),$$

$$\mu_2(g, f) = I(g) - I(f),$$

which implies L^2 closeness [$\mu_2(g, f)$ is small] if $I(g) - I(f)$ is small [$I(g) \geq I(f)$]. Similarly for the L^∞ case,

(22) $\Rightarrow \quad \sup_x |f(\mathbf{x}) - g(\mathbf{x})| = I(g) - I(f) \quad$ [since $I(f) = \delta$].

Again, the smaller the value of $I(g)[I(g) \geq I(f)]$, the closer $g(\mathbf{x})$ is to $f(\mathbf{x})$ in the L^∞ sense.

B. Appendix 2

The proof follows three intermediate steps.

(a) Let $I_{\text{emp}}^{(l)}(g) = \max_{i=1,\dots,l} |y_i - g(\mathbf{x}_i)|$. Then

$$I_{\text{emp}}^{(l)}(g) \leq I_{\text{emp}}^{(m)}(g) \text{ for } m > l \text{ and } \forall g \in G, \tag{23}$$

which essentially states that every new point reveals for every function in G an empirical error that is equal or worse than the one already encountered. Equation (23) also conforms with the fact that $I_{\text{emp}}(g) < I(g)$ and is the basis of convergence of $I_{\text{emp}}(g)$ to $I(g)$.

(b) (23) $\Rightarrow \quad I(g) - I_{\text{emp}}^{(l)}(g) \geq I(g) - I_{\text{emp}}^{(m)}(g), \quad m > l,$

$$\sup_g \left[I(g) - I_{\text{emp}}^{(l)}(g) \right] \geq \sup_g \left[I(g) - I_{\text{emp}}^{(m)}(g) \right],$$

$$\limsup_{l \to \infty} \sup_g \left[I(g) - I_{\text{emp}}^{(l)}(g) \right] = 0.$$

This basically proves uniform convergence of $I_{emp}^{(l)}(g)$ to $I(g)$ and that, in turn, means

$$\forall k > 0, \exists N > 0 \quad \text{such that} \quad \sup_g \left[I(g) - I_{emp}^{(l)}(g) \right] < \kappa \quad \forall l > N. \tag{24}$$

(c) Let $\varepsilon > 0$ be a given positive number and $G_\varepsilon = \{g \in G | I(g) < \varepsilon\}$, which we assume to be nonempty. Let also $g^* = \arg\min I(g)$. Obviously, $g^* \in G$ and $I(g^*) < \varepsilon$. If, in Eq. (24), we set $\kappa = \varepsilon - I(g^*)$, then

$$\exists N > 0 \quad \text{such that} \quad I(g) - I_{emp}^{(m)}(g) < \varepsilon - I(g^*) \quad \forall m > N$$

$$I_{emp}^{(m)}(g) > I(g) - \varepsilon + I(g^*) \tag{25}$$

We want to prove that, if this is the case, then only solutions with $I(g) < \varepsilon$ will be produced by the minimization of the empirical risk, and convergence in this weak sense will be guaranteed. Let g' be a function such that $I(g') > \varepsilon$. Then from Eq. (25)

$$I_{emp}^{(m)}(g) > I(g') - \varepsilon + I(g^*) > I(g^*),$$

but

$$I(g^*) > I_{emp}^{(m)}(g^*), \quad \text{so} \quad I_{emp}^{(m)}(g') > I_{emp}^{(m)}(g^*),$$

and, therefore, g' can never be the solution to the minimization of $I_{emp}^{(m)}(g)$ since there will always be at least be at least the best approximation g^*, with guaranteed lower value of the empirical risk.

C. Appendix 3

We first claim that if in some subspace $j < j^*$, $I_{emp}(g) > \varepsilon \; \forall g \in S_j$, there does not exist $g \in S_j$ so that $I(g) < \varepsilon$. That is straightforward since $I_{emp}(g) < I(g)$. By forcing a threshold on I_{emp}, we are guaranteed to find a data point for which no solution in S_j can satisfy the bound. The algorithm will then look for the solution at the immediate subspace S_{j+1} where, by Assumption 1, a better solution exists. Following this procedure, eventually subspace S_j^* is reached. Its existence is guaranteed by Assumption 2. The algorithm will never move to higher subspaces $j > j^*$ since $I_{emp}(g) < I(g) < \varepsilon \; \forall g \in G_\varepsilon$, and the requirement $I_{emp}(g) < \varepsilon$ will always be satisfied.

References

Bakshi, B., and Stephanopoulos, G., Wave-Net: A multiresolution, hierarchical neural network with localized learning. *AIChE J*. **39**, 57 (1993).

Bakshi, B., Koulouris, A., and Stepanopoulos, G., Learning at multiple resolutions: Wavelets as basis functions in artificial neural networks and inductive decision trees. *In* "Wavelet Applications in Chemical Engineering" (R. L. Motard and B. Joseph, eds.) Kluwer Academic Publishers, Dordrecht/Norwell, MA, p. 139 (1994).

Barron, A. R. Approximation and estimation bounds for artificial neural networks. *Mach. Learn*. **14**, 115 (1994).

Barron, A. R., and Barron, R. L., Statistical learning networks: A unifying view. *In* "Symposium on the Interface: Statistics and Computing Science." p. 192. Reston, VA, 1988.

Barto, A. G., Connectionist learning for control. *In* "Neural Networks for Control." (W. T. Miller, R. S. Sutton and P. J. Werbos, eds.) p. 5. MIT Press, Cambridge, MA, 1991.

Bhat, N. V., and McAvoy, T. J., Use of neural nets for dynamic modeling and control of chemical process systems. *Comput. Chem. Eng*. **14**, 573 (1990).

Bhat, N. V., and McAvoy, T. J., Determining the structure for neural models by network stripping. *Comput. Chem. Eng*. **16**, 271 (1992).

Cohen, A., Biorthogonal wavelets. *In* "Wavelets—A Tutorial in Theory and Applications," (C. K. Chui, ed.), Academic Press, San Diego, CA, p. 123. 1992.

Daubechies, I., "Ten Lectures on Wavelets." SIAM Philadelphia, 1992.

Feauveau, J. C., Nonorthogonal multiresolution analysis using wavelets. *In* "Wavelets—A Tutorial in Theory and Applications" (C. K. Chui, ed.), Academic Press, San Diego, CA, p. 153. 1992.

Girosi, F., "Rates of Convergence of Approximation by Translates and Dilates," AI Lab Memo, Massachusetts Institute of Technology, Cambridge, MA, 1993.

Girosi, F., and Anzellotti, G., Rates of convergence for radial basis functions and neural networks. "Artificial Neural Networks with Applications in Speech and Vision," (R. J. Mammone, ed.), p. 97. Chapman & Hall, London, 1993.

Hartman, E., Keeler, K., and Kowalski, J. K., Layered Neural Networks with Gaussian hidden units as universal approximators. *Neural Comput*. **2**, 210 (1990).

Hernandez, E., and Arkun, Y., A study of the control relevant properties of backpropagation neural net models of nonlinear dynamical systems. *Comput. Chem. Eng*. **16**, 227 (1992).

Hornik, K., Stinchcombe, M., and White, H., Multi-layer feedforward networks are universal approximators. *Neural Networks* **2**, 359 (1989).

Hoskins, J. C., and Himmelblau, D. M., Artificial neural network models of knowledge representation in chemical engineering. *Comput. Chem. Eng*. **12**, 881 (1988).

Kearns, M., and Vazirani, U., "An Introduction to Computational Learning Theory." MIT Press, Cambridge, MA. 1994.

Kon, M., and Raphael, L., "Convergence Rates of Wavelet Expansions," preprint 1993.

Koulouris, A., and Stepanopoulos, G., On-line empirical learning of process dynamics with Wave-Nets, submitted to *Comput. Chem. Eng.* (1995).

Kovačević, J., and Vetterli, M., Nonseparable multiresolutional perfect reconstruction banks and wavelet bases for R^n. *IEEE Trans. Inf. Theory*, **38**, 533 (1992).

Kramer, M. A., Thompson, M. L. and Bhagat, P. M., Embedding theoretical models in neural networks. *Proc. Am. Control Conf*. 475 (1992).

Kreinovich, V. Y., Arbitrary nonlinearity is sufficient to represent all functions by neural networks: A theorem. *Neural Networks* **4**, 381 (1991).

Lee, M., and Park, S., A new scheme combining neural feedforward control with model predictive control. *AIChE J*., **38**, 193 (1992).

Leonard, J. A., and Kramer, M. A., Radial basis function networks for classifying process faults. *IEEE Control Syst.* **11,** pp. 31–38, (1991).

Ljung, L., "System Identification: Theory for the User." Prentice-Hall Englewood Cliffs, NJ. (1987).

Mallat, S. G., A theory for multiresolution signal decomposition: The wavelet representation. *IEEE Trans. Pattern Anal. Mach. Intell.* **PAMI-11**, 674 (1989).

Mavrovouniotis, M. L., and Chang, S., Hierarchical Neural Networks. *Comput. Chem. Eng.* **16**, 347 (1992).

Moody, J., and Darken, C. J., Fast learning in networks of locally-tuned processing units. *Neural Comput.* **1**, 281 (1989).

Murty, K. G., "Linear Programming." Wiley, New York, 1983.

Narendra, K. S., and Parthasarathy, K., Identification and control of dynamical systems using neural networks. *IEEE Trans. Neural Networks* **1**, (1990).

Pati, Y. C., Wavelets and time-frequency methods in linear systems and neural networks. Ph.D. Thesis, University of Maryland, College Park (1992).

Platt, J., A resource-allocating network for function interpolation. *Neural Comput.* **3**, 213 (1991)

Poggio, T., and Girosi, F., "A Theory of Networks for Approximation and Learning," AI Lab. Memo. No. 1140. Massachusetts Institute of Technology, Cambridge, MA, 1989.

Psichogios, D. C., and Ungar, L. H., Direct and indirect model based control using artificial neural networks, *Ind. Eng. Chem. Res.* **30**, 2564 (1991).

Psichogios, D. C., and Ungar, L. H., A hybrid neural network-first principles approach to process modeling. *AIChE J.* **38**, 1499 (1992).

Rengaswamy, R., and Venkatasubramaniam, V., Extraction of qualitative trends from noisy process data using neural networks. *AIChE Annu. Meet. Los Angeles* (1991).

Rumelhart, D. E., McClelland, J. L., and the PDP Research Group, "Parallel Distributing Processing." MIT Press, Cambridge, MA, 1986.

Strang, G., Wavelets and dilation equations: A brief introduction. *SIAM Rev.* **31**, 614 (1989).

Ungar, L. H., Powell, B. A., and Kamens, S. N., Adaptive Networks for fault diagnosis and process control. *Comput. Chem. Eng.* **14**, 561 (1990).

Vapnik, V., "Estimation of Dependences Based on Empirical Data." Springer-Verlag, Berlin, 1982.

Ydstie, B. E., Forecasting and control using adaptive connectionist networks. *Comput. Chem. Eng.* **14**, 583 (1990).

Zhang, Q., and Benveniste, A., Wavelet networks. *IEEE Trans. Neural Networks* **3**, 889 (1992).

REASONING IN TIME: MODELING, ANALYSIS, AND PATTERN RECOGNITION OF TEMPORAL PROCESS TRENDS

Bhavik R. Bakshi

Department of Chemical Engineering
Ohio State University
Columbus, Ohio 43210

George Stephanopoulos

Laboratory for Intelligent Systems in Process Engineering
Department of Chemical Engineering
Massachusetts Institute of Technology
Cambridge, Massachusetts 02139

I. Introduction	487
A. The Content of Process Trends: Local in Time and Multiscale	488
B. The Ad Hoc Treatment of Process Trends	490
C. Recognition of Temporal Patterns in Process Trends	492
D. Compression of Process Data	493
E. Overview of the Chapter's Structure	494
II. Formal Representation of Process Trends	495
A. The Definition of a Trend	496
B. Trends and Scale-Space Filtering	500
III. Wavelet Decomposition: Extraction of Trends at Multiple Scales	507
A. The Theory of Wavelet Decomposition	508
B. Extraction of Multiscale Temporal Trends	516
IV. Compression of Process Data through Feature Extraction and Functional Approximation	527
A. Data Compression through Orthonormal Wavelets	527
B. Compression through Feature Extraction	530
C. Practical Issues in Data Compression	530
D. An Illustrative Example	532
V. Recognition of Temporal Patterns for Diagnosis and Control	535
A. Generating Generalized Descriptions of Process Trends	538

B. Inductive Learning through Decision Trees 541
C. Pattern Recognition with Single Input Variable 543
D. Pattern Recognition with Multiple Input Variables 544
VI. Summary and Conclusions 545
References 546

The plain record of a variable's numerical values over time does not invoke appreciable levels of *cognitive* activity to a human. Although it can cause a fervor of numerical computations by a computer, the levels of cognitive appreciation of the variable's temporal behavior remain low. On the other hand, if one presents the human with a graphical depiction of the variable's temporal behavior, the level of cognition increases and a wave of reasoning activities is unleashed. Nevertheless, when the human is presented with scores of graphs, depicting the temporal behavior of interacting variables, that person's reasoning abilities are severely tested. In such case, the computer will happily continue crunching numbers without ever rising above the fray and thus developing a "mental" model, interpreting correctly the temporal interactions among the many variables.

Reasoning in time is very demanding, because time introduces a new dimension with significant levels of additional freedom and complexity. While the real-valued representation of variables in time is completely satisfactory for many engineering tasks (e.g., control, dynamic simulation, planning and scheduling of operations), it is very unsatisfactory for all those tasks, which require decision making via logical reasoning (e.g., diagnosis of process faults, recovery of operations from large unsolicited deviations, "supervised" execution of startup for shutdown operating procedures).

To improve the computer's ability to reason efficiently in time, we must first establish new forms for the representation of temporal behaviors. It is the purpose of this chapter to examine the engineering needs for temporal decisionmaking and to propose specific models that encapsulate the requisite temporal characteristics of individual variables and composite processes. Through a combination of analytical techniques, such as *scale-space filtering* and *wavelet-based*, *multiresolution decomposition of functions*, and modeling paradigms from *artificial intelligence* (AI), we have developed a concise framework that can be used to model, analyze, and synthesize the temporal trends of process operations. Within this framework, the modeling needs for logical reasoning in time can be fully satisfied, while maintaining consistency with the numerical tasks carried out at the same time. Thus, through the modeling paradigms of this chapter one may put together intelligent systems that use consistent representations for their logical reasoning and numerical tasks.

I. Introduction

Present-day computer-aided monitoring and control of chemical plants has caused an explosion in the amount of process information that can be conveyed to process operators and engineers. The real-time history of thousands of variables can be displayed and monitored. However, whereas a simple visual inspection of scores of displayed trends is sufficient to allow the operator confirm the process' status during normal, steady-state operations, when the process is in significant transience or crises have occurred the displayed trends and alarms can confound even the best of the operators. When the process variables change with different rates, or are affected by varying transportation lags, or inverse the response dynamics, or information is "suspicious" or lost because of sensors' malfunctioning, it is very difficult for a human operator to carry out routine tasks, such as the following:

- Distinguish normal from abnormal operating conditions (Bastl and Fenkel, 1980; Long and Kanazava, 1980).
- Identify the causes of process trends, e.g., external load disturbances, equipment faults, operational degradation, operator-induced mishandling.
- Evaluate current process trends and anticipate future operational states.
- Plan and schedule sequences of operating steps to bring the plant at the desired operating level, e.g., recover from safety fallback position, return to feasible operation after a fault.

The key recognitive skill required to carry out the above tasks is the formation of a "mental" model of the process operations that fits the current facts about the process and enables the operators to correctly assess process behavior and predict the effects of possible control actions. Correct "mental" models of process operations have allowed operators to overcome the weakness of "lost" sensors and conflicting trends, even under the pressure of an emergency (Dvorak, 1987), whereas most of the operational mishandlings are due to an erroneous perception as to what is going on in the process (O'Shima, 1983).

In order to develop intelligent, computer-aided systems with systematic and sound methodologies for the automatic creation of "mental" models of process operations, we need to resolve the following two and interrelated issues:

- What is the appropriate representational model for describing the "true" process trends, and how is it generated from the process data?
- How does one generate relationships among process trends in order to provide the desired "mental" model of process operations?

In the subsequent paragraphs and sections of this chapter we will see that these two issues impose requirements that transgress the abilities of simple "smoothing" filters and conventional regression techniques.

A. THE CONTENT OF PROCESS TRENDS: LOCAL IN TIME AND MULTISCALE

The time-dependent behavior of measured variables in a chemical process reflects the composite effect of many distinct contributions, coming from the underlying physicochemical phenomena and the status of processing equipment, sensors, and control valves. Thus, basic process dynamics, sensor noise, actuator dynamics, parameter drifts, equipment faults, external load disturbances, and operator-induced actions combine their contributions in some unknown way to form the temporal behavior of the measured operational data. As an example, consider the process signal shown in Fig. 1 (Cheung, 1992). It reflects the composite effect of contributions from five distinct sources; as a slow drift caused by a fouling process (source 1), equipment faults (sources 2 and 3), a periodic disturbance (source 4), and changing sensor noise (source 5). Ideally, we would like to have analytical techniques that can take the observed signal apart and render the exact temporal reproduction of each contributing signal. We know that this is theoretically impossible, but practically acceptable representations of the individual components is feasible and very valuable for the correct interpretation of process trends.

A systematic analysis of a process signal over (1) different segments of its time record and (2) various ranges of frequency (or *scale*) can provide a *local* (in time) and *multiscale* hierarchical description of the signal. Such description is needed if an intelligent computer-aided tool is to be constructed in order to (1) localize in time the "step" and "spike" from the equipment faults (Fig. 1), or the onset of change in sensor noise characteristics, and (2) extract the slow drift and the periodic load disturbance.

The engineering context of the need for multiscale representation of process trends can be best seen within the framework of the hierarchical

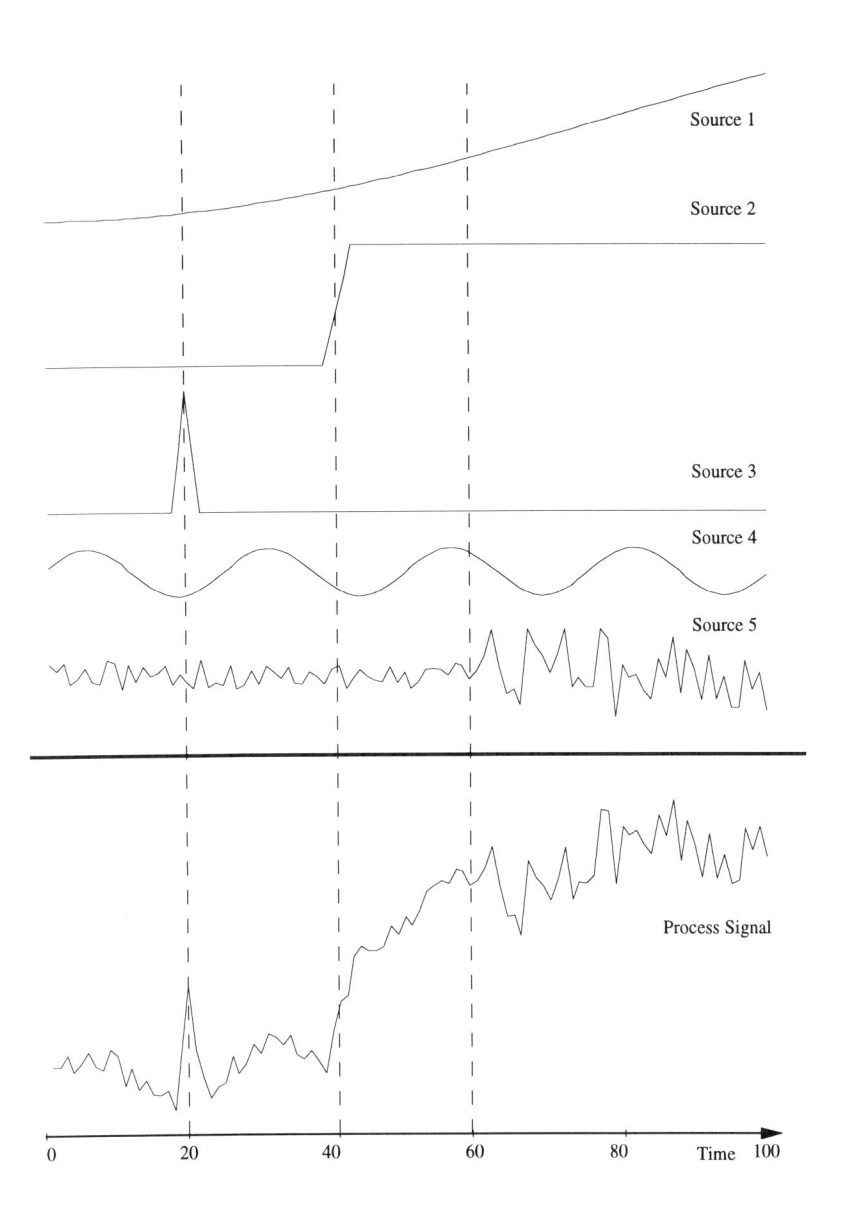

FIG. 1. A process signal and its component.

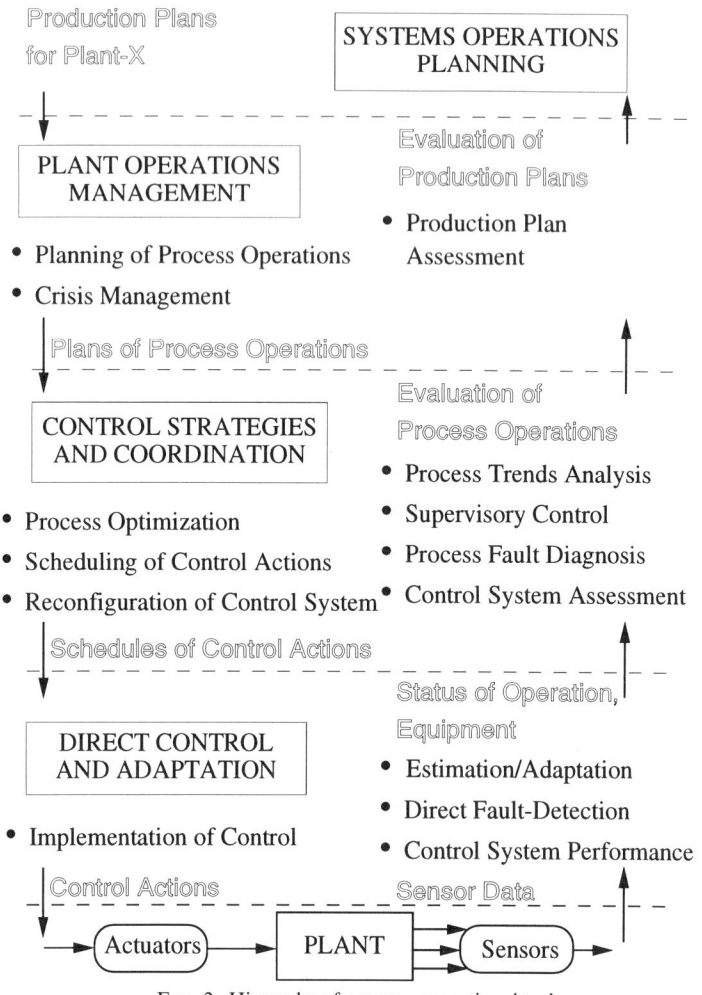

FIG. 2. Hierarchy of process operational tasks.

stratification of operational tasks, shown in Fig. 2 (Stephanopoulos, 1990). At the lowest level of abstraction, process data at a scale of seconds or minute are used to carry out a variety of numerical and logical tasks.

B. THE AD HOC TREATMENT OF PROCESS TRENDS

The term, *process trend*, undoubtedly carries an intuitive meaning about how process behavior changes over time. However, the exact meaning has never been formalized to allow one to articulate it in a general and concrete manner. For a long time, a process trend has been considered just as a smooth representation of a noisy signal, and consequently belonged to the province of digital signal processing community. However, the paradigms advanced are mostly ad hoc in character, and rarely can they sustain adequate fundamental physical justification. For example, during the design of digital filters, *the frequency response is selected* and the essential design task is to determine the coefficients of the filter that match the response. How do you come up, though, with the correct frequency response for the discovery of the "true" process trends in a signal? The answers have been essentially empirical and ad hoc, and have led to an explosive proliferation of design techniques. For example, the Fourier approach and over 100 windowing techniques (e.g., the uniform, von Hann, Hamming, and Kaiser windows) can be used for the design of nonrecursive filters, whereas the *impulse-invariant* method, the *bilinear transform*, and various computer-aided, trial-and-error techniques can be used for the design of recursive filters.

Furthermore, the current understanding of temporal process trends, produced by digital filters, is heavily founded on statistical considerations. Strong assumptions are often made about the statistical nature of the trend, in order to obtain "rigorously" optimal detection schemes. For example, the optimal Bayesian detection, the moving-average filters with adaptive controlled width, and the Wiener filters are only *theoretically optimal* (i.e., their practical implementation could yield grossly inadequate results) for (1) known signals with additive Gaussian noise (2) unknown signals with narrow band relative to the noise, and (3) random signals with known spectra, respectively (Van Trees, 1968; Papoulis, 1977). Although it is commonly agreed that a trend will appear if the random elements are removed from the signal, no filtering technique has incorporated a formal notion of the trend itself by eliciting its fundamental properties and its relation to the physicochemical process it describes.

Once a smooth signal has been constructed, how is the trend represented? Most of the available techniques do not provide a framework for the representation (and thus, interpretation) of trends, because their representations (in the frequency or time domains) do not include primitives that capture the salient features of a trend, such as continuity, discontinuity, linearity, extremity, singularity, and locality. In other words, most of the approaches used to "represent" process signals are in fact *data compaction techniques*, rather than trend representation approaches. Furthermore, whether an approach employs a frequency or a time-domain representation, it must make several major decisions before the data are compacted. For frequency-domain representations, assumptions about the origin of the trend must be made, e.g., whether it is stationary (i.e., frequency content independent of time), quasistationary (i.e., frequency content varies slowly with time), or nonstationary. For the curve-fitting representations in the time domain, the selection of the particular functional model and the thresholds for fitting errors are ad hoc decisions.

It is clear from the preceding discussion that *the deficiencies of the existing frequency- and time-domain representations of process trends stem from their procedural character* (Cheung and Stephanopoulos, 1990), i.e. they represent trends as the outputs of a computational process, which quite often bears no relationships to the process physics and chemistry. What is needed is a *declarative* representation, which can capture explicitly all the desirable characteristics of process trends.

C. Recognition of Temporal Patterns in Process Trends

The correct interpretation of measured process data is essential for the satisfactory execution of many computer-aided, intelligent decision support systems that modern processing plants require. In supervisory control, detection and diagnosis of faults, adaptive control, product quality control, and recovery from large operational deviations, determining the "mapping" from process trends to operational conditions is the pivotal task. Plant operators "skilled" in the extraction of real-time patterns of process data and the identification of *distinguishing features* in process trends, can form a "mental" model on the operational status and its anticipated evolution in time.

A formal induction of mappings from measured operating data to process conditions is composed of the following three tasks (Fig. 3):

Task 1. Extraction of pivotal, temporal features from process data.

Task 2. Inductive learning of the relationship between the features of process trends and process conditions.

Task 3. Adaptation of the relationship utilizing future operating data.

Linear, polynomial, or statistical discriminant functions (Fukunaga, 1990; Kramer, 1991; MacGregor *et al.*, 1991), or adaptive connectionist networks (Rumelhart *et al.*, 1986; Funahashi, 1989; Vaidyanathan and Venkatasubramanian, 1990; Bakshi and Stephanopoulos, 1993; third chapter of this volume, Koulouris *et al.*), combine tasks 1 and 2 into one and solve the corresponding problems simultaneously. These methodologies utilize a priori defined general functional relationships between the operating data and process conditions, and as such they *are not inductive*. Nearest-neigh-

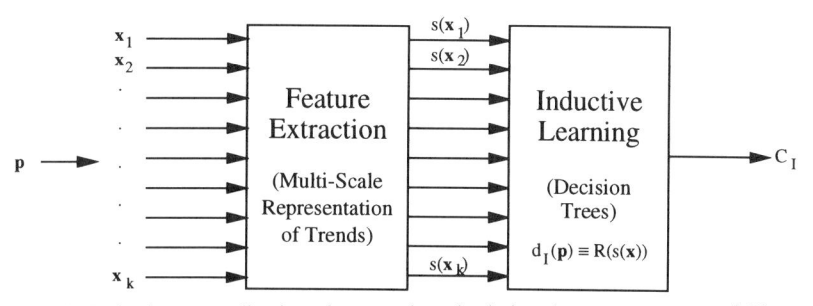

FIG. 3. Inductive generalization of pattern-based relations between process variables and operating conditions.

bor classifiers (Silverman, 1986; Kanal and Dattatreya, 1985), case-based analogical reasoners (Jantke, 1989), or inductive decision trees (Saraiva and Stephanopoulos, 1992), do not assume a functional model and produce truly inductive strategies.

The extraction, though, of the so-called pivotal features from operating data, encounters the same impediments that we discussed earlier on the subject of process trends representation: (1) localization in time of operating features and (2) the multiscale content of operating trends. It is clear, therefore, that any systematic and sound methodology for the identification of patterns between process data and operating conditions can be built only on formal and sound descriptions of process trends.

D. COMPRESSION OF PROCESS DATA

The expression "a mountain of data and an ant hill of knowledge" can be used to signify, besides its obvious message, the significance of the task of maintaining and sorting the vast amounts of data accumulated by present-day, digital process data acquisition systems. Efficient compression, storage, and recovery of historical process data is essential for many engineering tasks. Data compression methodologies must store with minimum distortion of qualitative and quantitative features, independently of whether these features represent behavior *localized in time* or a particular range of frequencies (i.e., *scale*). In other words, the successful compression of process data must necessarily go through an efficient representation of process trends (see Section I, A). Data compression has received surprisingly little attention by the chemical engineering practice, despite the clear articulation of the many benefits that can be drawn from it (Hale and Sellars, 1981; Bader and Tucker, 1987a, b). The *boxcar* and *backward slope* methods are fast, and produce piecewise linear interpolations of the data within predefined, acceptable error bounds, but are unable to capture small process transients and upsets. More techniques (Feehs and Arce, 1988; Bakshi and Stephanopoulos, 1995) have been inspired by technologies used for the compression of speech, images, and communications data, e.g., *vector quantization* (Gray, 1984), and *wavelet decomposition* theory (Mallat, 1989).

The primary objective of any data compression technique is to transform the data to a form that requires the smallest possible amount of storage space, while retaining all the relevant information. The desired qualities of a technique for efficient storage and retrieval of chemical process data are as follows:

1. Compacted data should require minimum storage space.
2. Compaction and retrieval should be fast, often in real time.
3. A clear and explicit measure of the quality of the signal, obtained after retrieval, should be available, to be used as a criterion for guiding the compression.
4. The retrieved signal should have minimum distortion and should contain all the desired features.
5. The compaction should be based on physically intuitive criteria and should require minimum a priori assumptions.

None of the practiced compression techniques satisfies all of these requirements. In addition, it should be remembered that compression of process data is not a task in isolation, but it is intimately related to the other two subjects of this chapter: (1) description of process trends and (2) recognition of temporal patterns in process trends. Consequently, we need to develop a common theoretical framework, which will provide a uniformly consistent basis for all three needs. This is the aim of the present chapter.

E. OVERVIEW OF THE CHAPTER'S STRUCTURE

Section II introduces the formal framework for the definition and description of process trends at all levels of detail: qualitative, order-of-magnitude, and analytic. A detour through the basic concepts of *scale-space filtering* is necessary in order to see the connection between the concept of process trends and the classical material on signal analysis. Within the framework of scale-space filtering we can then elucidate the notions of "episode," "scale," "local filtering," "structure of scale," "distinguished features," and others.

In Section III we introduce the theory of the multiresolution analysis of signals using *wavelet decomposition*, which is used to provide the scale-space image of a function with correct local characteristics. This localized description of a signal's features allows the correct extraction of distinguished attributes from a signal, a task that forms the basis for the inductive generation of pattern-based logical relationships among input and output variables, or the efficient compaction of process data. Of particular value is the construction of *translationally invariant* wavelet decompositions, which allow the correct formation of temporal patterns in process variables. Using translationally invariant decomposition of signals, Section III contains specific methodologies for the construction of the *wavelet interval–tree of scale*, the extraction of distinguished features and the generalization of process trends.

The ideas presented in Section III are used to develop a concise and efficient methodology for the compression of process data, which is presented in Section IV. Of particular importance here is the conceptual foundation of the data compression algorithm; instead of seeking noninterpretable, numerical compaction of data, it strives for an explicit retention of distinguished features in a signal. It is shown that this approach is both numerically efficient and amenable to explicit interpretations of historical process trends.

In Section V we discuss how the temporal distinguished features of many input and output signals can be correlated in propositional forms to provide logical rules for the diagnosis and control of processes, which are difficult to model. *Inductive decision trees* are used to capture the knowledge contained in previous operating data. The explicit description of process trends and the explicit statement of the knowledge "mined" from the data, overcome many of the real-world obstacles in applying these ideas at the manufacturing floor.

Throughout the remaining four sections of this chapter, we will provide illustrations on the use of the various techniques, using real-world case studies.

II. Formal Representation of Process Trends

Although the term "process trend" emulates a certain intuitive understanding in the minds of the speaker and the listener, this understanding may not be the same. Certainly, we do not have a clear, sound, and unambiguous definition of the term "trend," and this must be the first issue to be addressed.

A. The Definition of a Trend

In order to represent and reason with temporal information, we need to represent time explicitly and concisely. The discrete-time character of computer-aided data acquisition and control dictates that time should be represented as a sequence of strictly increasing time points:

$$t = \{t_{-\infty} \cdots t_i \cdots t_j \cdots t_\infty\},$$

where

$$t_i < t_j \Leftrightarrow i < j \quad \text{for all integer } i, j.$$

Furthermore, a *time interval*, I_{ij}, is defined to be an open interval of time as follows:

$$I_{ij} = (t_i, t_j), \quad \text{where} \quad t_i < t_j.$$

Thus, we may represent time as a sequence of open intervals separated by time points, or a sequence of time intervals. We will adopt the second interpretation, considering a time point as an interval of zero duration.

1. From the Quantitative to the Qualitative Representation of a Function

Consider a real-valued function, $x(t)$, with the following properties over a time interval $[a, b]$:

1. $x(t)$ is continuous over $[a, b]$ but is allowed to have a finite number of discontinuities in its value or/ and first derivative.
2. $x'(t)$ and $x''(t)$ are continuous in (a, b), and their one-sided limits exist at a and b.
3. $x(t)$ has a finite number of extrema and inflexion points in $[a, b]$.

Variables that satisfy the above requirements will be called *reasonable variables* (Cheung and Stephanopoulos, 1990). In defining the "reasonableness" of a function, we are concerned only with the properties of the function's value, first and second derivatives. Such definition is less restrictive (it does not require existence of all derivatives), but it is completely general and allows the characterization of a function at different levels (Cheung and Stephanopoulos, 1990). All the physical variables encountered in the operation of a plant are reasonable.

Since we elected to represent time as a sequence of time intervals, we consider that the *state* of a reasonable variable is completely known, if we know the value and the derivative of the variable over a time interval. As the duration of the defining time interval approaches zero, we take the following definition of the state (continuous) of the variable at a given time point (Cheung and Stephanopoulos, 1990):

a. State (Continuous). The state (continuous), $CS(x, t)$ of a reasonable variable, $x:[a, b] \in R \; \forall t \in [a, b]$, is the point value $PtVl(x, t)$ defined by a triplet as follows:

1. If x is continuous at t, then

$$CS(x, t) \equiv PtVl(x, t) = \langle x(t), x'(t), x''(t) \rangle.$$

2. If x is discontinuous at t, then

$$CS(x, t) \equiv PtVl(x, t)$$
$$= \langle \langle x_L(t), x_R(t) \rangle, \langle x_L'(t), x_R'(t) \rangle, \langle x_L''(t), x_R''(t) \rangle \rangle,$$

where the subscripts L and R denote the left- and right-side limits at the discontinuity.

Consequently, the *trend* (continuous) of a variable can be defined as follows

b. Trend (Continuous). The continuous trend of a reasonable variable, $x:[a, b] \to R$, is given by the continuous sequence of states over $[a, b]$.

As we go from continuous to discrete-time representations, the time is represented by a strictly increasing sequence of points. In such case, the state (discrete) of a reasonable function and the associated trend are given by the following definitions:

c. State (Discrete). The state (discrete), $DS(x, t)$ of a reasonable function, $x:[a, b] \to R$, over a set of strictly increasing time points $T = \{a = t_0, \ldots, t_j, \ldots, t_n = b\} \subseteq [a, b]$, is defined as follows:

1. If $t \in T$, then $DS(x, t) = PtVl(x, t)$.
2. If $t_i < t < t_{i+1}$ for $i = 0, 1, \ldots, n$, then $DS(x, t) = \langle PtVl(x, t_i), PtVl(x, t_{i+1}) \rangle$.

d. Trend (Discrete). The discrete trend of a reasonable variable, $x:[a, b] \to R$, is given by the set of discrete states corresponding to the strictly increasing time points of the time interval $[a, b]$.

Clearly, the quantitative description of the discrete state and of the discrete trend must be declaratively explicit, since we cannot perform differentiation at the single points defining the intervals of $[a, b]$. This is a strict requirement and can be theoretically met only if we know the underlying continuous function that provides the values of the derivatives at the time points of a discrete representation. The availability, though, of such a continuous function is based on a series of ad hoc decisions on the character and properties of the functions, and if one prefers to avoid them, then one must accept a series of approximations for the evaluation of first and second derivatives. These approximations provide a sequence of representations with increasing abstraction, leading, ultimately, to qualitative descriptions of the state and trend as follows (Cheung and Stephanopoulos, 1990):

e. State (Qualitative). Let $x:[a,b] \to R$ be a reasonable function. $QS(x,t)$, the qualitative state of x at $t \in [a,b]$, is defined as the triplet of qualitative values as follows:

$$QS(x,t) = \begin{cases} \text{undefined} & \text{if } x \text{ is discontinuous at } t \\ \langle [x(t)], [\partial x(t)], [\partial \partial x(t)] \rangle & \text{otherwise,} \end{cases}$$

where

$$[x(t)] = \begin{cases} + & \text{if } x(t) > 0, \\ 0 & \text{if } x(t) = 0, \\ - & \text{if } x(t) < 0; \end{cases}$$

$$[\partial x(t)] = \begin{cases} + & \text{if } x'(t) > 0, \\ 0 & \text{if } x'(t) = 0, \\ - & \text{if } x'(t) > 0; \end{cases}$$

$$[\partial \partial x(t)] = \begin{cases} + & \text{if } x''(t) > 0, \\ 0 & \text{if } x''(t) = 0, \\ - & \text{if } x''(t) > 0. \end{cases}$$

f. Trend (Qualitative). The qualitative trend of a reasonable variable, $x:[a,b] \to R$, is the continuous sequence of qualitative states over $[a,b]$.

2. Episodes and Trends

The practical value of the qualitative state and trend lies in the fact that both are very close to the intuitive notions employed by humans in interpreting the temporal behavior of signals. But humans capture the trend as a finite sequence of ordered segments with constant qualitative state over each segment. To emulate a similar notion we introduce the concept of *episode* through the following definition (Cheung and Stephanopoulos, 1990):

a. Episode. Let $x:[a,b] \to r$ be a reasonable function. For any time interval $I = (t_i, t_j) \subseteq [a, b]$ such that the qualitative state, $QS(x, t)$ is constant for $\forall t \in (t_i, t_j)$, an episode, E, of x over (t_i, t_j) is the pair, $E = \langle I, QS(x, I) \rangle$, with (1) I signifying the temporal extent of an episode and (2) $QS(, I) = QS(x, t)$ $\forall t \in (t_i, t_j)$ characterizing the constant qualitative state over I. Whenever two episodes, defined over adjacent time intervals, have the same qualitative state values, they can be combined to form an episode with broader temporal extent. On the other hand, if a qualitative state is constant over a time interval, then it is also constant over any of its time subintervals. These observations lead to the following definition of a *maximal episode*.

b. Maximal Episode. An episode, E_1, is maximal if there is no episode, E_2, such that (1) E_1 and E_2 have the same qualitative state and (2) $I_1 \subseteq I_2$, i.e., the temporal extent of E_1 is contained in the temporal extent of E_2.

From the definition of the maximal episode we conclude that the maximal episodes occur between adjacent time points, at which $x(t)$, $x'(t)$ or/ and $x''(t)$ change qualitative value. We will call these points *distinguished time points*, and from now on we will employ the following definition of a *trend*:

c. Trend. The trend of a reasonable function, $x:[a, b] \to R$ is a sequence of maximal episodes, defined over time intervals whose distinguished points are strictly ordered in time.

3. Triangular Episode: A Geometric Language to Describe Trends

If a trend is to be described by an ordered sequence of maximal episodes, then we only need to generate a language that contains the declarative value of an episode. Such a language was proposed by Cheung and Stephanopoulos (1990). It is composed of seven primitives (Fig. 4a) and possesses the following properties:

Completeness. Every trend can be represented by a legal sequence of triangular episodes (Fig. 4b).

Correctness. The recursive refinement of triangular episodes allows the description of a trend at any level of detail, converging to the real-valued description of a signal.

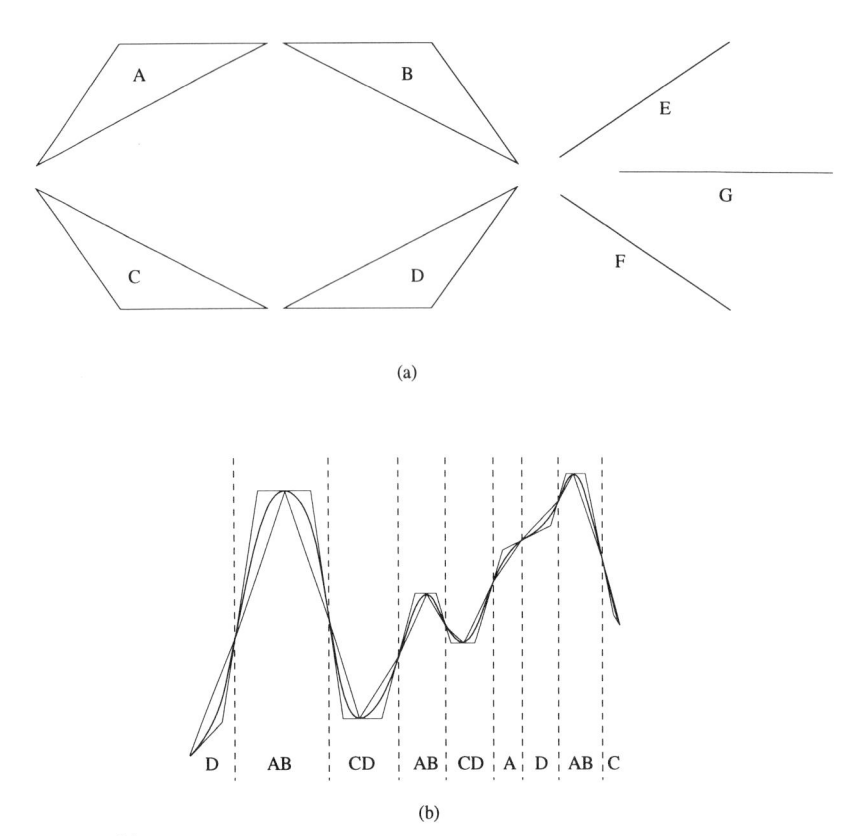

FIG. 4. (a) The seven primitives of the triangular description of trends; (b) sequence of triangular episodes describing a specific "trend" of a signal.

Robustness. The relative ordering of the triangular episodes in a trend is invariant to scaling of both the time axis and the function value. It is also invariant to any linear transformation (e.g., rotation, translation). Finally it is quite robust to uncertainties in the real value of the signal (e.g., noise), provided that the extent of a maximal episode is much larger than the period of noise.

B. Trends and Scale-Space Filtering

Consider the continuous function shown in Fig. 5a. Points 1 through 13 are the inflextion points and constitute a subset of all the distinguished points over the indicated period of time, and define the following

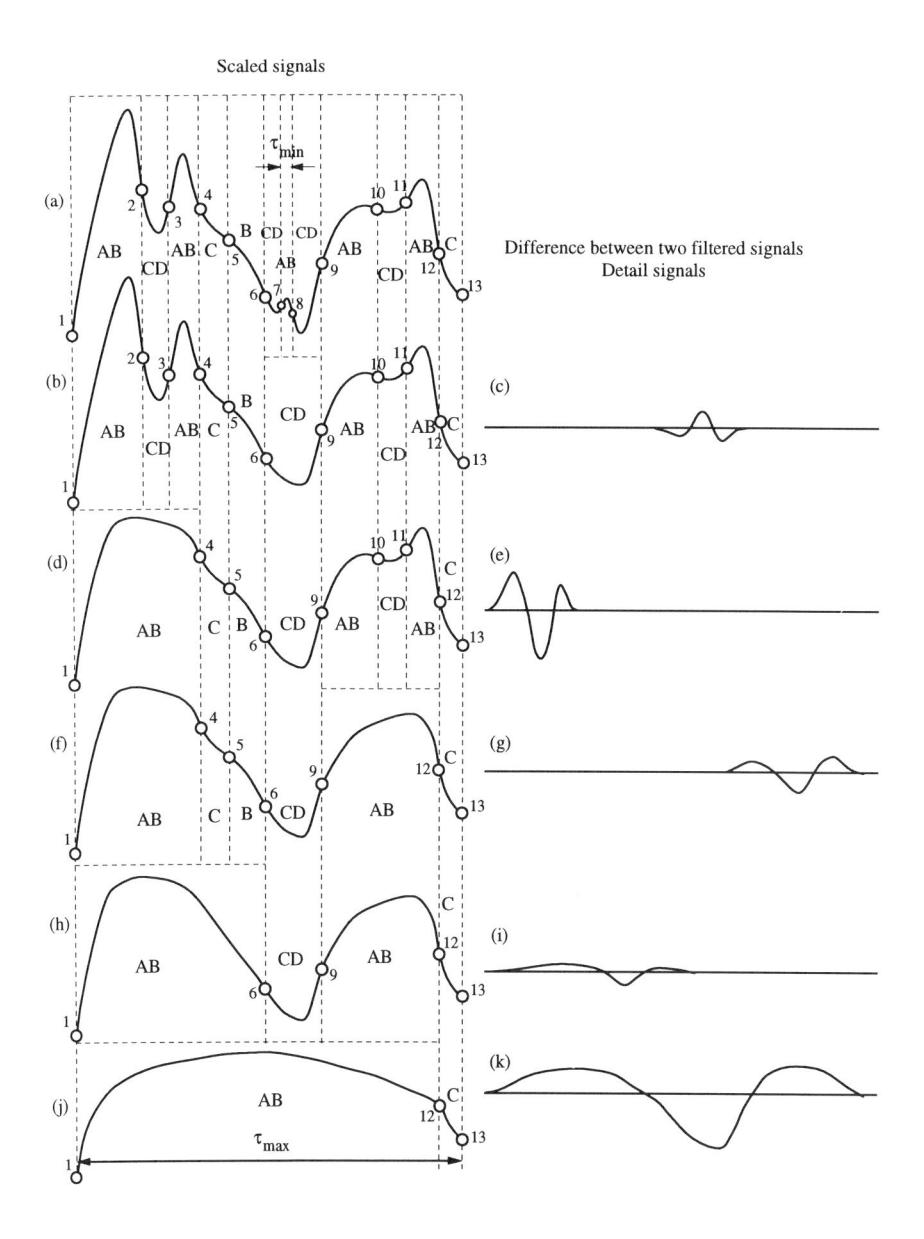

FIG. 5. Multiscale representation of process trends.

sequence of groupings of the triangular episodes:

$$AB(1,2) - CD(2,3) - AB(3,4) - C(4,5) - B(5,6)$$
$$- CD(6,7) - AB(7,8) - CD(8,9) - AB(9,10)$$
$$- CD(10,11) - AB(11,12) - C(12,13).$$

Let us filter the signal of Fig. 5a with a compact filter of variable width of compactness. The first distinguished feature to disappear from the trend of Fig. 5a is the sequence, $CD(6,7) - AB(7,8) - CD(8,9)$ (defined by the inflexion points, 6–9), which is replaced by the $CD(6,9)$ grouping of triangular episodes (see Fig. 5b). The trend of the filtered function (Fig. 5b) differs from the trend of the original one (Fig. 5a) over the time segment defined by the inflexion points 6–9. In an analogous manner, as the width of compactness of the local filter increases, additional features of the trend are replaced with more abstract descriptions. Examples of this process from Fig. 5 are the following:

(a) $\{AB(1,2) - CD(2,3) - AB(3,4)\}$ replaced by $\{AB(1,4)\}$
(Fig. 5d)
(b) $\{AB(9,10) - CD(10,11) - AB(11,12)\}$ replaced by $\{AB(9,12)\}$
(Fig. 5f)
(c) $\{AB(1,4) - C(4,5) - B(5,6)\}$ replaced by $\{AB(1,6)\}$
(Fig. 5h)
(d) $\{AB(1,6) - CD(6,9) - AB(9,12)\}$ replaced by $\{AB(1,12)\}$
(Fig. 5j)

From Fig. 5 we conclude that the original function can be represented by six (6) distinct trends (Fig. 5a, b, d, f, h, j), *each with its own sequence of triangular episodes*. Each successive trend of Fig. 5 contains information at a coarser resolution (scale). The differences among two successive trends are shown in Figs. 5c, 5e, 5g, 5i, and 5k, assuming for presentation purposes perfectly local filters.

The procedure described above is a pictorial approximation of a process called *scale-space filtering* of a function, proposed by Witkin (1983). The surface (e.g., Fig. 6) swept out by a filtered signal as the Gaussian filter's standard deviation is varied, is called *scale-space image* of the signal and is given by

$$F(t,\sigma) = f(t)*g(t,\sigma) = \int_{-\infty}^{+\infty} f(u) \left\{ \frac{1}{\sigma\sqrt{2\pi}} \exp\left[\frac{-(t-u)^2}{2\sigma^2} \right] \right\} du, \quad (1)$$

and gives the time-dependent behavior of the signal, $f(t)$, at different scales. As σ increases certain distinguishing features disappear, replaced

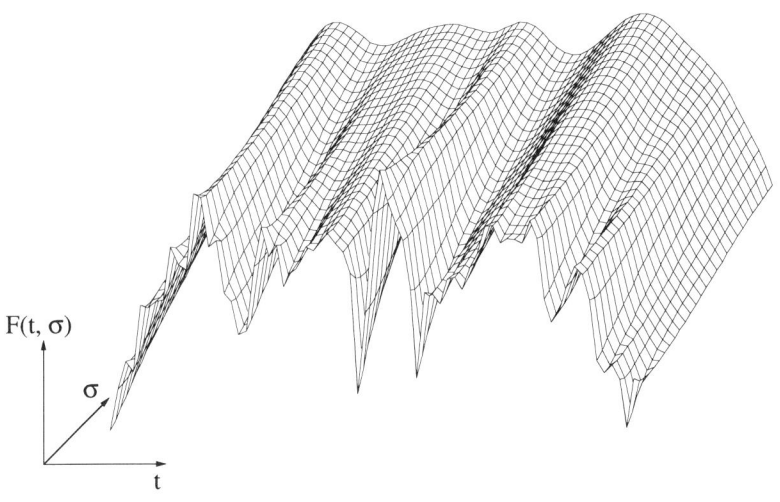

FIG. 6. Scale-space image of a function $F(t)$.

by coarser characteristics, very much in the spirit described by the pictorial examples of Fig. 5.

1. Structure of Scale

A close examination of the scale-space image of Fig. 6 reveals some interesting features:

1. As σ increases, pairs of inflexion points disappear and give rise to a smoother representation of the signal.
2. As σ decreases, a smooth segment of the signal goes through a singular point and gives rise to a "ripple" with the generation of a pair of inflexion points.
3. Some of the inflexion points persist over a large range of σ values, while others disappear. Clearly, the former correspond to *dominant features of the signal* while the latter indicate weak features, e.g., noise, which disappear readily with filtering.
4. The position of the inflexion points at higher values of σ, i.e., higher scales, shifts as a result of the increasingly global effect of the Gaussian filter.
5. If we connect the positions of the same inflexion points over various values of σ by straight lines, we create an *interval tree of scales*, as shown in Fig. 7 for the signals of Fig. 5. The interval tree allows us to generate two very important pieces of information about the trends of a measured variable:

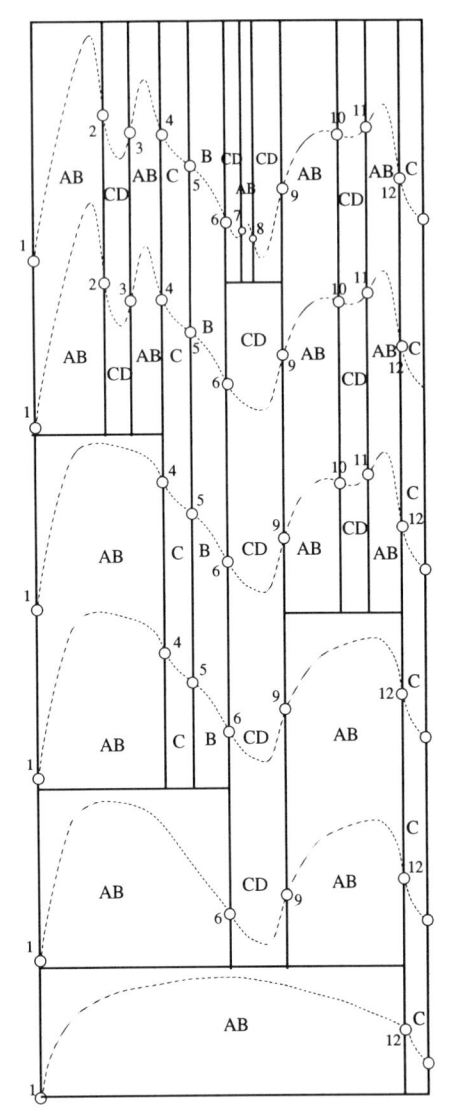

FIG. 7. The interval tree of scales for the signal of Fig. 5.

a. Number of Distinct Trends at Multiple Scales. The disappearance of inflexion points through filtering, or their generation through refinement, changes the sequence of triangular episodes needed for the description of a measured variable. This, in turn, implies that a new trend has been identified. Having the interval tree of scales, we can easily generate the complete set of trends that can be used to describe a measured signal at various scales of abstraction (or, detail). For example, the signal of Fig. 5 can be described by any of six (6) distinct trends.

b. *Dominant Features in a Signal.* The inflexion points which persist over extended ranges of σ (or scale), bound dominant features of a signal. For example, consider the signal of Fig. 5. The pairs of inflexion points $(1, 6)$ and $(9, 12)$ bound the features, $AB(1, 6)$ and $AB(9, 12)$ (see Fig. 5h), respectively, and have "survived" four levels of filtering. They correspond to dominant features of the original signal and as such characterize profoundly the signal itself.

2. Scaling Episodes and the Second-Order Zero Crossings

Scale-space filtering provides a multiscale description of a signal's trends in terms of its inflexion points (second-order zero crossings). The only legal sequences of triangles between two adjacent inflexion points are (in terms of triangular episodes):

AB
CD
A (if it is preceded and succeeded by D episodes)
B (if it is preceded and succeeded by C episodes)
C (if it is preceded and succeeded by B episodes)
D (if it is preceded and succeeded by A episodes)

Any of these legal sequences will be called a *scaling episode*.

If a signal is represented by a sequence of triangular episodes, scale-space filtering manipulates the sequences of triangular episodes with very concrete mechanisms. Here is the complete list of syntactic manipulations carried out by scale-space filtering:

1. $AB - CD - AB \Rightarrow AB$
2. $CD - AB - CD \Rightarrow CD$
3. $AB - C - B \Rightarrow AB$
4. $CD - A - D \Rightarrow CD$
5. $D - A - D \Rightarrow D$
6. $C - B - C \Rightarrow C$
7. $B - C - B \Rightarrow B$
8. $A - D - A \Rightarrow A$

The reader should note the following properties:

(a) Filtering occurs over a segment of *three* scaling episodes, producing a segment with *one* scaling episode.
(b) The syntax of the resulting scaling episode is determined by the type of the first and last episodes in the sequence of the three scaling episodes, e.g., $\underline{A}B - CD - A\underline{B} \Rightarrow AB$, $\underline{B} - C - \underline{B} \Rightarrow BB \equiv B$.

So, within the context of scale-space filtering, it is more convenient to express *a trend as a sequence of scaling episodes*, rather than as a sequence of episodes.

The utility of representing trends in terms of scaling episodes, or equivalently in terms of *second-order zero crossings* (i.e., inflexion points) increases even more, if we can reconstruct a signal from these zero crossings at multiple scales. Marr and Hildreth (1980), using the Laplacian of a Gaussian had surmised that the reconstruction was stable and complete, but gave no proof. Yuille and Poggio (1984) proved that the scale map of almost all signals filtered by a Gaussian of varying σ determines the signal uniquely, up to a constant scaling. Hummel and Moniot (1989) on the other hand have shown that reconstruction using second-order zero crossings alone is unstable, but together with gradient values at the zero crossings is possible and stable, although it uses a lot of redundant information. Mallat and Zhong (1992) have developed an algo-algorithm that converges quickly to stable reconstructions using second-order zero crossings of a signal, derived from the wavelet decomposition of the signal. Nevertheless, the completeness and stability of the reconstruction from second-order zero crossings remains an open theoretical question.

3. Properties of Scale-Space Filtering

The interval tree of scale allows us to extract the temporal features contained in a signal and systematically establish the trends at different scales. Witkin (1983) defined *the stability of a temporal episode as the number of scales over which it persists*. The most conspicuous features of a signal are also the most stable. He also developed a heuristic procedure through which he would construct representations of a signal by maintaining the most stable episodes. The stability criterion, though is entirely heuristic, and other criteria may be desired. In Section III we will see that wavelet decomposition of signals offers a sound and much more powerful approach in constructing stable representations of dominant trends. For the time being, it is very important to identify several important properties and shortcomings of the traditional scale-space filtering.

a. Distortion of Signal's Features. As a signal is filtered through a Gaussian, its features become progressively quantitatively more and more distorted. This is a natural effect of the averaging of the smoothing process over increasingly longer segments of a signal. Such distortion inhibits the extraction of accurate distinguished features. Feature extraction by selecting stable episodes from the interval tree involves fitting piecewise quadratic segments to the raw data covered by the stable episodes. This procedure is ad hoc, and creates discontinuities at the inflexion points.

b. Creation of Fictitious Features. Gaussian filtering has two very important properties:

1. Inflexion points never disappear as we move from more abstract to more detailed descriptions.
2. Fictitious inflexion points are not generated as the scale of filtering increases.

Taken together, these properties guarantee that Gaussian filtering never generates fictitious features. For many other filters, this is not the case. It is important that any filter we use maintains the integrity of the original signal.

III. Wavelet Decomposition: Extraction of Trends at Multiple Scales

The wavelet decomposition of continuous and discrete functions has emerged as a powerful theoretical framework over the last 10 years, and has led to significant technical developments in data compression, speech and video recognition and data analysis, fusion of multirate data, filtering techniques, etc. Starting with Morlet's work (Goupillaud *et al.* 1984) on the analysis of seismic data, wavelet decomposition of signals has become extremely attractive for two main reasons: (1) it offers excellent localization of a signal's features in both time and frequency and (2) it is numerically far more efficient than other techniques such as fast Fourier transform of signals (Bakshi and Stephanopoulos, 1994a). It is for the first reason that we discuss in this section the use of wavelet decomposition as the most appropriate framework for the extraction and representation of trends of process operating data.

A. The Theory of Wavelet Decomposition

A family of wavelets is a family of functions with all its members derived from the translations (e.g., in time) and dilations of a single, mother function. If $\psi(t)$ is the mother *wavelet*, then all the members of the family are given by

$$\psi_{su}(t) \equiv \frac{1}{\sqrt{s}} \psi\left(\frac{t-u}{s}\right) \quad \text{for} \quad (s,u) \in R^2. \tag{2}$$

the parameters s and u that label the members of a wavelet family are known as the *dilation* and *translation* parameters, respectively. A wavelet function, $\psi(t)$, belongs to the set of square-integrable functions, i.e., $\psi(t) \in L^2(t)$, and satisfies the following admissibility condition

$$C_\psi = \int_0^\infty \frac{|\hat{\psi}(\omega)|}{\psi} d\omega < +\infty, \tag{3}$$

with $\hat{\psi}(\omega)$ signifying the Fourier transform of $\psi(t)$. Condition (3) implies that as $\omega \to 0$, $\hat{\psi}(\omega)$ approaches zero faster and thus represents a *band-pass filter*. $\psi(t)$ is either compactly supported in time, or has some fast decay at infinity implying that condition (3) is equivalent to the vanishing of the first p moments (see Section III of third Chapter in this volume):

$$\int_{-\infty}^\infty t^m \psi(t)\, dt = 0, \quad m = 0, 1, 2, \ldots, p-1 \tag{3a}$$

where p is the order of approximation considered for $\psi(t)$. Figure 8 shows examples of various wavelet functions.

1. Resolution in Time and Frequency

In the time domain, a wavelet placed at a translation point, u, has a standard derivation, σ_u, around this point. Similarly, in the frequency domain a wavelet is centered at a given frequency, ω (determined by the value of the dilation parameter) and has a standard deviation, σ_ω, around the specific frequency. The values of σ_u and σ_ω are given by

$$\sigma_u^2 = -\int_{-\infty}^{+\infty} t^2 |\psi(t)|^2\, dt \qquad \sigma_\omega^2 = \int_{-\infty}^{+\infty} \omega^2 |\hat{\psi}(\omega)|^2\, d\omega \tag{4}$$

As the dilation parameter, s, increases, the wavelets have significant values over a broader time segment and the resulting resolutions in the time and frequency domains change as $s\sigma_u$ and σ_ω/s, respectively. Thus, at large

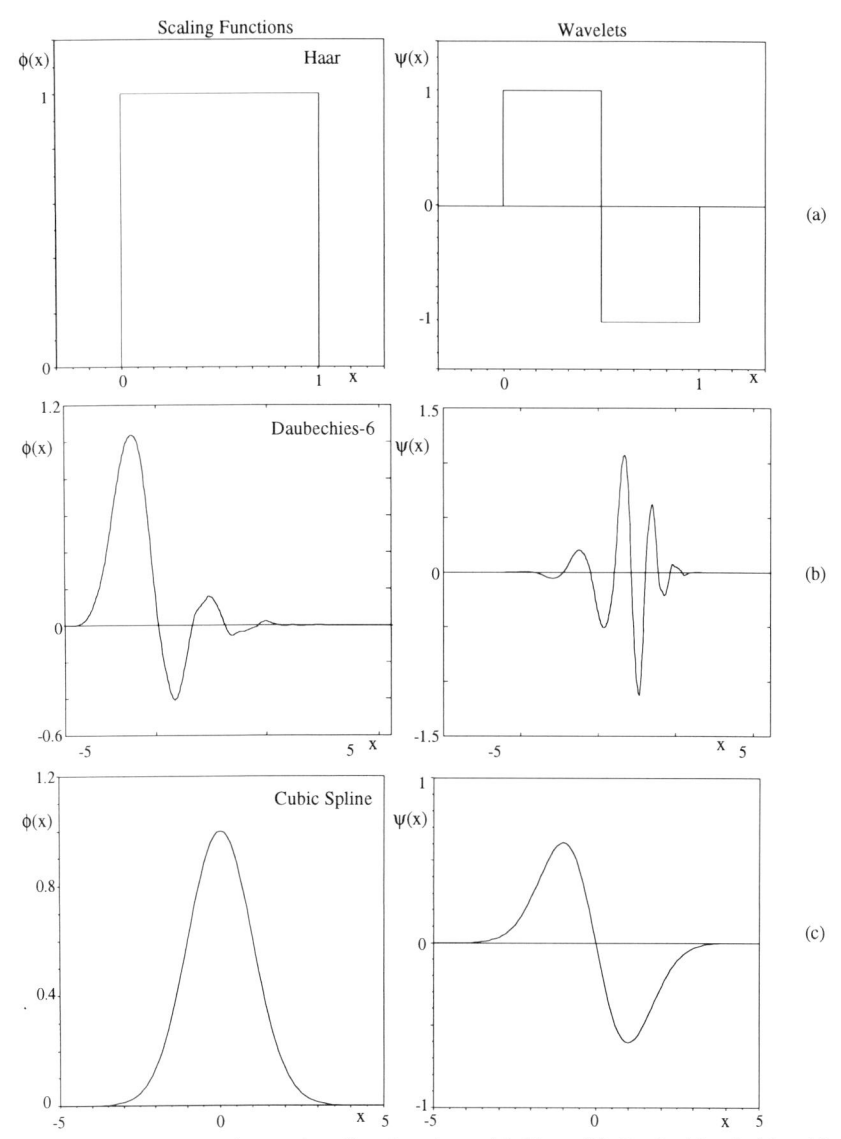

FIG. 8. Typical wavelets and scaling functions: (a) Haar, (b) Daubechies-6, (c) cubic spline.

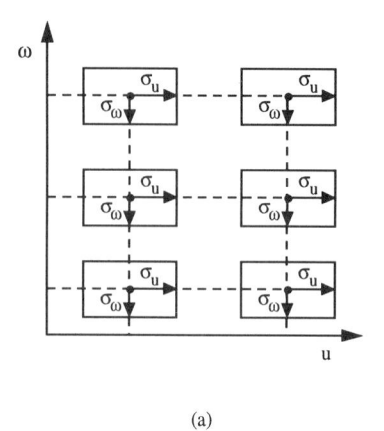

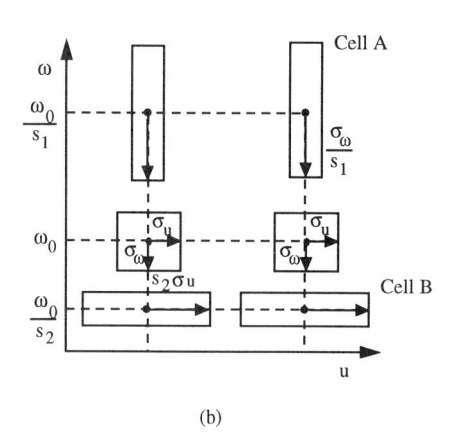

FIG. 9. Resolution in scale space of (a) window Fourier transform and (b) wavelet transform.

dilations (i.e., *scales*) the resolution in the time domain becomes coarser, while in the frequency domain, it becomes finer. The dimensions of the resolution cell (Fig. 9b) in the *scale-space* is equal to

$$[u_0 - s\sigma_u, u_0 + s\sigma_u] \times \left[\frac{\omega_0}{s} - \frac{\sigma_\omega}{s}, \frac{\omega_0}{s} + \frac{\sigma_\omega}{s}\right],$$

where u_0 is the translation point of a wavelet and consequently the center of its energy's concentration in the time domain, and ω_0/s is the frequency where the wavelet's energy is concentrated in the frequency domain. The variable dimensions of the resolution cell (Fig. 9b) are characteristic of the wavelet's ability to provide a variable-window framework for the localization in both time and frequency (i.e., scale), of the features contained in a signal. By comparison, the resolution cell of the windowed Fourier transform (Fig. 9a) maintains constant dimensions and thus produces inefficient representation of the signal's features.

Decomposition and Reconstruction of Functions

Let $F(t)$ be a square-integrable function, i.e., $F(t) \in L^2(t)$. The projection of $F(t)$ on a wavelet, $\psi_{su}(t)$, at dilation s and translation point, u, is given by

$$\alpha_{su} \equiv [W_{su}F(t)] = \int_{-\infty}^{+\infty} F(t)\psi_{su}(t)\, dt. \tag{5}$$

Using these projections over all wavelets with $(s, u) \in R^2$, we can reconstruct the function by the following equation:

$$F(t) = \int_{-\infty}^{+\infty}\int_{-\infty}^{+\infty} \alpha_{su}\psi_{su}(t)\, ds\, du. \tag{6}$$

If we consider a wavelet with compact support, then the projection given by Eq. (5) reflects the information content of $F(t)$ only in the resolution cell of the corresponding wavelet. For example, projections of $F(t)$ on wavelets with large dilation, s, represent the large-scale (i.e., low-frequency) components of $F(t)$. Consequently, the reconstruction of $F(t)$ through Eq. (6) indicates that we can put the functions together from "pieces," each of which represents the content of a specific segment of $F(t)$ over time [determined by $(u_0 - s\sigma_u, u_0 + s\sigma_u)$], and in terms of frequencies in a specific range [determined by $(\omega_0/s) - (\sigma_\omega/s), (\omega_0/s) + (\sigma_\omega/s)$]. For example, if a function contains a sharp feature, e.g., spike, then the bulk of the informational content carried by the spike will be reflected by the projection of the function on a "narrow" wavelet, whose resolution cell is narrow in u and long in ω (e.g., cell A in Fig. 9b). On the other hand, the content of a function reflected by a slowly rising trend will be reflected by the projection of the function on a "wide" wavelet with a resolution cell such as that of cell B in Fig. 9b.

3. Discretization of Scale

When we analyze the scale-space image of a function (see Section II, B) in order to extract the trends of process variables, we are interested only in a finite number of distinct trends, as they are defined by the interval tree of scale.

Usually, the dilation parameter s is discretized along the dyadic sequence,

$$s = 2^m \quad m \in Z,$$

where Z is the space of integer numbers. The resulting family of wavelets with a discretized dilation parameter is represented as follows:

$$\psi_{2^m u}(t) = \frac{1}{\sqrt{2^m}} \psi\left(\frac{t-u}{2^m}\right), \quad m \in Z. \tag{2a}$$

The projection of $F(t)$ to all wavelets of the above form with $m \in Z$ and $u \in R$, yields the so-called *dyadic wavelet transform* of (t), with the following components:

$$\alpha_{2^m u} \equiv [W_{2^m u} F(t)] = \int_{-\infty}^{+\infty} F(t) \left\{ \frac{1}{\sqrt{2^m}} \psi\left(\frac{t-u}{2^m}\right) \right\} dt,$$

$$m \in Z, \ u \in R. \tag{5a}$$

The discretization of the dilation parameter need not be dyadic. Discretization schemes with integer or noninteger factors are possible and have been suggested. The reconstruction of $F(t)$ is given by

$$F(t) = \sum_{m=-\infty}^{+\infty} \int_{-\infty}^{+\infty} \alpha_{2^m u} \psi_{2^m u}(t) \, du. \tag{6a}$$

To ensure that no information is lost on $\hat{F}(\omega)$ as the dilation is discretized, the scale factors 2^m for $m \in Z$ must cover the whole frequency axis. This can be accomplished by requiring the wavelets to satisfy the following condition:

$$\sum_{m=-\infty}^{+\infty} |\hat{\psi}(2^m \omega)|^2 = 1. \tag{7}$$

Equation (6a) implies that the scale (dilation) parameter, m, is required to vary from $-\infty$ to $+\infty$. In practice, though, a process variable is measured at a finite resolution (sampling time), and only a finite number of distinct scales are of interest for the solution of engineering problems. Let $m = 0$ signify the finest temporal scale (i.e., the sampling interval at which a variable is measured) and $m = L$ be coarsest desired scale. To capture the information contained at scales $m > L$, we define a *scaling function*, $\phi(t)$, whose Fourier transform is related to that of the wavelet, $\psi(t)$, by

$$|\hat{\phi}(\omega)|^2 = \sum_{m=1}^{+\infty} |\hat{\psi}(2^m \omega)|^2. \tag{8}$$

Since $\hat{\psi}(\omega)$ must satisfy condition (3), then from Eqs. (7) and (8) we conclude that

$$\lim_{\omega \to 0} |\hat{\phi}(\omega)| = 1.$$

In other words, the energy of $\phi(t)$ can be seen as a lowpass filter.

Let us now create a family of scaling functions through the dilation and translation of $\phi(t)$

$$\phi_{2^m u}(t) = \frac{1}{\sqrt{2^m}} \phi\left(\frac{t-u}{2^m}\right), \quad m \in Z, \quad u \in R \quad (9)$$

Consequently, the projection of $F(t)$ on $\phi_{2^m u}(t)$ produces a function whose content is completely stripped of high frequencies and includes information only at scales higher than m. As m increases, more and more details are removed from $F(t)$.

Through the employment of the scaling function we can reconstruct $F(t)$ using a finite number of scales, as follows:

$$F(t) = \sum_{m=0}^{L} \int_{-\infty}^{+\infty} \alpha_{2^m u} \psi_{2^m u}(t) \, du + \int_{-\infty}^{+\infty} \beta_{2^L u} \phi_{2^L u}(t) \, du, \quad (10)$$

where, $\beta_{2^L u}$ is the projection of $F(t)$ on the scaling function, $\phi_{2^L u}$, and is given by

$$\beta_{2^L u} \equiv [S_{2^L u} F(t)] = \int_{-\infty}^{+\infty} F(t) \phi_{2^L u}(t) \, dt. \quad (11)$$

4. Dyadic Discretization of Time

Having completed the decomposition and reconstruction of a function at a finite number of discrete values of scale, let us turn our attention to the discretization of the translation parameter, u, dictated by the discrete-time character of all measured process variables. The classical approach, suggested by Meyer (1985–1986), is to discretize time over dyadic intervals, using the sampling interval, τ, as the base. Thus, the translation parameter, u, can be expressed as

$$u = k(2^m \tau), \quad \text{with} \quad (m, k) \in Z^2, \quad (12)$$

where k is the translation parameter. At the initial dilation level $m = 0$, and $u = k\tau$, $k = 0, 1, 2, \ldots$, i.e., the translation points coincide with the sampling instants, i.e., $0, \tau, 2\tau, \ldots$. At the next dilation level, $u = (2\tau)k$, i.e., the distance between two adjacent wavelets double, and the translation points are $0, 2\tau, 4\tau, \ldots$.

As a result of the dyadic discretization in dilation and translation, the members of the wavelet family are given by

$$\psi_{mk}(t) \equiv \frac{1}{\sqrt{2^m}} \psi\left(\frac{t}{2^m} - k\tau\right), \quad (m,k) \in Z^2.$$

A discrete-time signal is decomposed into the following set of projections:

$$\alpha_{mk} \equiv \left[W_{2^m, k(2^m\tau)} F(t)\right] = \sum_{k=-\infty}^{+\infty} F(t)\psi_{mk}, \quad 1 \leq m \leq L, \quad (5b)$$

$$\beta_{Lk} \equiv \left[S_{2^L, k(2^L\tau)} F(t)\right] = \sum_{k=-\infty}^{+\infty} F(t)\phi_{Lk}, \quad (11a)$$

from which it can be completely reconstructed:

$$F(t) = \sum_{m=1}^{L} \sum_{k=-\infty}^{+\infty} \alpha_{mk}\psi_{mk} + \sum_{k=-\infty}^{+\infty} \beta_{Lk}\phi_{Lk}. \quad (6b)$$

For practical purposes, the wavelet decomposition can only be applied to a finite record of discrete-time signals. If N is the number of samples in the record, and $\tau = 1$, then the maximum value of the translation parameter can be found from Eq. (12), by setting $u = N$, and is equal to $k_{max} = N/2^m$. Consequently, the decomposition and reconstruction relations [Eqs. (5b), (11a), (6b)] take the following form:

$$\alpha_{mk} \equiv \left[W_{2^m, k2^m} F(t)\right] = \sum_{k=1}^{k_{max}} F(t)\psi_{mk}, \quad 1 \leq m \leq L, \quad (13)$$

$$\beta_{Lk} \equiv S_{2^m, k2^m} F(t) = \sum_{k=1}^{k_{max}} F(t)\phi_{Lk}, \quad (14)$$

$$F(t) = \sum_{m=1}^{L} \sum_{k=1}^{k_{max}} \alpha_{mk}\psi_{mk} + \sum_{k=1}^{k_{max}} \beta_{Lk}\phi_{Lk}. \quad (15)$$

Equation (15) provides a discrete, complete and nonredundant representation of the function $F(t)$, since it requires the computation of N coefficients, if N is the number of discrete-time samples describing $F(t)$.

5. Uniform Discretization of Time

An alternative approach to the discretion of the translation parameter u involves uniform sampling of the measured signal at all scales, i.e., $u = k\tau$, with $k \in Z$. The resulting decomposition algorithm is of complexity $O(N \log N)$, and the associated reconstruction requires the computation of $N \log N$ coefficients, i.e., it contains redundant information.

Nevertheless, uniform discretization of time at all scales leads to representations that are highly suitable for feature extraction and pattern recognition. More on this subject in a subsequent paragraph.

6. Practical Considerations

Let us now see how the theory of the wavelet-based decomposition and reconstruction of discrete-time functions can be converted into an efficient numerical algorithm for the multiscale analysis of signals. From Eq. (6b) it is easy to see that, given a discrete-time signal, $F_0(t)$ we have

$$F_0(t) = D_1(t) + F_1(t), \qquad (16)$$

where

$$F_1(t) = \sum_{k=-\infty}^{+\infty} \beta_{1k}\phi_{1k}(t) \quad \text{and} \quad D_1(t) = \sum_{k=-\infty}^{+\infty} \alpha_{1k}\psi_{1k}(t). \qquad (17)$$

$F_1(t)$ is called the *scaled signal* and is derived from the filtering of $F_0(t)$ with the lowpass scaling function. It represents a smoother version of $F_0(t)$. $D_1(t)$ is called the *detail signal* and is derived from the filtering of $F_0(t)$ with the bandpass wavelet functions. It represents the information that was filtered out of $F_0(t)$ in producing $F_1(t)$.

Generalizing Eqs. (16) and (17), we take

$$F_m(t) = D_{m+1}(t) + F_{m+1}(t), \qquad m \in Z, \qquad (18)$$

with

$$F_{m+1}(t) = F_m(t) * H_m \quad \text{and} \quad D_{m+1}(t) = F_m(t) * G_m, \qquad (19)$$

where the operator, *, signifies the convolution operation. Filter H_m is a lowpass filter, emulating the effects of the scaling function, and G_m is a bandpass filter affecting the influence of the wavelet function. Equations (18) and (19) imply that the wavelet-based decomposition and reconstruction of a discrete-time signal can be carried out through a cascade of convolutions with filters H and G (see Fig. 10). Figure 11 shows the scaled and detail signals generated from the wavelet decomposition of a given signal using dyadic and uniform sampling. The detailed methodological aspects of the cascaded convolution of signals, described above, can be found in Mallat (1989) for the case of dyadic sampling of time and in Mallat and Zhong (1992) for uniform sampling.

For real-world signals with finite records, the convolution processes leading to decomposition and reconstruction of a signal, require data in regions beyond the signal's endpoints. Assuming a mirror image of the

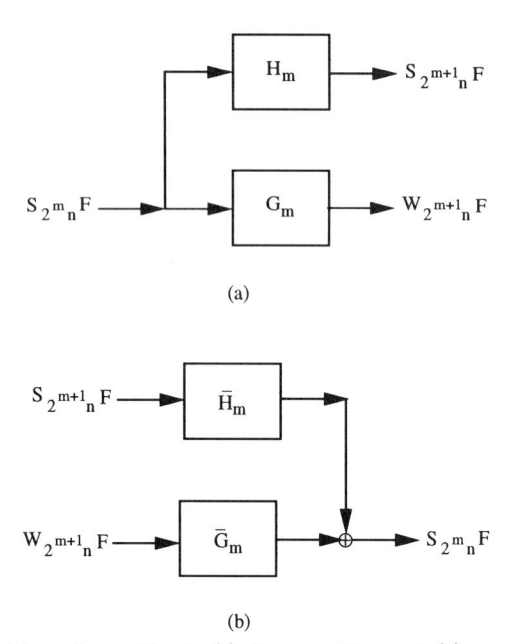

FIG. 10. Methodology for multiscale (a) decomposition and (b) reconstruction, using wavelets, with uniform sampling $(m, n) \in Z^2$.

original signal beyond both ends (Mallat, 1989), causes numerical inaccuracies in the computation of the wavelet coefficients. Bakshi and Stephanopoulos (1995) have studied various approaches for handling endpoint effects and they have suggested certain practical solutions in this problem.

B. Extraction of Multiscale Temporal Trends

In Section II we defined the trend of a measured variable as a strictly ordered sequence of scaling episodes. Since each scaling episode is defined by its bounding inflexion points, it is clear that the extraction of trends necessitates the localization of inflexion points of the measured variable at various scales of the scale-space image. Finally, the interval tree of scale (see Section II) indicates that there is a finite number of distinct sequences of inflexion points, implying a finite number of distinct trends. The question that we will try to answer in this section is, "How can you use the wavelet-based decomposition of signals in order to identify the distinct sequences of inflexion points and thus of the signal's trends?"

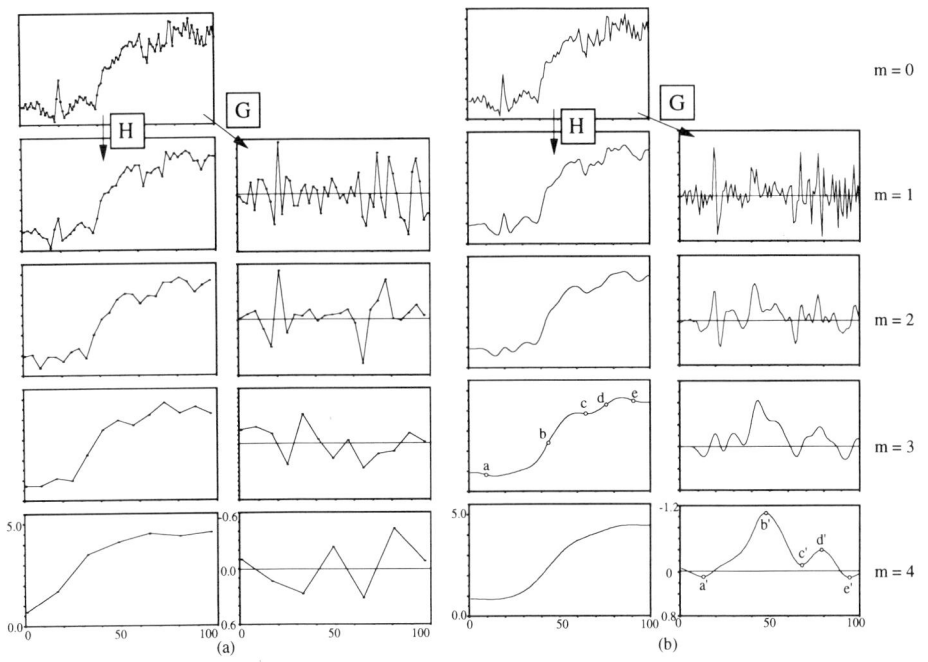

FIG. 11. Wavelet decomposition (a) dyadic sampling using Daubechies-6 wavelet; (b) uniform sampling using cubic spline wavelet.

1. Translationally Invariant Representation of Measured Variables

Measured variables may have identical features in different segments of the time record. Any analysis method should be able to identify and extract such common features, independently of the time segment in which they appear. Consequently, the wavelet decomposition of a time-translated signal should produce wavelet coefficients that are translated in time but equal in magnitude to the wavelet coefficients of the original (untranslated) signal. For example, if $F(t) \in L^2$ is the original signal and $F_\theta(t)$ is the same function translated in time by θ units, i.e., $F_\theta(t) = F(t - \theta)$, then

$$[W_{2^m u} F_\theta(t)] = [W_{2^m u} F(t - \theta)], \quad m \in Z, \quad u \in R. \quad (20)$$

Since the translation variable, u, is also discretized (dyadically, or uniformly at each scale), relationship (20) will hold only if the signal is translated in time by a period that is an integer multiple of the discretiza-

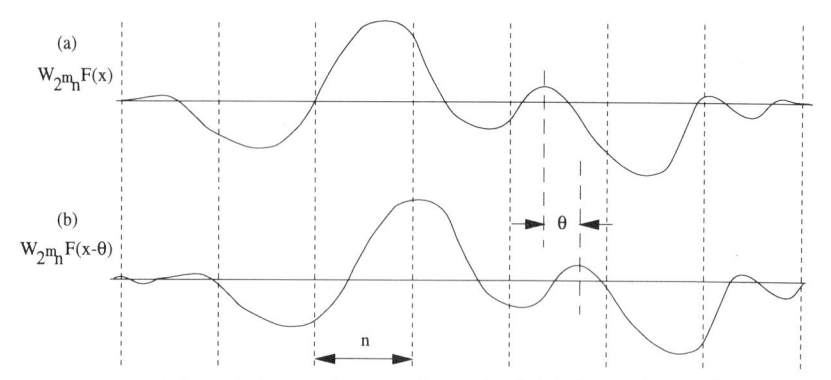

FIG. 12. Translation of the wavelet transform of $F(x)$ by θ, with sampling rate, n (Mallat, 1991).

tion interval at all scales. For example, in the case of dyadic sampling of time, translational invariance of the wavelet transform requires that $\theta = r(2^m \tau)$, $r \in Z$, whereas uniform sampling necessitates that $\theta = r\tau$, $r \in Z$, where τ is the sampling period of the original signal.

Unfortunately, the requirements for translational invariance of the wavelet decomposition are difficult to satisfy. Consequently, for either discretization scheme, comparison of the wavelet coefficients for two signals may mislead us into thinking that the two trends are different, when in fact one is simply a translation of the other.

On the other hand, the value of (zero-order) zero crossings and local extrema (first-order zero crossings) of a dyadic wavelet transform, $W_{2^m_u} F(t)$, do translate when the original signal is translated (Fig. 12). Therefore, we can generate translationally invariant representations of a signal if we employ zero-order and first-order zero crossings of the signal's wavelet transform.

2. Detection of Inflexion Points

Within the framework of scale-space filtering, inflexion points of $F(t)$ appear as extrema in $\partial F(t)/\partial t$ and zero crossings in $\partial^2 F(t)/\partial t^2$. Thus, filtering a signal by the Laplacian (second derivative) of a Gaussian will generate the inflexion points at various scales (Marr and Hildreth, 1980). In the same spirit, if the wavelet is chosen to be the first derivative of a scaling function, i.e., $\psi(t) = d\phi(t)/dt$, then from Eqs. (5a) and (11) we can easily show (Mallat and Zhong, 1992) that

$$[W_{2^m u} F(t)] = 2^m \frac{d}{dt}[S_{2^m u} F(t)]. \tag{21}$$

A wavelet defined as above is called a *first-order wavelet*. From Eq. (21) we conclude that the extrema points of the first-order wavelet transform provide the position of the inflexion points of the scaled signal at any level of scale. Similarly, if $\psi(t) = d^2\phi(t)/dt^2$, then the zero crossings of the wavelet transform correspond to the inflexion points of the original signal smoothed (i.e., scaled) by the scaling function, $\psi(t)$ (Mallat, 1991).

An obvious choice for the scaling function is the Gaussian, but it is not a compactly supported function, and the corresponding filter requires a large number of coefficients. The scaling function and associated wavelet chosen in this chapter to support the extraction of process trends were suggested by Mallat and Zhong (1992). The scaling function is a cubic spline with compact support of size 4 (i.e., the corresponding filter requires four coefficients), whereas the associated wavelet is not compactly supported but decreases exponentially at infinity (Fig. 8c). Using these functions, we can generate efficient representations of signals. The scaled signals at various scales are equivalent to the cubic spline fits of the data at the various scales. Unfortunately, since the scaling function is not a Gaussian, we cannot guarantee that spurious oscillations with fictitious inflexion points will not be generated at coarser scales. Nevertheless, such fictitious inflexion points, if generated, are short-lived, i.e., disappear quickly with increasing scale, and thus they will never be stable and distinguished features of any process trend. It is also worth noting that the variation in the magnitude of local extrema of the wavelet decomposition across several scales provides useful information about the nature of the inflexion points, and thus offers a measure of the signal's regularity.

3. The Wavelet Interval–Tree of Scale

Consider the given signal and its wavelet decomposition, as shown in Fig. 11. Since the wavelet is a first-order wavelet, the inflexion points of the original function appear as extrema of the detail signal at the corresponding scale (Fig. 11). If we connect the extrema of the detailed signal across various scales, we generate a structure similar to the interval tree of scale, discussed in Section II, B. We will call this tree, the *wavelet interval–tree of scale* (Fig. 13). It gives the evolution of the features of a signal in the scale space, and can be used to generate the trends of measured variables with various combinations of features. The wavelet

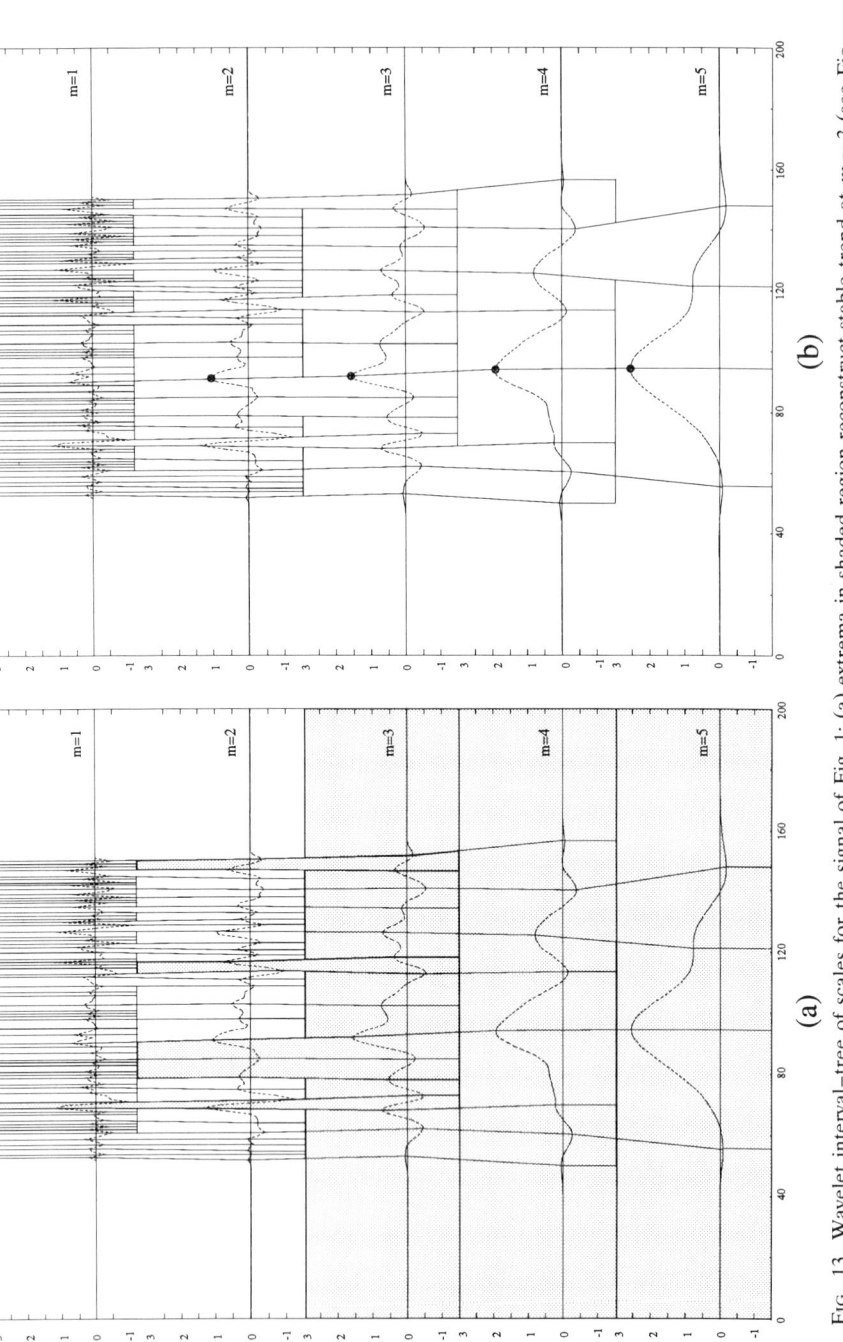

FIG. 13. Wavelet interval–tree of scales for the signal of Fig. 1: (a) extrema in shaded region reconstruct stable trend at $m = 3$ (see Fig. 15b); (b) encircled extrema reconstruct trend with accurate step change (see Fig. 15d).

interval–tree of scale presents certain advantages over the classical scale-space filtering, which are very important in the reconstruction of process trends. These advantages are (Bakshi and Stephanopoulos, 1994a):

1. The wavelet interval–tree of scale is constructed from $\log_2 N$ distinct representations, where N is the number of points in the record of measured data. This is a far more efficient representation than that of scale-space filtering with continuous variation of Gaussian σ.
2. In scale-space filtering, trends consisting of combinations of features at various scales are constructed by approximating the stable features through piecewise continuous segments, like parabolas. Such an approach leads to discontinuities in the trends. On the contrary, the trends generated from the wavelet interval-tree of scale with a mixture of features from different scales, are continuous up to the degree of continuity of the scaling function, and are based on formal and sound analysis.
3. The detail signals at the various scales of the tree provide the features of the trend that are distinct to the corresponding range of scales.
4. The evolution of the inflexion points, as given by the local extrema of the detail signals, also characterize the regularity of the original signal's inflexion points.

4. *The Algorithm for the Extraction of Trends*

Let us recall that *a trend is a strictly ordered sequence of scaling episodes*, whose bounding inflexion points can be identified by the local extrema of the detail signals, as discussed in the previous paragraph. Clearly, the more stable a scaling episode (over a range of scales), the more distinguished is the corresponding feature in describing the trend. Consider, for example, the noise-corrupted pulse shown in Fig. 14 along with the associated detail signals, resulting from its decomposition with a first-order wavelet. The extrema, A and B, of the detail signal at the smallest scale persist at all scales, indicating the presence of a very stable scaling episode. This scaling episode is bounded by the inflexion points on the left and right edges of the pulse and reveals the most dominant feature of the original signal, i.e., the pulse itself. On the contrary, the inflexion points corresponding to the fluctuations of the noise in the original signal disappear very quickly and do not represent distinguished scaling episodes, i.e., features. The range of each extremum point is equal to the range of the G filter and is equal to $2^{m+1}q + 1$, where m is the scale of the corresponding detail signal and $2q + 1$ is the length of the filter. By

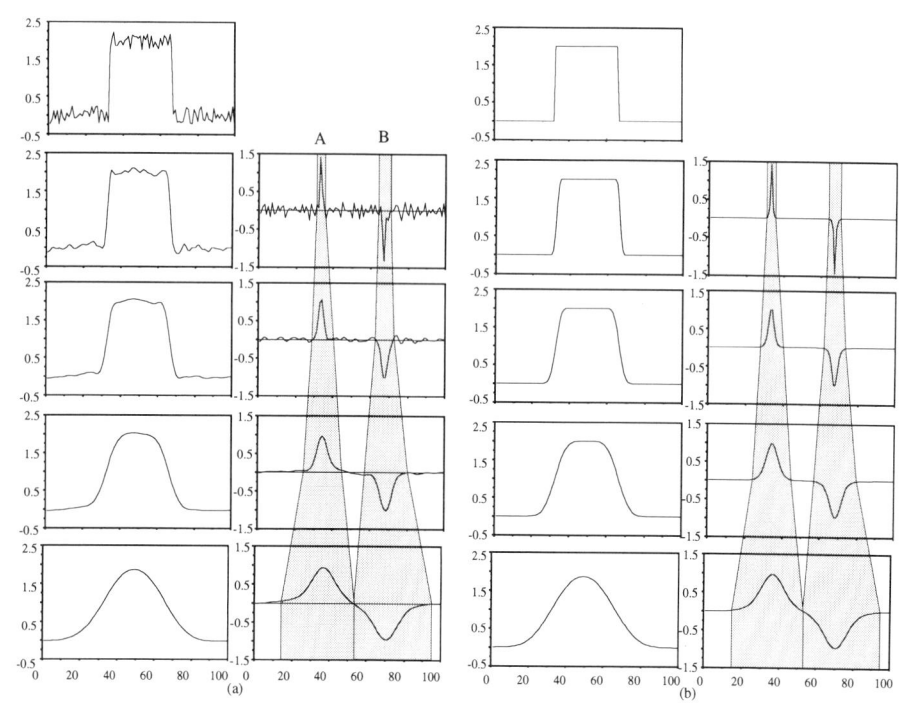

FIG. 14. Extracting distinguishing features from noise pulse signal. Wavelet coefficients in shaded regions represent stable extrema. (a) Wavelet decomposition of noisy pulse signal; (b) wavelet decomposition of pulse signal. (Reprinted from Bakshi and Stephanopoulos, "Representation of process trends, Part III. *Computers and Chemical Engineering*, **18**(4), p. 267, Copyright (1994), with kind permission from Elsevier Science Ltd., The Boulevard, Langford Lane, Kidlington OX5 1GB, UK.)

utilizing the wavelet coefficients of the stable extrema we can reconstruct a signal that contains the distinguished features at several scales. In Fig. 14, by utilizing the wavelet coefficients of the shaded region, we can completely reconstruct the pulse with minimum quantitative distortion.

The algorithm for extracting the most stable description of trends at each level of scale, proceeds as follows:

Step 1. Generate the finite, discrete dyadic wavelet transform of data using Mallat and Zhong's (1992) cubic spline wavelet (Fig. 8c).
Step 2. Generate the wavelet interval–tree of scale.
Step 3. For each scaling level, m, and for each scaling episode, (a) determine the range of stability and (b) collect the wavelet coefficients at all scales in the range of the episode's stability.

Step 4. Reconstruct the signal using the last scaled signal and the wavelet coefficients collected in step 3.

Once the stable reconstruction of a signal has been accomplished, its subsequent representation can be made at any level of detail, i.e. qualitative, semi-quantitative, or fully real-valued quantitative. The triangular episodes (described in Section I, A) can be constructed to offer an explicit, declarative description of process trends.

5. Illustrative Examples

a. Example 1: Generating Multiscale Descriptions. Consider the process signal and its wavelet decomposition shown in Fig. 11. Quantitatively accurate representations at each of the dyadic scales may be generated by selecting the wavelet transform extrema for the most stable episodes at each scale. For example, the extrema selected for reconstructing the process trend at scale, $m = 3$, are shown shaded in Fig. 13a. Note that wavelet coefficients at scales < 3 are needed to provide a stable representation. The reconstructed signal is shown in Fig. 15b, and is qualitatively equivalent to the scaled signal at $m = 2$, but is quantitatively much more accurate, since features such as the spike at $t = 20$ are undistorted. Similarly, the stable process trend at scale $m = 4$ is shown in Fig. 15c. The step change at $t = 40$ is not brought out accurately in the trends in Figs. 15b and 15c, due to the presence of other extrema in the range of the extremum corresponding to the step change. Most of the extrema in its range are much smaller than the extremum at $t = 40$. An empirical criterion, based on the relative magnitudes of wavelet transform extrema may be used for selecting large extrema, resulting in the reconstructed signal shown in Fig. 15d. The extremum at $t \approx 40$, at scales $m = 2-5$ used for this reconstruction, as shown in Fig. 13b. The step change is brought out quite accurately in this trend at $m = 5$.

Another powerful heuristic for extracting process trends is Witkin's stability criterion (Witkin, 1983). Starting with the episodes at the desired scale, lower-scale descriptions are considered only if the lower-scale episodes have a higher mean stability than their parent episode. The episodes selected using this criterion at scale, $m = 5$, and the reconstructed signal are shown in Fig. 16. The reconstructed signal contains all the conspicuous features in the raw data, without quantitative distortion. Witkin's stability criterion is very effective for extracting the most relevant features from a process signal.

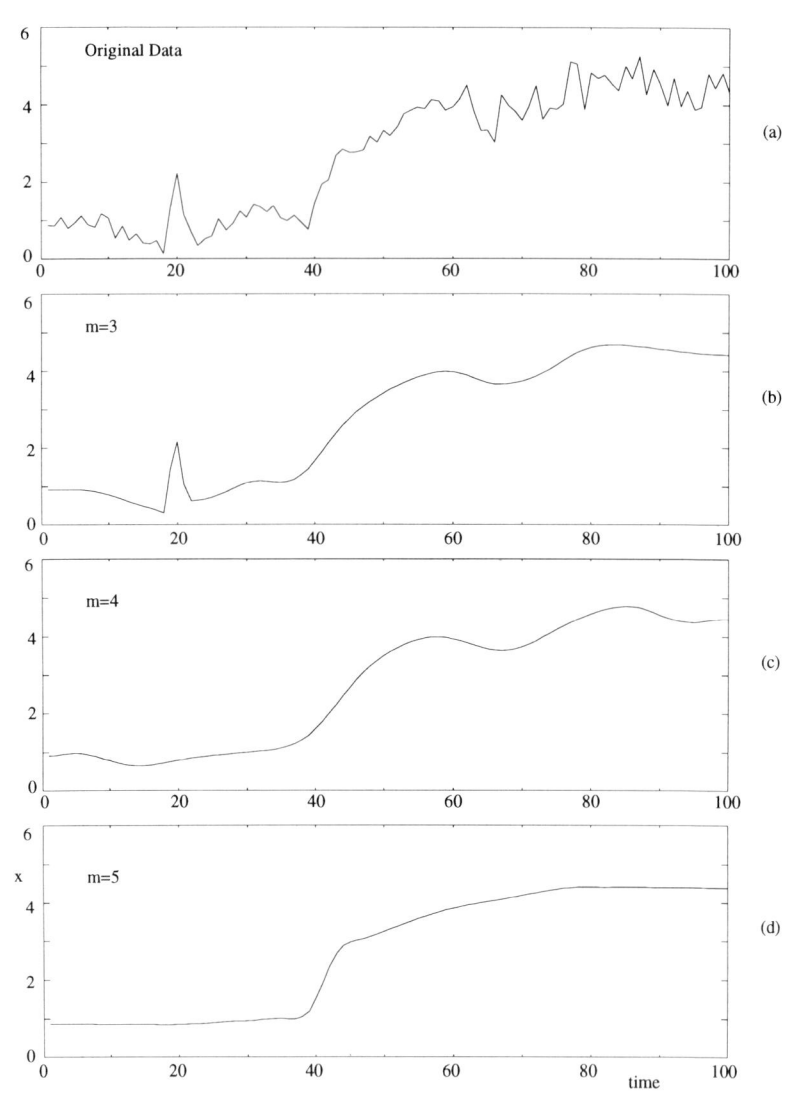

FIG. 15. Process trends of the signal in Fig. 1, extracted at various scales: (a) original data; (b) stable trend at $m = 3$ (see Fig. 13); (c) stable trend at $m = 4$; (d) stable trend at $m = 5$, neglecting small extrema.

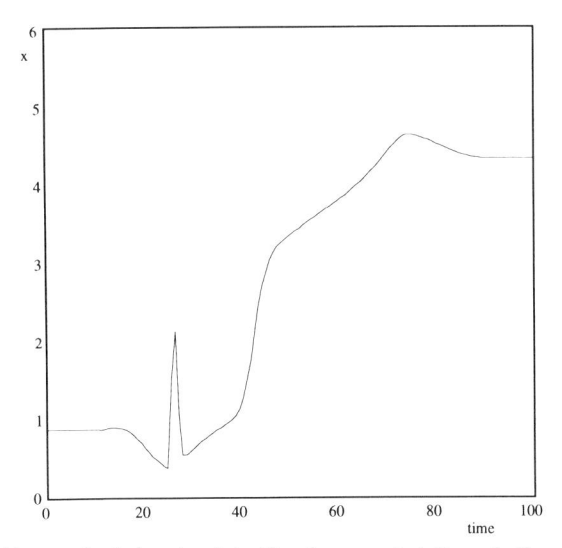

FIG. 16. Stable trend of the signal in Fig. 1, generated through the use of Witkin's stability criterion, and the wavelet coefficients shown in Fig. 13b.

b. Example 2. Generalization of process trends. Comparison of process signals obtained from several examples representing different process conditions is essential for evaluating the relevance of the extracted features. The representation of process trends at multiple scales provides a convenient, hierarchical technique for evaluating the features. Consider the signals shown in Fig. 17a obtained from three different batches of the fed-batch fermentation process described in Section V. At the level of the raw data, it is difficult to compare the three signals. On representing the signal at multiple scales, the differences in the signals disappear, and at a coarse-enough scale, the signals become qualitatively identical, as shown in Figs. 17–17g. The discovery of a generalized description provides a means of comparing the qualitative and quantitative features contained in the signals, and allows matching of features in the trends, facilitating easy extraction of qualitative differences. If the information in the signals at coarse scales is inadequate for distinguishing between the signals, then information in trends at finer scales is considered. As lower scales are considered, only the matched features need to be compared with each other. This provides a natural decomposition to the learning problem, and simplifies it significantly. The utility of generalized descriptions of process signals will be exploited further in the learning of input/output mappings, as described in Section V. The process trends obtained by applying Witkin's stability criterion at the coarsest scale are shown in Fig. 17h. These process trends are also quantitatively very similar, and the extracted features are physically meaningful and thus interpretable, as we will see in the illustrations of Section V (Figs. 21–23).

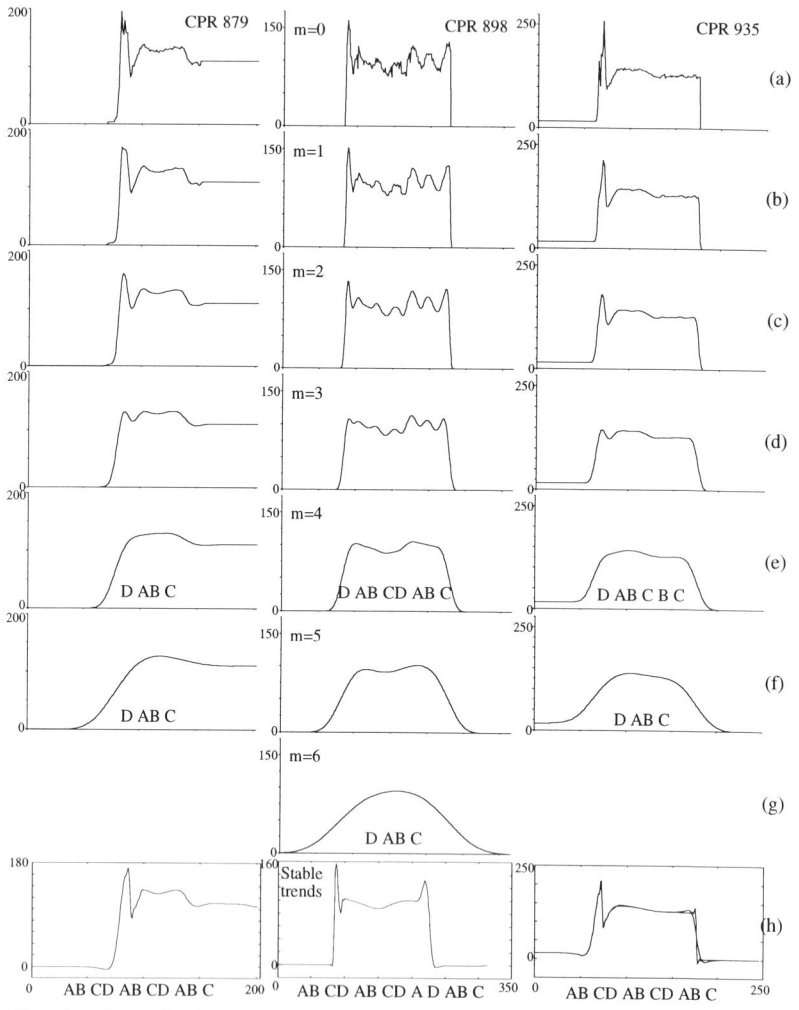

FIG. 17. Generalization of process trends for three distinct records; (a) raw data; (b)–(g) scaled signals; (h) stable trends.

IV. Compression of Process Data through Feature Extraction and Functional Approximation

With increasing computerization and improvements in sensor technology, it is relatively easy and inexpensive to collect large quantities of process data. Since measured data are useful for performing a variety of analytical and decision support engineering tasks, it is essential to store the data in historical records for future use. Efficient storage techniques are needed for two primary reasons: (1) to reduce the space required for the historical records and (2) to retrieve the data in a manner than renders the data easily interpretable for the execution of engineering tasks. In this section we will examine how both of these needs can be satisfied within the same theoretical framework of wavelet decomposition for the representation of measured signals.

A. Data Compression through Orthonormal Wavelets

Data compaction involves representation of the measured signal as an approximation that requires less storage space, at the cost of losing some information from the original signal. The data compaction problem may be stated as an approximation problem (Bakshi and Stephanopoulos, 1995) as follows:

1. Definition: Data Compaction Problem

Determine the approximate representation $\tilde{F}$ of a discrete-time function, $F(t)$ so as to either (1) minimize the error of the approximation given by

$$e_{L_p} = \frac{\|F - \tilde{F}\|_{L_p(I)}}{N} = \frac{\left(\int_I |F(t) - \tilde{F}(t)|^p \, dt\right)^{1/p}}{N}, \qquad 0 < p < \infty,$$

where N is the number of data points contained in the original signal, for a given compression ratio, $C_R = (N/\tilde{N})$, where $\tilde{N}$ is the number of points stored in the compacted representation; or (2) maximize the compression rate, C_R, for a given error of approximation, $e_{L_p}^2$.

A popular technique for approximation consists of representing the data as a weighted sum of a set of basis functions. As described in Section III, A, wavelets form a convenient set of basis functions to represent signals consisting of a variety of features. A signal decomposed on an orthonormal basis, using dyadic discretization of the translation and dilation parameters may be represented in terms of its wavelet coefficients and the coefficients of the last scaled signal, as given by Eq. (15). Since the wavelets at all translations and dilations, and scaling functions at a given dilation are orthonormal, the total energy in the signal is equal to the sum of the square of the coefficients:

$$\|F(t)\|_{L_2}^2 = \sum_{m=1}^{L}\sum_{k=1}^{k_{max}} \alpha_{mk}^2 + \sum_{k=1}^{k_{max}} \beta_{Lk}^2.$$

Compression may be achieved if some regions of the time–frequency space in which the data are decomposed do not contain much information. The square of each wavelet coefficient is proportional to the least-squares error of approximation incurred by neglecting that coefficient in the reconstruction;

$$e_N^2 = \frac{\|F(t) - \tilde{F}(t)\|^2}{N^2} = \frac{\sum_{m=1}^{L}\sum_{k=1}^{k_{max}} \alpha_{mk,\text{neglected}}^2 + \sum_{k=1}^{k_{max}} \beta_{Lk,\text{neglected}}^2}{N^2}, \tag{22}$$

where $\alpha_{mk,\text{neglected}}$ and $\beta_{mk,\text{neglected}}$ are the coefficients that are not stored in the compacted representation. Similarly, the local error in a region consisting of $(2r + 1)$ points, bounded by the interval $[l - r, l + r]$ is given by

$$e_{2r+1}^2 = \frac{\|F(t) - \tilde{F}(t)\|^2}{(2r+1)^2} = \frac{\sum_{m=1}^{L}\sum_{k=l-r}^{l+r} \alpha_{mk,\text{neglected}}^2 + \sum_{k=l-r}^{l+r} \beta_{Lk,\text{neglected}}^2}{(2r+1)^2}. \tag{23}$$

As $r \to 0$, the error of approximation tends to the L^∞ norm. Equations (22) and (23) are useful for data compaction with orthonormal wavelets.

Several complete orthonormal bases may be used for data compression. The selection of the best basis may be performed by utilizing several different criteria suggested by Coifman, Wickerhauser, and coworkers (Coifman and Wickerhauser, 1992; Wickerhauser, 1991). Some of the most interesting basis selection criteria include those discussed in the following paragraphs.

a. *Number above a Threshold.* The set of wavelets with the minimum number of coefficients above a threshold, ε, is selected as the best basis. This gives the best basis to represent a signal to precision ε. This measure is similar in principle to the error measure for the box car and backward slope methods.

b. *Entropy.* The statistical thermodynamical entropy is given by

$$H(\alpha) = -\sum_j p_j \log p_j,$$

where

$$p_j = \frac{|\alpha_j|^2}{\|\alpha_j\|^2} \quad \text{and} \quad p \log p = 0 \quad \text{for} \quad p = 0.$$

Coefficients of the wavelet expansion are indicated by α_j. Entropy is not an additive cost function, but it can be written as

$$H(\alpha) = \|\alpha\|^{-2}\lambda(\alpha) + \log\|\alpha\|^2,$$

where

$$\lambda(\alpha) = -\sum_j |\alpha_j|^2 \log|\alpha_j|^2$$

is additive, and can be minimized. The entropy measure provides a physically meaningful criterion for selecting the appropriate coefficients for data compression because $\exp[H(\alpha)]$ is proportional to the number of coefficients needed to represent the signal to a fixed mean square error.

c. *Number Capturing Given Percentage of Signal Energy.* This measure explicitly selects the smallest number of coefficients necessary to represent the signal with a given least-squares error. The cost function is the number of coefficients, N_e, with the largest absolute value that capture a percentage, e, of signal energy. These N_e coefficients represent the signal with the least-squares error, e, in the local region covered by the wavelet in the most compact form:

$$\sum_{j=1}^{N_e} \alpha_j^2 \geq \frac{e}{100} \sum_{j=1}^{N} \alpha_j^2,$$

where N is the total number of coefficients in the given wavelet decomposition. This measure evaluates the number of coefficients necessary to approximate the signal with a given least-squares error of approximation.

B. COMPRESSION THROUGH FEATURE EXTRACTION

Compression of process data through feature extraction requires

1. Techniques for representing features in the process data in an explicit manner to allow selection of the relevant features.
2. Criteria for determining what features in a process signal are relevant and worth storing, and what features may be lost due to compression.

Multiresolution representation of a process signal allows satisfaction of both these requirements. Decomposition of a signal using derivative wavelets and uniform sampling of the translation parameter provides a technique for representing the dominant features in a signal at various scales, as discussed in Section III, B. The relevance of features in the process signal may be determined based on their persistence over multiple scales. These properties of derivative wavelets are based on the work of Mallat and Zhong (1992), and have been exploited for extracting features from process data by Bakshi and Stephanopoulos (1994a) and were discussed in Section III, B.

C. PRACTICAL ISSUES IN DATA COMPRESSION

The practical, implementational issues that arise during the utilization of the data compression techniques, presented in the previous two subsections, are discussed in the following paragraphs.

1. Compression in Real Time

The speed with which the data need to be compressed depends on the stage of data acquisition at which compression is desired. In intelligent sensors it may be necessary to do some preliminary data compression as the data are collected. Often data are collected for several days or weeks without any compression, and then stored into the company data archives. These data may be retrieved at a later stage for studying various aspects of the process operation.

In order to compress the measured data through a wavelet-based technique, it is necessary to perform a series of convolutions on the data. Because of the finite size of the convolution filters, the data may be decomposed only after enough data has been collected so as to allow convolution and decomposition on a wavelet basis. Therefore, point-by-point data compression as done by the boxcar or backward slope methods is not possible using wavelets. Usually, a window of data of length 2^m, $m \in Z$, is collected before decomposition and selection of the appropriate features for storage are carried out. The optimal window size depends on the nature of the signal, the compression parameter used, and the decomposing wavelet. The window size should be greater than the duration of the longest discardable or irrelevant even in the signal. For a measured variable, the best basis may be different for compressing data in different windows. Usually, the time–frequency characteristics of the signal from a given measured variable do not vary significantly with time. Therefore, a best basis may be selected from a few sets of data and then used for new data. The feasibility of the selected best basis may be evaluated at regular intervals, if necessary.

2. Inaccuracies Due to End Effects

A practical problem in decomposing and reconstructing signals using wavelets is the errors introduced due to boundary effects. Both wavelet decomposition and reconstruction with limited amounts of data require assumptions about the signal's behavior beyond its endpoints. Usually, the signal is assumed to have its mirror image beyond its boundaries. This assumption often introduces an unacceptable error in the reconstructed signal, particularly near its endpoints. For processes, where the signal is compressed by analyzing windows of finite sizes, this error may be eliminated by augmenting the signal beyond its endpoints by segments of constant value. Thus, for a signal of length 128 points, the decomposition is performed on a signal that is augmented by constant segments of length equal to 64 points on either end. The resulting decomposition has $L + 1$, i.e., one additional scale that is created because of the augmentation. For data compression, this additional scale is disregarded, and the scaling function and wavelet coefficients are selected from L scales only. This simple procedure results in highly accurate reconstruction by elimination of the boundary effects. The quality of the compression is also unaffected since the wavelet coefficients for the augmented portions are constant or zero. The computational complexity of decomposition and reconstruction of the augmented signal is $O(2N)$, and still linear in the length of the original signal.

3. Selecting the Mother Wavelet and Compression Criteria

Several types of wavelets have been developed, but no formal criteria exist for selecting the mother wavelet for compressing a given signal. Some qualitative criteria and experience-based heuristics are normally used. In approximating a signal by wavelet or derivative wavelet transform extrema, the smoothness of the reconstructed signal depends directly on the nature of the basis functions. Several orthonormal wavelets with different degrees of smoothness have been designed. The orthonormal wavelets with compact support designed by Daubechies (1988) are reasonably smooth for orders greater than 6. The Haar wavelets provide piecewise constant approximation which may be adequate for some process signals, such as those of manipulated variables. Among derivative wavelets, quadratic and cubic wavelets are described by Mallat and Zhong (1992). The first derivative of a Gaussian is an infinitely differentiable wavelet. A priori knowledge of the nature of the signal may be used to select the mother wavelet based on its smoothness.

The accuracy of the error equations (Eqs. (22) and (23)] also depends on the selected wavelet. A short and compactly supported wavelet such as the Haar wavelet provides the most accurate satisfaction of the error estimate. For longer wavelets, numerical inaccuracies are introduced in the error equations due to end effects. For wavelets that are not compactly supported, such as the Battle–Lemarie family of wavelets, the truncation of the filters contributes to the error of approximation in the reconstructed signal, resulting in a lower compression ratio for the same approximation error.

D. An Illustrative Example

A typical example of half a day's process data from a distillation tower, and its wavelet decomposition using an orthonormal wavelet (Daubechies, 1988), with dyadic sampling are shown in Fig. 18. The raw process data represent pressure variation measured every minute (Takei, 1991). The raw signal is augmented by a constant segment of half the length of the original signal to avoid boundary effects. The augmentation results in an additional level of the decomposition, which is disregarded in the selection of coefficients, and the signal reconstruction. The wavelet coefficients are normalized to be proportional to their contribution to the overall signal. From Fig. 18 it is clear that many of the wavelet coefficients have very small values, which may be neglected without significant loss of information. The reconstructed signal with compression ratios of 2, 4, and 24 are shown in Fig. 19. The performance of orthonormal wavelet-based data compaction is compared with that of the conventional techniques such as backward slope and boxcar methods. As shown in Fig. 20, the wavelet-based method outperforms both the boxcar, and the backward slope methods. The quality of the reconstructed signal is significantly better for the wavelet-based method for similar compression ratios. Process data compaction using biorthogonal wavelets, wavelet packets and wavelet trans-

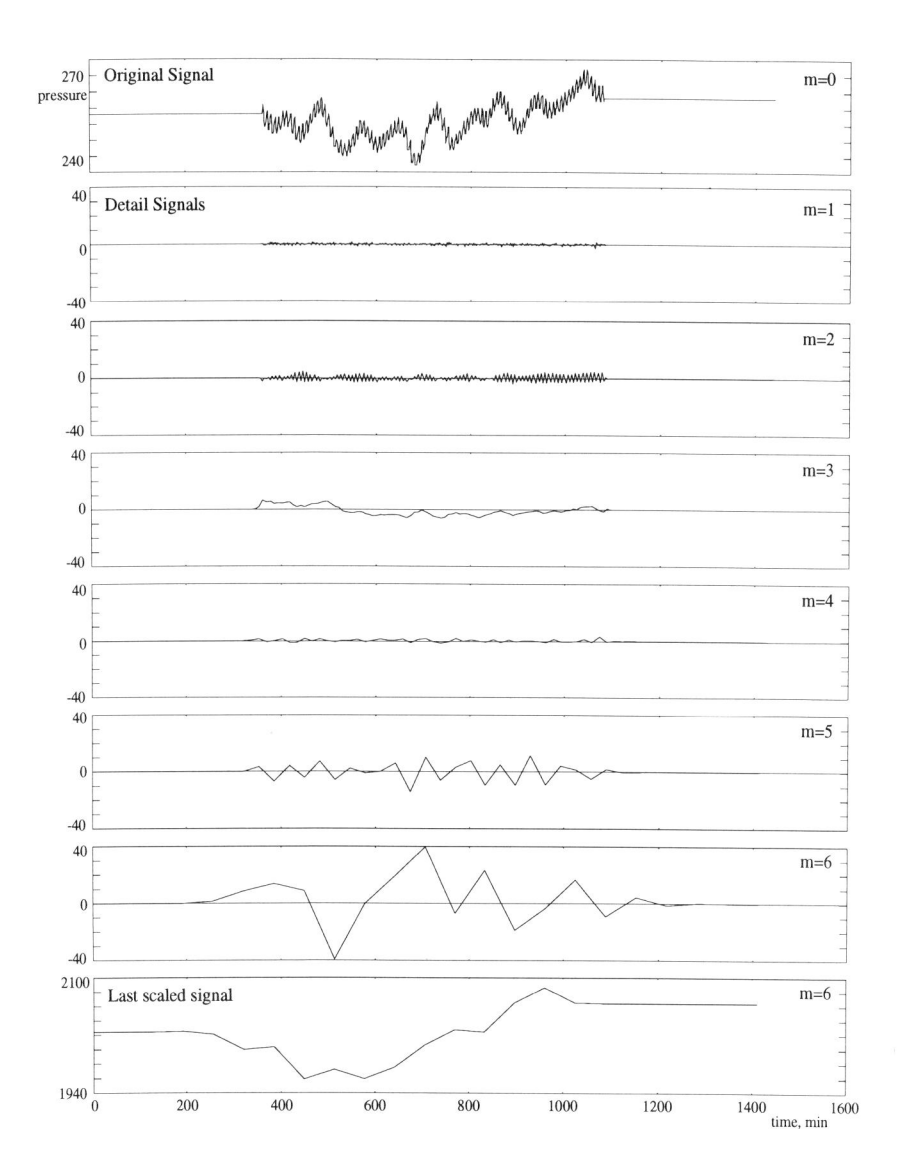

FIG. 18. Wavelet decomposition of a pressure signal, using Daubechies-6 wavelet.

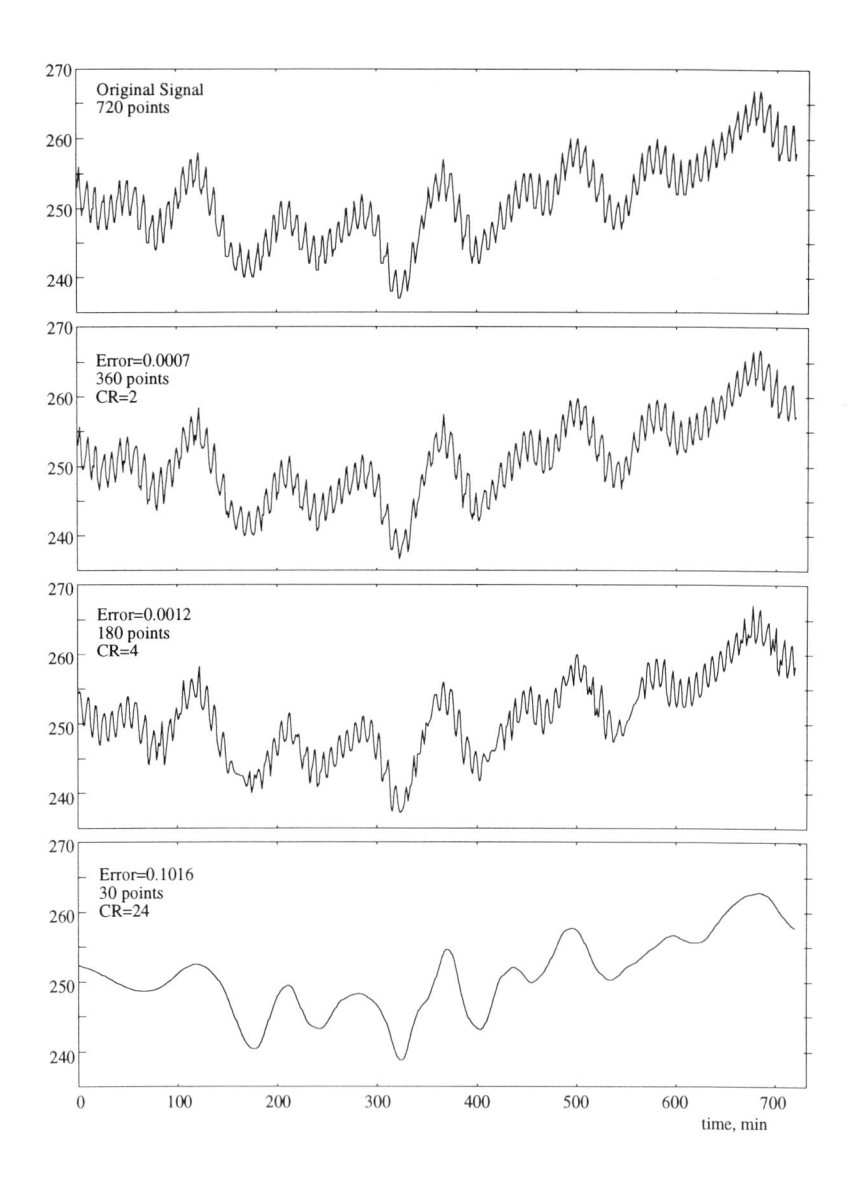

FIG. 19. Reconstruction of compressed signal from the wavelet decomposition of Fig. 18.

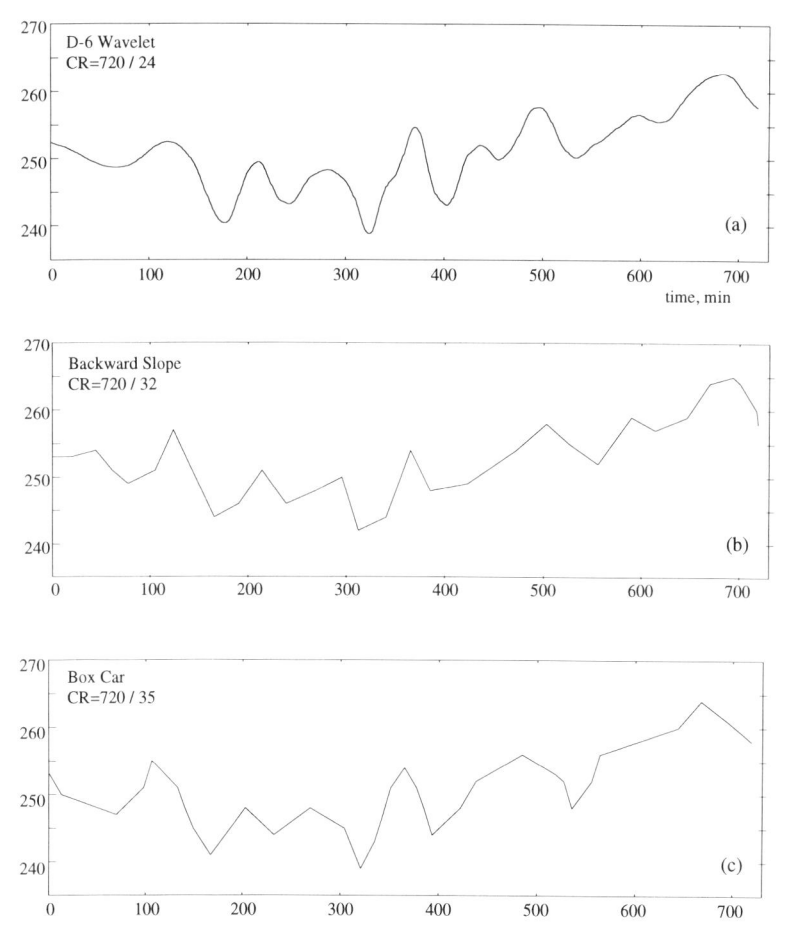

FIG. 20. Performance of data compression techniques: (a) orthonormal wavelet; (b) backward slope; (c) boxcar.

form extrema, and several more examples are presented in Bakshi and Stephanopoulos (1995).

V. Recognition of Temporal Patterns for Diagnosis and Control

The pivotal component of many engineering tasks—e.g., process fault diagnosis and product quality control—is the recognition of certain distinguishing temporal patterns (i.e., features) during the operation of the plants, and the development of associations between process operating patterns and process conditions. This two-step process is depicted in Fig. 3. Pattern recognition is the process through which a given pattern, $\mathbf{p}_i$, is assigned to the correct output class, C_I. A *pattern*, $\mathbf{p}$, is designated as the N-dimensional vector of inputs $(x_1, x_2, \ldots, x_N)$ in the input space, which may be partitioned into regions indicated by $(C_1, C_2, \ldots, C_K)$. The inputs may represent the value of manipulated and measured process variables, measured external disturbances, violation of output constraints, and/or set points at a given time point, or over a time interval. The feature extraction phase transforms the pattern $\mathbf{p}$, to a feature space, S_x, where only the most relevant parts of the inputs are retained. The extracted features may include the values of computed variables such as derivatives, integrals, or averages over time of the measured variables. Any pattern $\mathbf{p}_i = (x_1, x_2, \ldots, x_N)_i^T$ corresponds to a particular class of operating situations, such as sensor or actuator failure, process equipment failure, process parameter changes, activation of unmodeled dynamics, and effect of unmodeled disturbances. During inductive learning, the feature space S_x is partitioned into K mutually exclusive regions, $S_x^{(I)}$, with $I = 1, 2, \ldots, K$. Thus

$$S_x^{(I)} \cap S_x^{(J)} = 0, \quad I, J = 1, 2, \ldots, K, \text{ but } I \neq J,$$

and

$$\bigcup_{I=1}^{K} S_x^{(I)} = S_x.$$

This mapping from S_x to the classes C_I is determined by the discriminant functions that define the boundaries of regions $S_x^{(I)}$, $I = 1, 2, \ldots, K$ in S_x. Let $d_I(\mathbf{p})$ be the discriminant function associated with the Ith class of operating situations, where $I = 1, 2, \ldots, K$. Then, a pattern of measurements $\mathbf{p}$, implies the operating situation C_I iff

$$d_I(\mathbf{p}) > d_J(\mathbf{p}) \quad \text{for all } J \neq I.$$

Also, the boundary of the region $S_x^{(I)}$ with another region $S_x^{(J)}$ is given by

$$d_I(\mathbf{p}) - d_J(\mathbf{p}) = 0.$$

The inductive learning process determines the discriminant functions, using prior examples of $(\mathbf{p}, C_I)$ associations.

For solving the pattern recognition problem encountered in the operation of chemical processes, the analysis of measured process data and extraction of process trends at multiple scales constitutes the feature extraction, whereas induction via decision trees is used for inductive learning. A formal framework for the multiscale analysis of process data and the extraction of qualitative and quantitative features at various scales, based on the mathematical theory of wavelets, has already been developed (see Section III, B). The formal and efficient framework of wavelet decomposition provides a translationally invariant representation, and enables the extraction of qualitative and quantitative features at various scales and temporal locations with minimum distortion.

Having represented measured process data at multiple scales, we can develop consistent models for the various operational tasks. Learning input/output mappings between features in measured data and process conditions involves determining features that are most relevant to the process conditions. Inductive learning via decision trees provides explicit input/output mappings by identifying the most relevant qualitative and quantitative features from the measured variables. The mapping is easily expressed as *if-then rules* and may even be physically interpretable.

Several techniques from statistics, such as partial least-squares regression, and from artificial intelligence, such as artificial neural networks have been used to learn empirical input/output relationships. Two of the most significant disadvantages of these approaches are the following:

1. If the input to the learning procedure consists of raw process data from, e.g., several production batches, and the corresponding product yields, the learning technique has to overcome the "curse of dimensionality." The raw data consists of information, much of which may be irrelevant to the learning task, since the sensor data are measured at a scale much smaller than that of the events that may be relevant to the process yield. Such extraneous information increases the complexity of the learning process and may necessitate a large number of training examples to achieve the desired error rate.
2. The model learned is usually a "blackbox" and does not provide any insight into the physical phenomena and events influencing the process outputs.

These disadvantages are overcome by the methodology we will describe in the subsequent paragraph developed by Bakshi and Stephanopoulos. Effects of the curse of dimensionality may be decreased by using the hierarchical representation of process data, described in Section III. Such a multiscale representation of process data permits hierarchical development of the empirical model, by increasing the amount of input information in a stepwise and controlled manner. An explicit model between the features in the process trends, and the process conditions may be learned by *inductive learning* using *decision trees* (Quinlan, 1986; Breiman *et al.*, 1984), as described later on in this section.

A. GENERATING GENERALIZED DESCRIPTIONS OF PROCESS TRENDS

Consider a measured operating variable, $x(t)$, and its M distinct measurement records, $[x(t)]_i$, $i = 1, 2, \ldots, M$ over the same range of time. Using the multiscale decomposition of measured variables, discussed in Section III, we can represent each measurement record, $[x(t)]_i$, $i = 1, 2, \ldots, M$ by a finite state of trends, where each trend is a pattern of triangular episodes;

$$[p]_i^k = \{T_1, T_2, \ldots, T_{m_{i,k}}\}_i^k, \quad k = 1, 2, \ldots, l_i,$$

where superscript k indicates the kth representation of the record at some temporal scale, and l_i is the number of distinct ranges of scale generated by the wavelet interval-tree of scale. $T_1, T_2, \ldots, T_m$ denote the primitive triangles, describing the monotonic temporal behavior of $[x(t)]_i$ over a certain period of time, and could be any of the seven types shown in Fig. 4a;

$$T \in \{A, B, C, D, E, F, G\}.$$

1. Qualitatively Equivalent Patterns

Two patterns are qualitatively equivalent if and only if their corresponding sequences of triangular episodes are qualitatively equivalent episode by episode; i.e., the condition

$$\{T_1, T_2, \ldots, T_{m_{i,k}}\}_i^k \langle QE \rangle \{T_1, T_2, \ldots, T_{m_{i,j}}\}_i^j$$

implies that

$$\{T_r\}_i^k \langle QE \rangle \{T_r\}_i^j, \quad r = 1, 2, \ldots, m_{i,k}; \quad m_{i,k} = m_{i,j},$$

where $\langle QE \rangle$ denotes the qualitative equivalence.

2. Generalized Description of Trends

Consider M distinct batch records of the same measured variable. Let $[p]_i^k$; $i = 1, 2, \ldots, M$ be the appropriate representation of the ith record at some scale of abstraction. If $[p]_1^{k_1} \langle QE \rangle [p]_2^{k_2} \langle QE \rangle \ldots [p]_i^{k_i} \langle QE \rangle \ldots [p]_M^{k_M}$, then the common qualitative trend of all records is called the *generalized description* of trends contained in M records of a measured variable.

Figure 17 shows the raw data and the scaled signals for three records (CPR 879, 898, and 935) of carbon dioxide (CO_2) production rate (CPR) from a fed-batch fermentor. The description of the three records at the coarsest scale is $\{DABC\}$ and is identical for the three batches, thus leading to a general description of the variation of CPR. Representation at scales coarser than that of the generalized description will also be qualitatively equivalent. Figure 21 shows the generalized description of six (6) operating variables. These generalized descriptions were generated

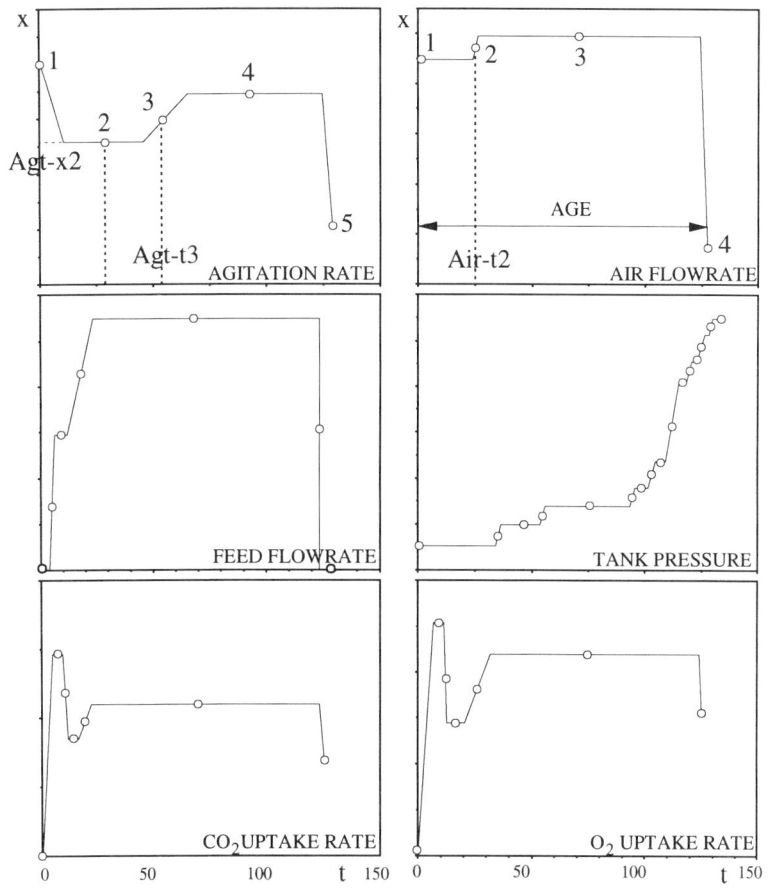

FIG. 21. Generalized description of fed-batch fermentation process data.

from 32 records of each variable, using Witkin's stability criterion to extract the most stable trends.

3. Pattern Matching of Multiscale Descriptions

Extraction of the most relevant parts of a process trend requires pattern matching of features that arise from the same physical phenomena, and are qualitatively identical. Matching of qualitatively identical features allows comparison of the qualitative and quantitative features at each scale in an organized, hierarchical manner, and helps fight the curse of dimensionality.

Consider the process data shown in Fig. 17a. These data represent the CPR for the fed-batch fermentation process for three batches giving different yields, as shown in the figure. The raw data are quite different, both in their qualitative and quantitative features. At coarser representations, the three signals start looking similar, and finally, at the coarsest scale, they are qualitatively identical, as shown in Figs. 17f and 17g, resulting in a unique generalized description: $\{D - AB - C\}$, and matching qualitatively identical features is straightforward.

The descriptions obtained using Witkin's stability criterion are shown in Fig. 17h. Pattern matching of the features in these descriptions results in one of the following matches:

Match 1: $AB - CD - AB - CD - A$ $\quad B - C \quad$ CPR 879, CPR 935
$\quad\quad\quad\; AB - CD - AB - CD - A - (D - A)$
$\quad\quad\quad\quad\quad\quad\quad\quad\quad\quad\quad\quad\quad\quad\quad\quad B - C \quad$ CPR 898

Match 2: $AB - CD - AB - C \quad\quad\quad D - AB - C \quad$ CPR 879 CPR 935
$\quad\quad\quad\; AB - CD - AB - C(D - A) - D - AB - C \quad$ CPR 898

The qualitative feature, $\{D - A\}$, distinguishes the variable CPR 898 from the other two, and should be evaluated for its relevance to solving the classification problem. The pattern matching task may not be straightforward, especially at lower scales, due to greater detail in the trends. Heuristics may then be used to organize the matches in terms of their likelihood of being physically meaningful. Often, domain-specific information about the process is available for matching features in trends from different examples. For example, information about the sequence and type of features in a trend under normal operation, may be available. Such information simplifies the pattern matching problem, and eliminates several infeasible matches. In the absence of domain-specific information, all matches are equally likely, and need to be evaluated for their relevance to solving the learning problem. For more details on the technical aspects of pattern matching see Bakshi and Stephanopoulos (1994b).

TABLE I
DATA FOR ILLUSTRATING INDUCTION USING DECISION TREES[a]

Example #	Input Features		Output Features	
	Pressure	Temperature	Color	Production Quality
1	N	99	N	Good
2	N	105	A	Bad
3	H	108	N	Good
4	L	92	N	Bad
5	L	106	N	Bad
6	N	106	N	Good
7	L	104	A	Bad
8	N	95	A	Bad

[a] Key: H—high; L—low; N—normal; A—abnormal. *Source*: reproduced from Bakshi and Stephanopoulos (1994b), by permission.

B. INDUCTIVE LEARNING THROUGH DECISION TREES

Once the several records of a process variable have been generalized into a pattern, as indicated in the previous paragraph, we need a mechanism to induce relationships among features of the generalized descriptions. In this section we will discuss the virtues of inductive learning through decision trees.

Inductive learning by decision trees is a popular machine learning technique, particularly for solving classification problems, and was developed by Quinlan (1986). A decision tree depicting the input/output mapping learned from the data in Table I is shown in Fig. 22. The input information consists of pressure, temperature, and color measurements of

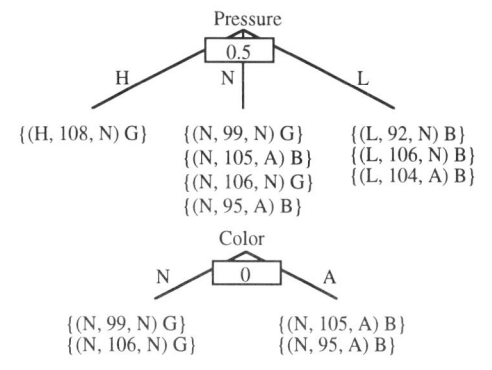

FIG. 22. Decision tree for data in Table I.

a chemical process. The output variable is the product quality corresponding to each set of measured variables. The decision tree in Fig. 22 provides the following information for obtaining product of good quality:

1. If pressure is normal and color is normal, then quality is good.
2. If pressure is low, then quality is bad.
3. If pressure is normal and color is abnormal, then quality is bad.

The model induced via the decision tree is not a blackbox, and provides explicit and interpretable rules for solving the pattern classification problem. The most relevant variables are also clearly identified. For example, for the data in Table I, the value of the temperature are not necessary for obtaining good or bad quality, as is clearly indicated by the decision tree in Fig. 22.

The procedure for generating a decision tree consists of selecting the variable that gives the best classification, as the root node. Each variable is evaluated for its ability to classify the training data using an information theoretic measure of entropy. Consider a data set with K classes, C_I, $I = 1, 2, \ldots, K$. Let M be the total number of training examples, and let M_{C_I} be the number of training examples in class C_I. The information content of the training data is calculated by Shannon's entropy:

$$I(M_{C_1}, M_{C_2}, \ldots, M_{C_K}) = -\sum_{I=1}^{n} \frac{M_{C_I}}{M} \log_2 \left(\frac{M_{C_I}}{M} \right). \quad (24)$$

Equation (24) provides a measure of the variety of classes contained in the data set. If all examples belong to the same class, then the entropy is zero. Smaller entropy implies less variety of classes (more order) in the data set. If the data set is split into groups, G_1 and G_2, with M_{C_I, G_J} being the number of examples belonging to class, C_I, that are present in group, G_J, then the total information content is

$$E(G_1, G_2) = \frac{M_{G_1}}{M} I(M_{C_1, G_1}, \ldots, M_{C_K, G_1}) + \frac{M_{G_2}}{M} I(M_{C_1, G_2}, \ldots, M_{C_K, G_2}). \quad (25)$$

Equations (24) and (25) are adequate for designing decision trees. The feature that minimizes the information content is selected as a node. This procedure is repeated for every leaf node until adequate classification is obtained. Techniques for preventing overfitting of training data, such as cross validation are then applied.

Induction via decision trees is a greedy procedure and does not guarantee optimality of the mapping, but works well in practice, as illustrated by successful applications in several areas. The attractive features of learning by decision trees are listed below.

1. The model is easy to understand, and interpret physically. Concise rules may be developed.
2. Only the most relevant features are selected as nodes. Redundant, or unnecessary features are clearly identified through the maximization of entropy change.
3. Both continuous and discrete-valued features can be handled in a uniform framework. This permits easy combination of quantitative and structural methods.
4. No a priori assumptions about the distribution of data or class probability are required.
5. The technique is robust to noisy examples.

The discriminant hypersurface is approximated in a piecewise constant manner. This may result in decision trees that are very large and complicated, and deriving meaningful rules may not be very easy. This problem is alleviated by the hierarchical learning procedure, since the number of features evaluated is increased gradually, only if necessary. Techniques for overcoming some of the disadvantages of decision trees are described by Saraiva and Stephanopoulos (1992).

C. Pattern Recognition with Single Input Variable

Consider the three distinct records (examples) of measurements for the same operating variable, shown in Fig. 23. Let the records $[x(t)]_1$ and $[x(t)]_2$ correspond to operating conditions of class C_A, while the third, $[x(t)]_3$ corresponds to operating conditions of class C_B. At the lowest scale, the high-frequency components of the three records make all of them look very different. The first scale at which the first two records have the same qualitative description, yields the following common trend: $AB - C - B - C$. On the other hand, the most stable representation of the third record is $AB - C - B - CD - AB - C$. In this particular example, syntactic representation of trends has been sufficient to provide complete classification of the three records. Using inductive learning through decision trees leads to the following two cases for classifying trends of class C_A

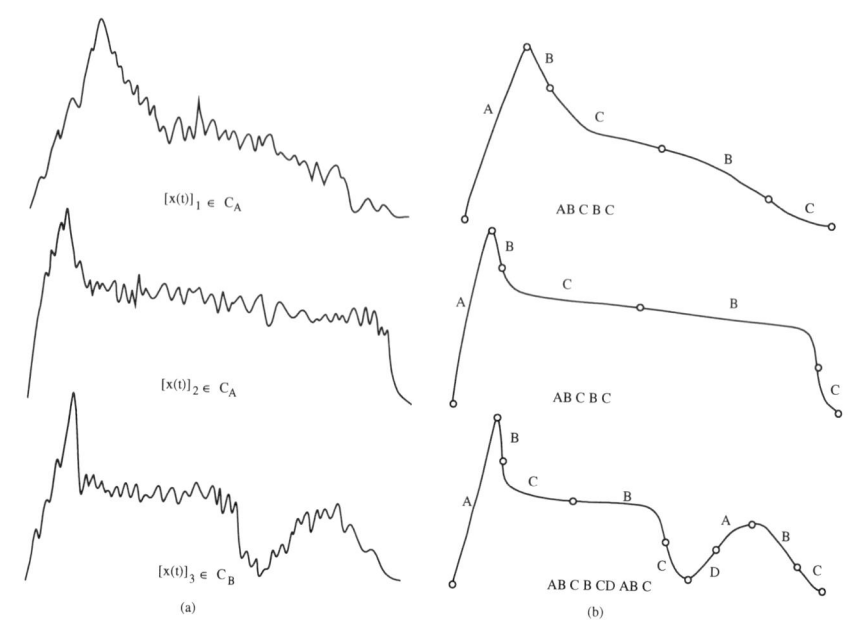

Fig. 23. (a) The raw data of three distinct records and (b) their corresponding syntactic generalizations. (Reprinted from Bakshi and Stephanopoulos, Representation of Process Trends, Part IV. *Computers and Chemical Engineering*, **18**(4), p. 303, Copyright (1994), with kind permission from Elsevier Science Ltd., The Boulevard, Langford Lane, Kidlington OX5 1GB, UK).

from those of class C_B:

Case 1. The syntactic difference is the string $(D - AB - C)$ at the end of the trend, since

$$AB - C - B - C \qquad \text{(description of first two records)}$$
$$AB - C - B - C(D - AB - C) \qquad \text{(description of third record)}$$

Case 2. The syntactic difference is the string $(B - CD - A)$ in the middle of the trend, since

$$AB - C - \qquad\qquad B - C \qquad \text{(for first two records)}$$
$$AB - C - (B - CD - A)B - C \qquad \text{(for third record)}$$

D. Pattern Recognition with Multiple Input Variables

The inductive classification of multiple-dimensional trends involves the mapping between the distinguishing features of several input-variables and the corresponding classes of a single output. It is carried out in two phases, as follows:

1. *Phase 1: Generate Generalized Descriptions and Extract Distinguishing Features for Each Variable*

The procedure for generating generalized trends was described earlier. The features of the generalized trends are given by the triangular episodes at any level of required detail, i.e., qualitative, semiquantitative, real-valued analytic.

2. *Phase 2: Learn Mapping between Extracted Features and Output Classes*

Once the distinguishing features for each input variable have been extracted through the generalization of descriptions of the available records, these features become the inputs of the inductive learning procedure through decision trees.

Bakshi and Stephanopoulos (1994b) have applied the above procedure to a fed-batch fermentation process. The problem involved 41 sets of batch records on 24 measured variables. Of these variables only very few were found by the decision tree to be relevant, and yield rules such as the following for guiding the diagnosis or control of a fermentor.

Diagnostic Rule-1 If the duration of the high bottom dissolved oxygen (BDO) phase is > 3.6 h and the level of CO_2 generation during the production phase is > 4.3 units, then the quality of the fermentation is excellent.

Control rule-1. To achieve fermention of excellent quality, keep (a) the agitation rate in the first scaling episode (growth phase) > 37 units and (b) the duration of the first episode in the air flowrate ≤ 25 h.

For the detailed discussion on the inductive learning of diagnostic and control rules around the fed-batch fermentor system, the reader should refer to the work of Bakshi and Stephanopoulos (1994b).

VI. Summary and Conclusions

The wavelet decomposition of measured data provides a natural framework for the extraction of temporal features, which characterize operating process variables and their trends. Such characterization, local in frequency and time, provides valuable insight as to what is going on in a process, and thus it can explicitly support a number of engineering methodologies, such as data compression, diagnosis of process upsets, and pattern recognition for quality control. Nevertheless, a number of interesting issues arise that can shape future developments in the area process operations and control.

1. *Multiscale modeling of process operations.* The description of process variables at different scales of abstraction implies that one could create models at several scales of time in such a way that these models communicate with each other and thus are inherently consistent with each other. The development of multiscale models is extremely important and constitutes the pivotal issue that must be resolved before the long-sought integration of operational tasks (e.g., planning, scheduling, control) can be placed on a firm foundation.

2. *Multiscale process identification and control.* Most of the insightful analytical results in systems identification and control have been derived in the frequency domain. The design and implementation, though, of identification and control algorithms occurs in the time domain, where little of the analytical results in truly operational. The time-frequency decomposition of process models would seem to offer a natural bridge, which would allow the use of analytical results in the time-domain deployment of multiscale, model-based estimation and control.

3. *Integration of process operational tasks.* The industrial deployment of computer-aided systems that can integrate planning-scheduling-diagnosis-control rests on two pillars; multiscale process models (see paragraph 1, above) and multiscale depiction of process data. Although the wavelet decomposition offers the theoretical answer to the second, the industrial implementation requires the solution of the following problems: (a) integration of quantitative, qualitative and semiquantitative descriptions of process operations through the establishment of an "appropriate" language; and (b) integration of planning, scheduling, diagnosis, and control methodologies around the common language. These issues require creative modeling of process data around the wavelet decomposition and need to be addressed soon by researchers.

References

Bader, F. P., and Tucker, T. W., Data compression applied to a chemical plant using a distributed historian station. *ISA Trans.* **26**(4), 9–14 (1987a).
Bader, F. P., and Tucker, T. W., Real-time data compression improves plant performance assessment. *InTech.* **34**, 53–56 (1987b).

Bakshi, B. R., and Stephanopoulos, G., Wave-net: A multi resolution, hierarchical neural network with localized learning. *AIChE J.* **39**(1), 57–81 (1993).

Bakshi, B. R., and Stephanopoulos, G., Representation of process trends. Part III. Multi-scale extraction of trends from process data. *Comput. Chem. Eng.* **18**, 267 (1994a).

Bakshi, B. R., and Stephanopoulos, G., Representation of process trends. Part IV. Induction of real-time patterns from operating data for diagnosis and supervisory control. *Comput. Chem. Eng.* **18** 303 (1994b).

Bakshi, B. R., and Stephanopoulos, G., Compression of chemical process data through functional approximation and feature extraction. *AIChE J.*, accepted for publication (1995).

Bastl, W., and Fenkel, L., Disturbance analysis systems. In "Human Diagnosis of System Failures," (Rasmussen and Rouse, eds.) Nato Symp. Denmark, Plenum, New York, 1980.

Breiman, L., Friedman, J. H., Olshen, R. A., and Stone, C. J., "Classification and Regression Trees." Wadsworth, Belmont, CA (1984).

Cheung, J. T.-Y., Representation and extraction of trends from process data. Sc.D. Thesis, Massachusetts Institute of Technology, Dept. Chem. Eng., Cambridge, MA (1992).

Cheung, J. T.-Y., and Stephanopoulos, G., Representation of process trends. Part I. A formal representation framework. *Comput. Chem. Eng.* **14**, 495–510 (1990).

Coifman, R. R., and Wickerhauser, M. V., Entropy-based algorithms for best basis selection. *IEEE Trans. Inf. Theory* **38**(2), 713–718 (1992).

Daubechies, I., Orthonormal Bases of Compactly Supported Wavelets, *Comm. Pure Appl. Math.*, **XLI**, 909–996 (1988).

Dvorak, D. L., "Expert Operations Systems," Tech. Rep. Dept. of Computer Science, University of Texas at Austin, 1987.

Feehs, R. J., and Arce, G. R., "Vector Quantization for Data Compression of Trend Recordings," Tech. Rep. 88-11-1, University of Delaware, Dept. Elect. Eng., Newark, 1988.

Fukunaga, K., "Introduction to Statistical Pattern Recognition." Academic Press, Boston, 1990.

Funahashi, K. I. On the approximate realization of continuous mappings by neural networks. *Neural Networks* **2** 183–192 (1989).

Goupillaud, P., Grossmann, A., and Morlet, J., Cycle-octave and related transforms in seismic signal analysis. *Geoexploration* **23**, 85–102 (1984).

Gray, R. M., Vector quantization. *IEEE ASSP Mag.*, April, pp. 4–29 (1984).

Hale, J. C., and Sellars, H. L., Historical data recording for process computers. *Chem. Eng. Prog.*, November, pp. 38–43 (1981).

Hummel, R., and Moniot, R., Reconstructions from zero crossings in scale space. *IEEE Trans. Acoust., Speech, Signal Processing* **ASSP 37**(12), 2111–2130 (1989).

Jantke, K., "Analogical and Inductive Inference." Springer-Verlag, Berlin, 1989.

Kanal, L. N., and Dattatreya, G. R., Problem-solving methods for pattern recognition. In Handbook of Pattern Recognition and Image Processing," (T. Y. Young, and K.-S. Fu, eds.) Academic Press, New York, 1985.

Kramer, M. A., Nonlinear principal component analysis using autoassociative neural networks. *AIChE J.* **37**, 233–243 (1991).

Long, A. B., and Kanazava, R. M., "Summary and Evaluation of Scoping and Feasibility Studies for Disturbance Analysis and Surveillance Systems," EPRI Report NP-1684. Electr. Power Res. Inst., Palo Alto, CA, 1980.

MacGregor, J. F., Marlin, T. E., Kresta, J. V., and Skagerberg, B., Multivariate statistics methods in process analysis and control. In "Chemical Process Control, CPCIV," (Y. Arkun and W.H. Ray, eds.). CACHE, AIChE Publishers, New York, 1991.

Mallat, S. G., A theory for multiresolution signal decomposition: The wavelet representation. *IEEE Trans. Pattern Anal. Mach. Intell.* **PAMI-11**(7), 674–693 (1989).

Mallat, S. G., Zero crossing of a wavelet transform. *IEEE Trans. Inf. Theory* **IT-37**(4), 1019–1033 (1991).

Mallat, S., and Zhong, S., Characterization of signals from multiscale edges, *IEEE Trans. Pattern Anal. Mach. Intell.* **PAMI-14**(7), 710–732 (1992).

Marr, D., and Hildreth, E., Theory of edge detection. *Proc. R. Soc. London B Ser.* **207**, 187–217 (1980).

Meyer, Y., Principle d'incertitude, bases hilbertiennes et algebres d'operateurs. *Bourbaki Sem.* No. 662 (1985–1986).

O'Shima, E., Computer-aided plant operation. *Comput. Chem. Eng.* **7**, 311 (1983).

Papoulis, A., "Signal analysis." McGraw-Hill, New York, 1977.

Quinlan, J. R., Induction of decision trees. *Mach. Learn.* **1**(1), 81–106 (1986).

Rumelhart, D. E., McClelland, J. L., "Parallel Distributed Processing," Vol. 1. MIT Press, Cambridge, MA, 1986.

Saraiva, P., and Stephanopoulos, G., Continuous process improvement through inductive and analogical learning. *AIChE J* **38**(2), 161–183 (1992).

Silverman, B. W., "Density Estimation for Statistics and Data Analysis." Chapman & Hall, New York, 1986.

Stephanopoulos, G., Artificial intelligence... 'What will its contributions be to process control?' *In* "The Second Shell Process Control Workshop," (D.M. Prett, C. E. Garcia, and B. L. Ramaker, eds.) Butterworth, Stoneham, MA, 1990.

Takei, S., Multiresolution analysis of data in process operations and control. M.S. Thesis, Massachusetts Institute of Technology, Dept. Chem. Eng., Cambridge, MA, 1991.

Vaidyanathan, R., and Venkatasubramanian, V., Process fault detection and diagnosis using neural networks: II. Dynamic processes. *AIChE Ann. Meet.*, Chicago, IL (1990).

Van Trees, H., "Detection, Estimation and Modulation Theory." Wiley, New York, 1968.

Wickerhauser, M. V., "INRIA Lectures on Wavelet Packet Algorithms." Yale University, New Haven, CT, 1991.

Witkin, A. P., Scale space filtering: A new approach to multi-scale description. *In* "Image Understanding" (S. Ullman and W. Richard, eds.), pp. 79–95. Ablex, Norwood, NJ, 1983.

Yuille, A.L., and Poggio, T., Fingerprints theorems. *Proc. Natl. Conf. on Artif. Intell.*, pp. 362–365 (1984).

INTELLIGENCE IN NUMERICAL COMPUTING: IMPROVING BATCH SCHEDULING ALGORITHMS THROUGH EXPLANATION-BASED LEARNING

Matthew J. Realff[1]

School of Chemical Engineering
Georgia Institute of Technology
Atlanta, Georgia 30332

I. Introduction	550
A. Flowshop Problem	552
B. Characteristics of Solution Methodology	553
II. Formal Description of Branch-and-Bound Framework	555
A. Solution Space Representation—Discrete Decision Process	555
B. The Branch-and-Bound Strategy	557
C. Specification of Branch-and-Bound Algorithm	563
D. Relative Efficiency of Branch-and-Bound Algorithms	564
E. Branching as State Updating	566
F. Flowshop Lower-Bounding Scheme	568
III. The Use of Problem-Solving Experience in Synthesizing New Control Knowledge	570
A. An Instance of a Flowshop Scheduling Problem	570
B. Definition and Analysis of Problem-Solving Experience	573
C. Logical Analysis of Problem-Solving Experience	578
D. Sufficient Theories for State-Space Formulation	579
IV. Representation	581
A. Representation for Problem Solving	583
B. Representation for Problem Analysis	588
V. Learning	593
A. Explanation-Based Learning	594
B. Explanation	598
C. Generalization of Explanations	601
VI. Conclusions	607
References	608

---

[1] The work reported in this chapter was carried out while the author was a Ph.D. student in the Laboratory for Intelligent Systems in Process Engineering, Department of Chemical Engineering, Massachusetts Institute of Technology, Cambridge, MA 02139.

Learning comes from reflection on accumulated experience and the identification of patterns found among the elements of previous experience. All numerical algorithms used in scientific and engineering computing are based on the same paradigm: *Execute a predetermined sequence of calculational tasks and produce a numerical answer*. The implementation of the specific numerical algorithm is oblivious to the experience gained during the solution of a specific problem, and the next time a different, or even the same, problem is solved through the execution of exactly the same sequence of calculational steps. The numerical algorithm makes no attempt to reflect on the structure and patterns of the results it produced, or reason about the structure of the calculations it has performed. This chapter shows that this need not be the case. By allowing an algorithm to reflect on and reason with aspects of the problems it solves and its *own structure of computational tasks, the algorithm can learn* how to carry out its tasks more efficiently. This form of *intelligent numerical computing* represents a new paradigm, which will dominate the future of scientific and engineering computing. But, in order to unlock the computer's potential for the implementation of truly intelligent numerical algorithms, the *procedural* depiction of a numerical algorithm must be replaced by a *declarative* representation of the algorithmic logic. Such a requirement upsets an established tradition and imposes new educational challenges which most educators and educational curricula have not, as yet, even recognized. This chapter shows how one can take a branch-and-bound algorithm, used to identify optimal schedules of batch operations, and endow it with the ability to learn to improve its own effectiveness in locating the optimal scheduling policies for flowshop problems. Given that most batch scheduling problems are NP-hard, it becomes clear how important it is to improve the effectiveness of algorithms for their solution. Using the framework of Ibaraki (1978) a branch-and-bound algorithm is declaratively modeled as a *discrete decision process*. Then, *explanation-based machine learning* strategies can be employed to uncover patterns of generic value in the experience gained by the branch-and-bound algorithm from solving specific instances of scheduling problems. The logic of the uncovered patterns (i.e., new knowledge) can be incorporated into the control strategy of the branch and bound algorithm when the next problem is to be solved.

I. Introduction

The main goal of this chapter is to introduce the concept of *intelligent numerical computing* within the context of solving optimization problems of relevance to chemical engineering. Before focusing on this area, it is worth noting that intelligent scientific computing has been the subject of research within the context of other problems such as bifurcation analysis and nonlinear dynamics, and chemical reaction kinetic schemes (Abelson et al., 1989; Yip, 1992), both of which are clearly subjects of interest to chemical engineers.

Our first task is to define intelligent numerical computing and to indicate what features it possesses to make it a distinct subclass of general numerical computing. For us, the essence of the distinction is the difference between *reasoning* and *calculation* (Simon, 1983). Implementations of numerical algorithms, such as the simplex method for linear programming (Dantzig, 1963), or gradient descent methods for more general nonlinear optimization problems (Avriel, 1976), take a problem formulated as a set of algebraic relationships and calculate an answer. Following performance of the calculation task, the implementation is completely oblivious to the *experience*, and the next time a different, or even the same, problem is solved exactly, the same set of steps will be performed. The implementation makes no attempt to reflect on its experience, or reason about the calculation it performed; it is completely oblivious to its own structure, which has not been declaratively represented, but that is procedurally embodied in the set of calculations performed. *Intelligent numerical computing* attempts to incorporate reasoning into the calculation procedure by allowing the computer to reason and reflect on various aspects of the problem, and its own structure. To do this we need to adopt new representations of information pertinent to problem solving, in particular symbolic representations, and to introduce new algorithms, which compute with symbolic as opposed to numerical information, such as natural deduction and resolution (Robinson, 1965) for logic-based reasoning.

To illustrate these concepts we will focus on a particular numerical algorithm, branch and bound (Nemhauser and Wolsey, 1988), which has been a workhorse for solving optimization problems with discrete structure, such as chemical batch scheduling (Kondili et al., 1993; Shah et al., 1993). This is a generic problem-solving strategy, and its successful application relies on the identification of effective *control information*, often in (1) the form of a lower-bound function, which characterizes how good an incomplete solution can be, and (2) as dominance and equivalence conditions, which use information about one partial solution to terminate another. In this chapter, we will present a methodology for *automating* the acquisition of this control knowledge for branch-and-bound algorithms, using flowshop scheduling as an illustration. Here, automation means that the *computer itself* acquires new control information, unaided by the human user, save for the specification of the problem. The acquisition is carried out by analyzing the experience gained by the computer during problem solving, illustrating the incorporation of one facet of reasoning, *learning from experience*.

The rest of the chapter will focus on solving four goals, engendered by the problem of automatically improving problem-solving experience.

1. Before we can learn new control information, we must formally specify what it is we are trying to learn and how what we learn is to be applied within the original problem-solving framework (Section II).
2. Having established the formal specification of the problem-solving framework, the next goal is to establish how the solution of example problems can be turned into relevant problem-solving experience (Section III).
3. To convert the problem-solving experience into a useful form, we need to be able to represent it to the computer, along with the other information necessary to reason about it efficiently (Section IV).
4. Finally, having set up the learning problem, we need to employ a learning method that will guarantee preservation of the correctness of the branch-and-bound algorithm and make useful additions to the control information we have about the problem (Section V).

These four goals are addressed sequentially in the next four sections. The flowshop problem will be used as an illustration throughout, because of its practical relevance, difficulty of solution, and yet relative simplicity of its mathematical formulation.

A. FLOWSHOP PROBLEM

Many chemical batch production facilities are dedicated to producing a set of products that require for their manufacturing a common set of unit operations. The unit operations are performed in the same sequence for each product. This type of production problem is often solved by configuring the available equipment so that each unit operation is carried out by a fixed set of equipment items that are disjoint from those used in any other operation. If one unit is assigned to each step, this is called a *flowshop* (Baker, 1974).

The flowshop problem has been widely studied in the fields of both operations research (Lagweg et al., 1978; Baker, 1975) and chemical engineering (Rajagopalan and Karimi, 1989; Wiede and Reklaitis, 1987). Since the purpose of this chapter is to illustrate a novel technique to synthesize new control knowledge for branch-and-bound algorithms, we will adopt the simplest, and most widely studied, form of the flowshop problem. The assumptions with respect to the problem structure are given below.

1. *Storage policy.* Much attention has been paid to exploring the implications of different storage policies, such as having no intermediate storage, finite intermediate storage, running the plant with a zero-wait policy, or combinations of the policies discussed above. We will assume unlimited intermediate storage is available to simplify the constraints between batches.
2. *Clean out, set up, and transfer policies.* In general, these operations could be dependent on the order in which the products are routed through the equipment. We will assume that these operations are *sequence independent* and can be factored into the processing times for each step.

With these simplifications, we can now formulate the mathematical model, which describes the flowshop. Let

p_{ik} ≡ the processing time of job i on machine k,
$c(\sigma, k)$ ≡ the end-time of the last job of sequence σ on machine k.

Then, to calculate the end-time of an operation of job a_i on machine k, when a_i has been scheduled after a sequence σ, we use the following equations:

Machine 1:

$$c(\sigma a_i, 1) = c(\sigma, 1) + p_{a_i 1}. \tag{1}$$

Machine k:

$$c(\sigma a_i, k) = \max\{c(\sigma, k), c(\sigma a_i, k-1)\} + p_{a_i k} \quad k = 2\ldots m. \tag{2}$$

To generate a specific instance of a flowshop problem we will assume that the plant produces a fixed set of products. In addition to allowing the type of product to vary, we will also allow the size of the batch to be one of a fixed set of sizes. Further details of the formulation are given in Section III, A.

B. CHARACTERISTICS OF SOLUTION METHODOLOGY

To solve the flowshop scheduling problem, or indeed most problems with significant discrete structure, we are forced to adopt some form of enumeration to find, and verify, optimal solutions (Garey and Johnson, 1979). To perform this enumeration, we must solve two distinct problems.

1. *Representation*. The solution space is composed of discrete combinatorial alternatives of batch production schedules. For example, in the permutation flowshop problem, where the batches are assumed to be executed in the same order on each unit, there are $N!$ number of solutions, where N is the number of batches. We must find a way to compactly represent this solution space, in such a way that significant portions of the space can be characterized with respect to our objective as either "poor" or "good" without explicitly enumerating them.
2. *Control strategy*. Having found an appropriate representation for the space of solutions, we must then take advantage of this representation to explore only as small a fraction of the space as possible. This involves the following:
   (a) Finding a method to systematically enumerate subsets of the solution space without explicitly enumerating their members.
   (b) Finding ways to prove that certain subsets of the space cannot contain optimal solutions, or if they do, that we have other subsets of the space that will produce equivalent solutions.

To solve the problems of representation and control, we will employ the framework of the branch-and-bound algorithm, which has been used to solve many types of combinatorial optimization problems, in chemical engineering, other domains of engineering, and a broad range of management problems. Specifically, we will use the framework proposed by Ibaraki (1978), which is characterized by the following features:

1. The combinatorial problem is represented by a *discrete decision process* (DDP) (Ibaraki, 1978) where the underlying information in the problem is captured by an explicit *state-space model* (Nilsson, 1980).
2. The control is divided into three parts.
   (a) Elimination of alternative solutions through the use of a *lower bound* on the value of the objective function.
   (b) Elimination of alternative solutions through *dominance* conditions.
   (c) Elimination of alternative solutions through *equivalence* conditions.

The next section will highlight these features of the branch-and-bound framework, within the context of the flowshop scheduling problem. Then we will give an abstract description of the algorithm, followed by the paraphrasing of an important theorem from Ibaraki (1978), which characterizes the relative efficiency of different branch-and-bound algorithms in terms of their control strategies.

II. Formal Description of Branch-and-Bound Framework

If an algorithm is to reflect on its own computational structure and thus *learn* how to improve its performance, then this algorithm needs to have a *declarative* representation of its components. Thus, a branch-and-bound algorithm used to generate optimal schedules of batch operations should possess declarative representations of (1) the schedules of batch operations (2) the predicates that determine the feasibility of a scheduling policy, (3) the objective function, and (4) the conditions that determine the control strategy of the branch-and-bound algorithm. The material in this section will provide such declarative representations for the various components of branch-and-bound algorithms, and will introduce a metric that can be used to measure improvements in the efficiency of such algorithms.

A. SOLUTION SPACE REPRESENTATION—DISCRETE DECISION PROCESS

The first step in solving a combinatorial optimization problem is to model the solution space itself. Such a model should be declarative in character, if it is to be independent of the characteristics of the specific algorithm that will be used to find the solution within the solution space. The model we have adopted for the scheduling of flowshop operations is the *discrete decision process* (DDP) introduced originally by Karp and Held (1967). As defined by Ibaraki (1978) a DDP, Y, is a triple (Σ, S, f) with its elements defined as follows:

Σ is a finite nonempty alphabet whose symbols are used to build the description of solutions. Let Σ^* denote the set of finite strings generated from the concatenation of symbols present in Σ.

S is a subset of Σ^* whose members satisfy a set of feasibility predicates over Σ^*. The members of Σ^* will represent the "feasible" solutions.

f is a cost function over S, mapping S into the set of real numbers.

For the flowshop problem, one choice of Σ is an alphabet with as many symbols as the number of distinct batches to be scheduled, i.e., one symbol for each batch, then, each discrete schedule of batches is represented by a finite string, σ, of symbols drawn from the alphabet, Σ. Only strings which contain each alphabetic element once and only once represent acceptable schedules of batch operations. We will use the notation σ to denote the set of symbols from Σ, which are contained in the solution string, Σ^*. Also, with $O(Y)$ we will denote the set of optimal solutions of the DDP, Y.

The DDP is a formalism more general than the customary formalisms of the combinatorial optimization problems and offers an excellent framework for unifying the formalization of very broad classes of such problems (Karp and Held, 1967; Ibaraki, 1978). It also exhibits many common features with the *state-space representation* of problems commonly employed by researchers in artificial intelligence (AI) (Kumar and Kanal, 1983; Nilsson, 1980). This is a feature that we will find very convenient in subsequent sections as we try to integrate machine learning algorithms, which use the state space representation, with branch-and-bound algorithms solving combinatorial optimization problems.

Thus, it can be effectively argued that the DDP formalism is superior to other formalisms since it can uniformly accommodate

1. A variety of problems seeking the set of feasible solutions, the set of optimal solutions, or one member of either set.
2. Various types of alphabets (used to describe solutions), predicates (used to describe feasibility of solutions), and cost functions.

1. Illustration: Flowshop Example

Let

$\Sigma \equiv \{\text{all batches}\}$.

$S \equiv$ Set of all σ_x, if $\{\sigma_x\}$ contains all symbols of Σ once and only once.

$f(x) \equiv C(\sigma_x, m)$, completion time of the last operation of the final batch.

In this alphabet, each batch is assigned its own symbol. The problem formulation allows for the same combination of product and size to be selected multiple times. Hence the schedules that have the same batch type in two or more different positions will be enumerated multiple times, even though they represent schedules which are indistinguishable from one another.

A more compact alphabet for this situation is one for which we create a unique symbol for each batch type, and allow the multiple occurrence of this symbol to stand for scheduling the batch type more than once. The feasibility predicate would be suitably modified to check to see when enough of the type had been added to a given branch. This gives the following DDP:

$\Sigma \equiv \{N$ distinct symbols: $N =$ all (product size) combinations$\}$.

$S \equiv$ Set of all σ_x, such that $\{\sigma_x\}$ contains all the symbols of Σ a number of times equal to the number of batches for each product-size combination that was selected for production.

$f \equiv C(\sigma_i, m)$ completion time of the last operation on the final machine.

The elimination of spurious equivalent solutions is important computationally, because either we will have to enumerate the spurious equivalent solutions, doubling the effort for each equivalent pair of solutions, or we will have to introduce rules that can detect the equivalence explicitly. For the purpose of illustrating the ideas of this chapter we will continue to use the naive alphabet, although the method is not restricted to such a choice.

A scheduling policy, x, is optimal if and only if the following conditions are satisfied:

Condition 1. Scheduling policy is feasible, i.e., $\sigma_x \in S$.

Condition 2. Scheduling policy has cost not higher than that of an other policy, y, i.e., $f(x) \leq f(y)$ for any y such that $\sigma_y \in S$.

B. THE BRANCH-AND-BOUND STRATEGY

1. Branching in General

The large size of the solution space for combinatorial optimization problems forces us to represent it *implicitly*. The branch-and-bound algorithm encodes the entire solution space in a *root node*, which is successively expanded into branching nodes. Each of these nodes represents a subset of the original solution space specialized to contain some particular element of the problem structure.

In the flowshop example, the subsets of the solution space consists of subsets of feasible schedules. We can organize these subsets in a variety of ways; for example, fixing any one position of the N available positions in the schedule to be a particular batch creates N subsets of size $(N-1)!$. Subsequently, as we fix more and more of the positions, the sets will include fewer and fewer possibilities, until all the positions are fixed, and we have a single element in the set corresponding to a single, feasible schedule.

2. Formal Statement of Branching

The branching, or specialization of the solution subsets, should obey certain constraints to avoid potential problems of inefficiency, nontermination, or incorrectly omitting solutions.

1. *Mutual exclusivity.* In branching from one node to its descendants, we should ensure that none of the subsets overlap with one another; otherwise we could potentially explore the same solution subsets in multiple branches of the tree. Formally, if X is the original solution space and x_i is the ith subset, then

$$x_i \cap x_j = \emptyset \; \forall i, j > i$$

2. *Inclusiveness.* In branching to the descendants, we should ensure that no solution of the parent set is omitted from the child subsets. If we fail to have inclusiveness, the procedure may not correctly find the optimum solution. Formally, this requirement implied that

$$\bigcup_i x_i = X.$$

We must now link these two properties, and the notion of branching to our problem representation, i.e., the DDP formalism. To do this, we introduce a variant on our earlier notation; let

$$Y(x) \equiv [\Sigma, S(x), f],$$

where $S(x)$ is the feasibility predicate over all strings that begin with the partial sequence of symbols, x.

If ε denotes the empty strings, then

$$Y(\varepsilon) \equiv (\Sigma, S, f)$$

and $Y(\varepsilon)$ is equivalent to the original problem, since $S(\varepsilon)$ includes all the original set of solutions. Thus $Y(\varepsilon)$ is the problem associated with the root node and branching from the root node creates problems $Y(a_1)$, $Y(a_2) \cdots Y(a_n)$, one for each of the n elements of the alphabet. Each problem, $Y(a_i)$ captures a subset of the original problem, specialized by the inclusion of the alphabet symbol. If $\Sigma = \{a_1, a_2, \ldots, a_n\}$, then a branch-and-bound algorithm decomposes $Y(\varepsilon)$ into $|\Sigma|$ problems, $\{Y(a_1), Y(a_2), \ldots, Y(a_n)\}$, where the solution strings are required to start with the respective alphabetic element. We will also extend the definition of the objective function values as follows:

$$f(x) = \begin{cases} \inf\{f(xy) : \sigma = xy \in S\} \\ \infty \text{ otherwise, i.e., } Y(x) \text{ infeasible} \end{cases}$$

According to this extension of the cost function definition, the following two conditions are always satisfied:

(a) $f(xy) \geq f(x)$ for $x, y \in \Sigma^*$.
(b) $f(x) = \min\{f(xa): a \in \Sigma\}$ for $x \notin S$.

3. Illustration: Flowshop Problem

With our alphabet equal to the set of batches to be scheduled, we will interpret ε, the empty string, to be the empty schedule. $Y(a_j)$ is then the problem $Y(\varepsilon)$ specialized to have schedules that "begin with" alphabet element a_j in the first position of the schedule. Thus, the discrete decision process $Y(a_j)$ is equivalent to solving the original scheduling problem with the first batch fixed to be a_j. The set $S(a_j)$ includes all feasible solutions that have a_j as their first batch, and, in our naive formulation, $S(a_j)$ differs from $S(\varepsilon)$ by prohibiting a_j to appear again in the string.

We can see that, in general, the branching process satisfies the property of inclusiveness, since we generate a subset for each possible specialization of the original set. We cannot, however, claim exclusiveness unless we are guaranteed that each string maps to a unique solution subset. For example, if an alphabet element corresponds to assigning a batch to a specific position, then the order in which batches are assigned their positions is irrelevant, hence, with that alphabet, $Y(a_i a_j) \equiv Y(a_j a_i)$.

Having defined the process of branching, we must now formally define the mechanisms for controlling the expansion of the subsets. The basic intuition behind each of these mechanisms is that we can "measure" the quality not just of a single feasible solution, but of the entire subset represented by the partial solution string.

4. Lower-Bound Function

Having formally defined the branching structure, we must now make explicit the mechanisms by which we can eliminate subsets of the solution space from further consideration. Ibaraki (1978) has stated three major mechanisms for controlling the evolution of the branch-and-bound search algorithms, by eliminating potential solution through

(a) Comparison with an estimate of the objective function's lower bound.
(b) Dominance conditions.
(c) Equivalence conditions.

In this section we will discuss the characteristics of the first mechanism, leaving the other two for the subsequent two sections.

In our scheduling example, a partial solution represents all the possible completions of a partial schedule. To estimate the quality of the partial solution, we could assume the best scenario, and assign it a lower-bound value based on the lowest possible value of its makespan. There are many ways to estimate the makespan; for example, we could simply ignore the remainder of the batches, and report the makespan of the current partial schedule, clearly a lower bound on the final makespan. Stated formally, the lower-bound function $g(x)$ satisfies the following requirements (Ibaraki, 1978):

(a) $g(x) \leq f(x)$
(b) $g(x) = f(x) \quad x \in S$
(c) $g(x) \leq g(xy) \quad$ for $x, y \in \Sigma^* \quad x \in \Sigma^*$

If $g(x)$ satisfies these conditions, we can use the following *lower-bound elimination* criterion to terminate the solution of a discrete decision process, $Y(y)$.

Condition: If $x \in S$ and $g(y) \geq f(x)$, then all solutions of $Y(y)$ can be discarded.

Clearly, since $g(y) \leq f(y)$, it follows that $f(y) \geq f(x)$, and hence the solution of $Y(y)$ cannot lead to a better objective function value.

The disadvantage of the lower-bound elimination criterion is that it cannot eliminate solutions with lower bounds better than the optimal solution objective function value.

Since

$$f(x^*) \leq f(x) \; \forall x \, x^* \in \text{optimal solution set}, O[Y(\varepsilon)]$$

and

$$\text{If } g(y) < f(x^*)$$
$$\text{then } g(y) < f(x) \; \forall x$$

and the condition for lower-bound elimination cannot be satisfied.

The efficiency of branch-and-bound algorithms is profoundly influenced by the *strength* of a lower-bounding scheme or function. In general, the closer a lower-bound function value of a node is to the true objective function value, the fewer nodes will be expanded, and the more efficient the algorithm will be. For example, in solving mixed-integer linear programs (MILP) via branch and bound, if the lower-bound scheme employed is a linear programming (LP) relaxation of the integer variables, then the efficiency of the solution technique is a strong function of how closely the

LP objective function value approximates the best integer solution value that could be generated by solutions emanating from that node.

If using a stronger lower-bound function leads to the expansion of fewer nodes, why not always use the strongest lower bound? There are two potential problems with this. First, in many cases, it is impossible to prove that one lower-bounding function, f_1, yields a higher value for minimization than another, f_2, for every node in the tree. Second, stronger lower bounds tend to require more computational effort to evaluate; thus, even though fewer nodes are evaluated, they take longer to evaluate. This tradeoff can lead to weaker lower bounds, yielding better overall computational efficiency.

In addition to the elimination of partial solutions on the basis of their lower-bound values, we can provide two mechanisms that operate directly on pairs of partial solutions. These two mechanisms are based on *dominance* and *equivalence conditions*. The utility of these conditions comes from the fact that we need not have found a feasible solution to use them, and that the lower-bound values of the eliminated solutions do not have to be higher than the objective function value of the optimal solution. This is particularly important in scheduling problems where one may have a large number of equivalent schedules due to the use of equipment with identical processing characteristics, and many batches with equivalent demands on the available resources.

5. Dominance Test

The intuitive notion behind a *dominance condition*, D, is that by comparing certain properties of partial solutions x and y, we will be able to determine that for every solution to the problem $Y(y)$ we will be able to find a solution to $Y(x)$ which has a better objective function value (Ibaraki, 1977). In the flowshop scheduling problem several dominance conditions, sometimes called elimination criteria, have been developed (Baker, 1975; Szwarc, 1971). We will state only the simplest:

Condition: If $\{x\} = \{y\} \land \forall i \; C(x,i) < C(y,i)$ then $x.D.y.$

For example, if the completion times of the partial schedule x on each machine i are less than those of the partial schedule y on each machine i, and x and y have scheduled the same batches, then, if we complete the partial schedules x and y in the same way, the completion time of any schedule emanating from x will be less than the completion time of the same schedule emanating from y; that is,

$$\forall u \; C(xu,i) < C(yu,i),$$

and thus, specifically

$$C(xu, m) < C(yu, m),$$

and the makespan of schedule, xu, will be less than that of yu.

Formally, we define the dominance condition by specifying a set of necessary and sufficient conditions, which it has to satisfy (Ibaraki, 1978):

Definition. A dominance relation, D, is a partial ordering of the partial solutions of the discrete decision processes in Σ^*, which satisfies the following three properties for any partial solutions, x and y:

1. Reflexivity $x.D.x$
2. Transitivity $x.D.y \wedge y.D.z \to x.D.z$
3. Antisymmetry $x.D.y \to \neg(y.D.x)$

Furthermore, a dominance relationship is constructed in such a way that it satisfies the following properties:

4. $x.D.y$ and $x \neq y \Rightarrow f(x) < f(y)$
5. $x.D.y$ and $x \neq y \Rightarrow g(x) < g(y)$
6. $x.D.y$ and $x \neq y \Rightarrow \forall u \in \Sigma^* \exists v \in \Sigma^*$ such that $xu.D.yv$ and $xv \neq yu$
7. For each $x \in \Sigma^*$ $\exists y \in \Sigma_D^*$ such that $y.D.x$

Note that the set Σ_D^* is defined such that it contains all solutions that have not been dominated by any other solution. Property 4 guarantees that we will not miss an element of the optimal set, since the objective function value of x is lower than that of y.

6. Equivalence Test

The final relationship between solution subsets, which we can use to curtail the enumeration of one subset, is an *equivalence condition*, EQ. Intuitively, equivalence between subsets means that for every solution in one subset, y, we can find a solution in the subset x, which will have the same objective function value, and that plays a similar role in the execution of the algorithm.

In the case of the flowshop example, we have the following equivalence condition

Condition: If $\{x\} = \{y\} \wedge \forall i\ C(x, i) = C(y, i)$ then $x.EQ.y$.

This condition implies that any complete schedule, xu, emanating from the partial schedule, x, will have the same end-times on all machines as the completions, yu, of y, not just for the completed schedule, but also for all intermediate partial schedules.

Formally, we define the equivalence condition by specifying a set of necessary and sufficient conditions, which it has to satisfy (Ibaraki, 1978).

Definition. An equivalence condition, EQ, between two partial solutions, $x \in \Sigma^*$ and $y \in \Sigma^*$ is a binary relationship, $x.EQ.y$, which has the following properties:

1. Reflexivity $x.EQ.x$
2. Transitivity $x.EQ.y \wedge y.EQ.z \rightarrow x.EQ.z$
3. Symmetry $x.EQ.y \rightarrow y.EQ.x$

and satisfies the following conditions.

Condition a:

$$x.EQ.y \Rightarrow \forall u \in \Sigma^* \begin{cases} xu \in S \Leftrightarrow yu \in S, \\ f(xu) = f(yu). \end{cases}$$

Condition b:

$$x.EQ.y \Rightarrow \forall u, w \in \Sigma^* \begin{cases} g(xu) = g(yu), \\ xu.EQ.w \Leftrightarrow yu.EQ.w, \\ xu.D.w \Leftrightarrow yu.D.w, \\ w.D.xu \Leftrightarrow w.D.yu. \end{cases}$$

These two definitions of dominance and equivalence have been stated in the context of seeking the optimal solution set $O(Y(\epsilon))$. If we are trying to seek only a single optimal solution, we can merge the definitions of dominance and equivalence into a single relationship, since we no longer need to retain solutions that might potentially have equal objective function values, but that are not strictly equivalent. In our flowshop example, this is equivalent to allowing the end-times of x be less than *or equal* to those of y, instead of strictly less than.

C. SPECIFICATION OF BRANCH-AND-BOUND ALGORITHM

The preceding definitions allow us to explicitly characterize a branch-and-bound algorithm by

1. Its representation as a discrete decision process, Y denoted by the triple of (Σ, S, f).
2. Its control information, which is represented by the lower-bound function, g, the dominance conditions, D, and the equivalence conditions, EQ.

To complete the specification of the algorithm, we require one additional decision parameter: *how to select the next problem $Y(x)$, which we will solve*, or equivalently, *which node in the branching structure to expand*. We will define a *search function*, s, which allows us to select a node from the currently unexpanded nodes for expansion. In this chapter, as in Ibaraki (1978), we consider only *best bound search*, where we select the node with the minimum $g(x)$ value for expansion. Thus our branch-and-bound algorithm, A, is explicitly specified by

$$A = (Y, g, D, EQ, s),$$

and procedurally is given by the following steps:

Step-1. **Initialize**
Let $A \leftarrow \{\epsilon\}$, $N \leftarrow \{\ \}$ $z \leftarrow \infty$ where $\{\epsilon\}$ denotes the set of null strings (solutions of length zero).

Step 2. **Search**
If $A = \emptyset$, exit; else $x \leftarrow s(A)$ goto Step-3.

Step-3 **Test for feasibility**
If $x \in S$, let $z \leftarrow min(z, f(x))$. Goto Step-4

Step-4 **Test through the lower bound**
If $g(x) > z$ goto Step-8; else goto Step-5.

Step-5 **Test for relative dominance**
If $y.D.x$ for some $y(\neq x) \in N$; goto Step-8; else goto Step-6.

Step-6 **Test for equivalence**
If $y.EQ.x$ for some $y \in N$ which has already been tested; goto Step-8; else goto Step-7

Step-7 **Decompose active node**
$A \leftarrow A \cup \{xa|a \in \Sigma\} - \{x\}$
$N \leftarrow N \cup \{xa|a \in \Sigma\}$; return to Step-2.

Step-8 **Terminate**
$A \leftarrow A - \{x\}$ and return to Step-2.

D. Relative Efficiency of Branch-and-Bound Algorithms

One measure of efficiency is the number of nodes of the branching tree that are expanded. Ibaraki (1978) has proved that the number of nodes expanded can be linked to the strength of the control knowledge, and proves that if one branch-and-bound algorithm has a better lower-bounding function, dominance and equivalence rules, then it will expand fewer nodes.

Let us briefly discuss the theoretical results providing the basis for the improved efficiency of branch-and-bound algorithms. Let $F = \{x \in \Sigma^* : g(x) \leq f(x^*)\}$ be the set of solutions that cannot be terminated by the lower-bound test. Then, the set L_A, defined by $L_A = F \cap \Sigma_D^*$, contains all the partial solutions, which can be terminated only by an equivalence relation. Recall that, by definition, no node in Σ_D^* can be terminated by a dominance rule.

Theorem 1

Let $A = (Y, g, D, EQ, s)$ be a branch-and-bound algorithm. The algorithm terminates after decomposing exactly $|L_A/EQ|$ nodes, provided that $|L_A/EQ| < \infty$, where L_A/EQ denotes the set equivalent classes of solutions induced by the equivalence conditions EQ.

For the proof of this theorem see Ibaraki (1978) as a result of Theorem 1, a natural measure of the efficiency of a branch-and-bound procedure is the number of the resulting equivalence classes under *EQ*. Furthermore, the following theorem (Ibaraki, 1978) allows a direct comparison of the efficiencies of two distinct branch-and-bound algorithms:

Theorem 2

Assume that the following two branch-and-bound algorithms

$$A = (Y, g, D, EQ, s) \quad \text{and} \quad A' = (Y, g', D', EQ', s')$$

terminate and satisfy the following conditions:

$g(x) \leq g'(x)$ for $x \in \Sigma^*$
$x.D.y \Rightarrow x.D'.y$ for $x, y \in \Sigma^*$
$x.EQ.y \Rightarrow x.EQ'.y$ for $x, y \in \Sigma^*$
Then

$$T'_A = |L_A/EQ'| < |L_A/EQ| = T_A,$$

i.e., algorithm A' is more efficient than algorithm A.

The importance of Theorem 2 is that it provides a *theoretical basis* for improving the control knowledge we have available to solve a given problem class. If we can derive new dominance and/or equivalence rules —or in the single optimal solution case, the enhanced dominance rules—we will be able to reduce the number of nodes enumerated. The focus of this chapter is the description of a methodology for achieving this goal, so that the *computer itself*, can carry out the process of acquiring new control information. The acquisition is based on the analysis of problem-solving activity, in the light of the formal specifications laid out for dominance and equivalence rules, and the generalization of this "experience" to apply it to new problems within the same problem class.

E. BRANCHING AS STATE UPDATING

Branching from one partial solution to another involves more than just concatenating an alphabet element to a string; it also changes the underlying *state* of the problem. In our flowshop example the state of the partial schedule is conveniently and parsimoniously represented by the start- and end-times of the last operation on each unit. If the problem involved sequence-dependent equipment cleaning, we would have to include information about the identity of the batches as well. The choice of state variables for a problem is critical in determining the ease with which additional state properties and new state properties are calculated. For example, we could record just the list of batches scheduled in a given state and each time we wanted to calculate a new start- or end-time recalculate it from the list of batches. This would be computationally expensive, but our state information would be less complex. Thus, what we choose to record in a state addresses the issue of time/space tradeoff explicitly. In general, the more state variables we have the easier it will be to update each one during branching, but the more we will have to compute and store.

In addition to having to assign state variables to the strings of the DDP, we also have to assign properties to the alphabet symbols. In our flowshop example, the alphabet symbols can be interpreted as batches to be executed with a series of processing times. Thus, if we use the notation, $\Theta(x)$, to denote the state of partial solution, x, then

$$Y(x) \to Y(xa_i) \qquad \Theta(xa_i) = H[\Theta(x), I(a_i)],$$

where I returns the interpretation of the alphabet symbol and H is a vector of functions h_j; $j = 1, 2, \ldots, k$, which compute the properties, p_j; $j = 1, 2, \ldots, k$, characterizing $\Theta(xa_i)$:

$$p_j(xa_i) = h_j[\Theta(x), I(a_i)].$$

The state updating functions combine information about the constraints on the state variables with the objective function minimization. The feasibility predicate forces the state variables to obey certain constraints, such as the nonoverlap of batches, forcing the start-times of successive operations to be greater than the end-times of the previous operation. The constraints do not force the start-times to be equal to the previous end-times; this comes as the effect of the objective function minimization. Consequently, in addition to the update functions, we also need to represent the underlying set of constraints:

Update functions:

$$s_{i,\sigma_{j+1}} = \max(e_{i,\sigma_j}, e_{i-1,\sigma_{j+1}}),$$
$$e_{i,\sigma_{j+1}} = s_{i,\sigma_{j+1}} + p_{ij}.$$

Constraints:

$$s_{i,\sigma_{j+1}} \geq e_{i,\sigma_j}, \tag{3}$$

$$s_{i,\sigma_{j+1}} \geq e_{i-1,\sigma_{j+1}}, \tag{4}$$

$$e_i, \sigma_{j+1} = s_{i,\sigma_{j+1}} + p_{ij}. \tag{5}$$

Figure 1 gives the structure of the constraint set. We can now classify these constraints into two classes. The first class of constraints includes the *intersituational constraints*, which have variables of the next state as their potential output variables, and the variables of the current state as their input variables. We can see that these correspond to constraints (3). We then have the *intrasituational constraints*, which relate variables of the next state together and correspond to constraints (4) and (5). We will similarly define those variables that appear in intersituational constraints to be *intersituational variables*, and those that appear in intrasituational constraints to be *intrasituational*. Note that these classes are not mutually exclusive. Thus, we will refine our classes to reflect whether the variable in a specific constraint is a potential output of that constraint or is an input to the constraint. In essence this input/output distinction separates the variables of the current state from those of the next state. In the flowshop example, the end-times of the current state are the input intersituational variables and the start-times of the next state are the output intersituational variables. The start-times also belong to the class of input intrasituational variables, and the end-times of the next state are the output intrasituational variables.

Our ability to make these distinctions rests on the fact that we know the direction that the branching generation imposes on the updating of the variables. If we were not solving the problem in such a way that all the variables are explicitly determined by the branching, then these distinctions would not be so clear. For example, if some variable values were the result of solving an auxiliary linear program that involved these constraints, we could not classify the variables this way.

568   MATTHEW J. REALFF

	$s_{i+1,1}$ $e_{i+1,1}$ $s_{i+1,2}$ $e_{i+1,2}$ $s_{i+1,3}$ $e_{i+1,3}$ $s_{i+1,4}$ $e_{i+1,4}$ $s_{i+1,5}$ $e_{i+1,5}$ $e_{i,1}$ $e_{i,2}$ $e_{i,3}$ $e_{i,4}$ $e_{i,5}$
Inter	× ×
Intra	× ×
Intra	× ×
Inter	× ×
Intra	× ×
Intra	× ×
Inter	× ×
Intra	× ×
Intra	× ×
Inter	× ×
Intra	× ×
Intra	× ×
Inter	× ×
Intra	× ×

Inter   Intersituational Constraints

Intra   Intrasituational Constraints

s   Output Intersituational Variables, Input Intrasituational Variables

$e_{i+1,j}$ Output Intrasituational Variables

$e_{i,j}$ Input Intersituational Variables

FIG. 1.

The purpose of defining these classes of constraints and variables is that this will enable us to conveniently express general notions of equivalence and dominance, without explicit reference to the flowshop domain, but at a more abstract level.

F. FLOWSHOP LOWER-BOUNDING SCHEME

To define the lower-bound function for flowshop scheduling, we will use the lower-bounding schemes proposed in Lagweg et al. (1978). The lower-bound schemes are organized into a hierarchy that reflects the *strength* of the lower-bounding scheme.

The *strength* of a lower-bounding scheme is a binary relation, over two lower-bounding schemes, which indicates that if $\Omega' \to_D \Omega$, then the lower-bounding scheme Ω' will always yield an equal or lower value for any σ, and hence will always lead to the expansion of fewer nodes, by the theorem of Section II, D.

We will use the simplest of lower-bounding schemes proposed in Lagweg et al. (1978), where all but one of the flowshop units are relaxed to be nonbottleneck machines, A nonbottleneck machine is a machine in one which the capacity constraint has been relaxed; i.e., it can process all the jobs simultaneously. Thus, to compute the lower bound, assuming that the bottleneck machine is u, and that we have completed the partial schedule σ, we need to schedule a single machine knowing that each unscheduled job, $i \in \phi$ will be released after a time $q_{i.u}$, representing the time it spends on the nonbottleneck machines up to u processed on u for a time p_{iu}, and then completed after a further period $q_{iu.}$, representing the time it spends on the nonbottleneck machines after u. We can calculate $q_{i.u}$ by adding the sum of the processing times of i on machines up to u to the partially completed schedule end-time where $q_{i.u} \equiv$ release time before reaching machine u and $q_{iu.} \equiv$ time spent on machines after u,

$$q_{i.u} = \max_l \left\{ C(\sigma, l) + \sum_{k=l}^{u-1} p_{ik} | l = 1 \ldots u \right\};$$

$q_{iu.}$ is simply the sum of the remaining processing time of i on the machines following u.

$$q_{iu} = \sum_{k=u+1}^{m} p_{ik}.$$

At this point the lower-bounding scheme consists of solving a single machine scheduling problem where for each job i, the release time is $q_{i.u}$, the due date is $-q_{iu.}$, and the processing time is p_{iu}. This nonbottleneck scheme can be further simplified in two steps. First, we can assume that $q_{i\alpha\beta} = \min_{i \in \phi} \{q_{\alpha\beta}\}$ for $\alpha\beta = .uu.$, avoiding the need to consider release times of $q_{i.u}$, and due dates of $-q_{iu.}$, which turns an NP-complete problem, into one solvable in polynomial time. If only one of these were to be relaxed, the schedule can still be found in polynomial time by Jackson's rule (Jackson, 1955). Second, we can avoid the computation of $q_{i.u}$ completely, by assuming that the maximum is obtained at $l = u$ for all values of i.

$$q_{i.u} = C(\sigma, u) \; \forall i.$$

Thus, our final lower-bound scheme

$$g(\sigma) = \max_u \left\{ C(\sigma, u) + \sum_{j \in \phi} p_{ju} + q_{u.} \right\},$$

where $q_{u.} = \min_j \{q_{ju.}\}$.

The reason we call this a scheme is because it is still parameterized by, u, the machine that we choose to retain as our bottleneck. The simplest choice of u is $u = m$, for which we get

$$g(\sigma) = C(\sigma, m) + \sum_{j \in \phi} p_{jm}.$$

The computational complexity of the lower bound is presented in Lagweg et al. (1978).

1. At the root node, we can calculate q_{iu}. for all (i, u) in $O(mn)$ steps, for $u = m$ we avoid this calculation.
2. At each node during the branching, we calculate the lower bound, by taking the minimum of the remaining unscheduled batches, q_{iu}.. This takes $o(|\phi|)$ for each machine or $O(m|\phi|)$ in all. The final calculation is then of order 1, provided we update and store $\sum_{j \in \phi} p_{ju}$ at each node. For $u = m$, the calculation is $O(1)$.

III. The Use of Problem-Solving Experience in Synthesizing New Control Knowledge

The purpose of this section is to illustrate the role that the problem-solving experience plays in improving problem-solving performance, via learning, and examine the question; *"exactly what is it that we are trying to learn?"*

A. AN INSTANCE OF A FLOWSHOP SCHEDULING PROBLEM

To make the ideas of this section more concrete we will use a specific example of a flowshop problem. The problem is a small one, only five batches will be considered, since this enables us to examine the enumeration tree by hand.

Consider the flowshop in Fig. 2. The unit operations being performed are mixing, reaction, distillation, cooling, and filtration. We have made the following assumptions about the way the flowshop operates:

1. Processing times will be similar for each product type on a given processing step, reflecting the fact that technologically identical operations have the same, or similar, equations governing their behavior, with only parametric variation between products.

2. Processing times have some proportionality to batch size. We have chosen to allow three batch sizes (500, 1000, 1500 mass units). The reactor

Flowshop Example Plant:

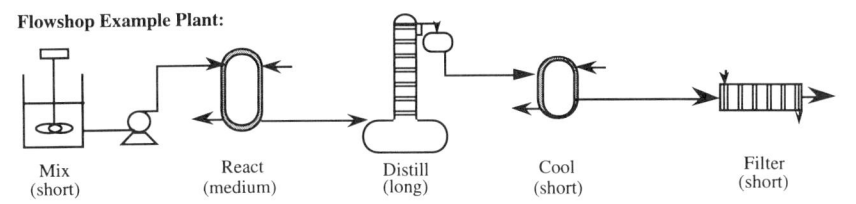

FIG. 2. Example configuration of unit operations in flowshop production.

processing time is the least sensitive to batch size, because the reaction time is assumed proportional to concentration and the increased time is due only to a constant pump rate assumption. The distillation processing time is the most sensitive, due to the assumption of constant boilup rate. The filtration period is roughly proportional to the size, because we have assumed a constant filter cake size and divided the larger batches to achieve this size.

3. The processing parameters have been chosen so that the distillation step is, in general, the longest step. In fact, processing periods for each operation have been roughly ordered as follows: D_1 (distillation) > R_1 (reaction) > F_1(filtration) > C_1(cooling) > M_1(mixing). Obviously, the size considerations may alter this ranking for batches of the same type but of different sizes.

The example problem was generated by picking the type and size at random, while ensuring that no size and type combination was repeated. This last requirement was imposed to avoid making the equivalence condition appear to perform better due to spurious equivalences.

In particular, consider the batches shown in Table I. The first two columns of the table indicate the product type and batch size. The next column gives the processing times for each operation. The fourth column gives a unique identifier to each batch composed of a prefix "B-P," the

TABLE I

PRODUCTS IN THE SCHEDULING FLOWSHOP

Product	Size	M_1	R_1	D_1	C_1	F_1	Code	Alphabet symbol
2	500	2	17	24	4	10	"B-P-2-0"	a_1
1	1000	4	20	26	8	12	"B-P-1-1"	a_2
3	1500	5	26	36	15	19	"B-P-3-2"	a_3
3	500	2	24	12	5	6	"B-P-3-3"	a_4
1	1500	6	21	40	11	19	"B-P-1-1"	a_5

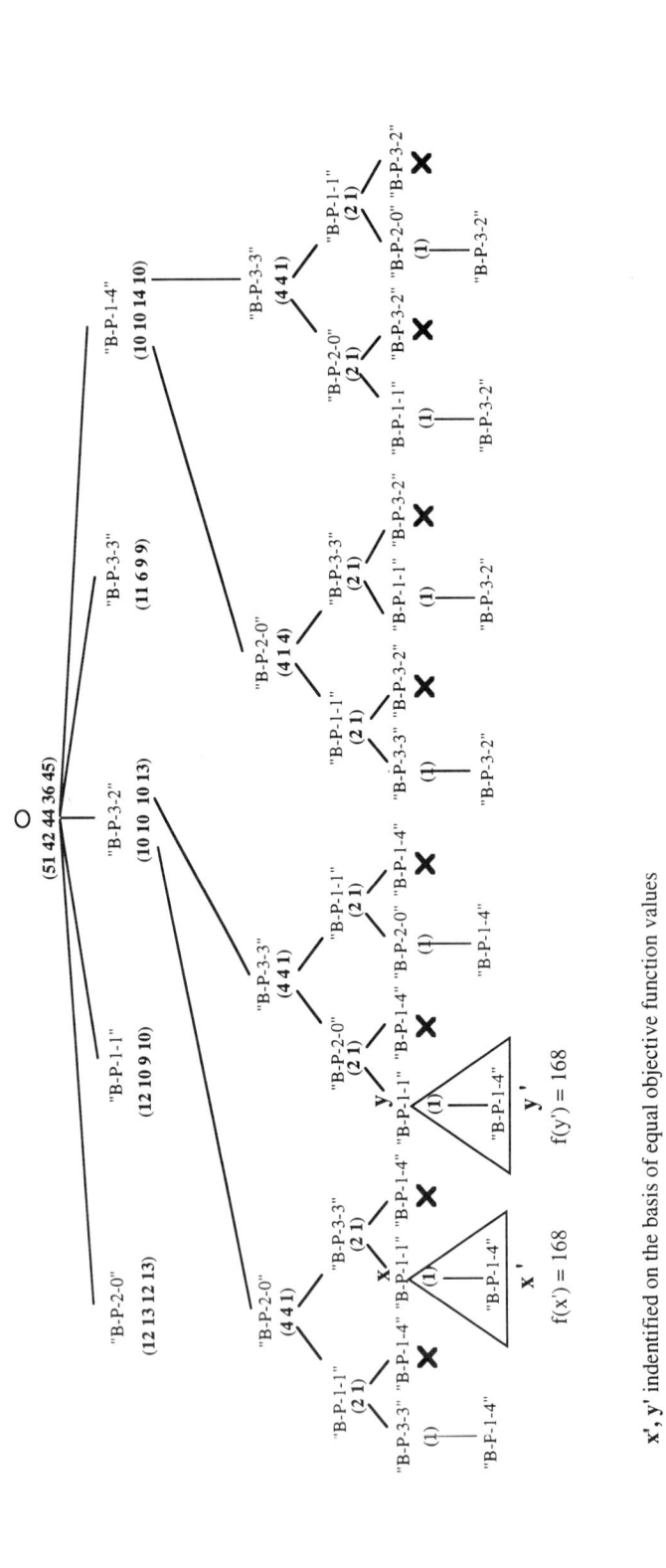

FIG. 3. Enumeration without any equivalence conditions; 218 nodes enumerated out of 326 possible.

x', y' indentified on the basis of equal objective function values

x,y identified by identical sets of alphabetic elements

product type and a counter. The fifth column lists the alphabet symbol associated with each batch. The total number of feasible schedules is 5! or 120. The total number of nodes in the tree is 326 (including the root node).

Utilizing only the simple lower bounding scheme with the bottleneck machine fixed to the last machine, we generate the branching tree given in Fig. 3, where much of the detail has been suppressed. The boldface numbers indicate the size of the subtrees (i.e., the number of nodes in the subtree) beneath the node.

We can now pose the following question:

> Do any of the nodes, which we have enumerated, satisfy the necessary and sufficient conditions for dominance or equivalence?

If we could answer the question affirmatively, we might be able to extract out the features of the nodes that cause them to participate in the relationship, and use these features to identify similar nodes in future problem solving activities. *This is the essence of the goal of learning*.

B. DEFINITION AND ANALYSIS OF PROBLEM-SOLVING EXPERIENCE

For branch-and-bound algorithms the notion of "problem-solving experience" is defined in term of the branching structure generated during the solution, and the various relationships between the nodes of the branching structure that are engendered by existing control knowledge.

1. Identification of Nodes for Dominance and Equivalence Relationships

To identify pairs of nodes that might be capable of being proved to participate in dominance or equivalence relationships, we will employ a strategy of *successive filtering* of the branching structure. We will concentrate on identifying equivalent nodes (for purposes of illustration), although a similar methodology applies to the case of nodes that participate in a dominance relationship.

The necessary and sufficient conditions for equivalence are repeated here for convenience.

Condition-a:

$$x.EQ.y \Rightarrow \forall u \in \Sigma^* \begin{cases} xu \in S \leftrightarrow yu \in S, \\ f(xu) = f(yu) \end{cases}$$

Condition-b:

$$x.EQ.y \Rightarrow \forall u, w \in \Sigma^* \begin{cases} g(xu) = g(yu), \\ xu.EQ.w \leftrightarrow yu.EQ.w, \\ xu.D.w \leftrightarrow yu.D.w, \\ w.D.xu \leftrightarrow w.D.yu \end{cases}.$$

First, let us concentrate on the meaning of *Condition-a*. Essentially, we are required to ensure that all the solutions, which lie below nodes x and y, have the same objective function value, and that x and y engender the same set of solutions. A problem involving a finite set of objects, each of which must be included in the final solution, is termed a *finite-set problem*. For finite-set problems, we can be confident that $x\omega \in S \leftrightarrow y\omega \in S$ if x and y contain the same set of alphabet symbols. In fact, if x and y do not contain the same set of alphabet symbols, they cannot be equivalent, the proof of which is trivial. Thus our first step in identifying candidates is to find two solutions $x', y' \in S$ that have equal objective function values. We then try to identify their ancestors x, y, such that the set of alphabetic symbols contained in x and y are equal. If in the process of finding this "common stem" we find that the stem is the same for both x', y', then we have the situation depicted in Fig. 4, and $x = y$. This is of no value; a solution is trivially equivalent to itself. We term this *ancestral-equality*. So far, all that has been established for x and y is that there exist two

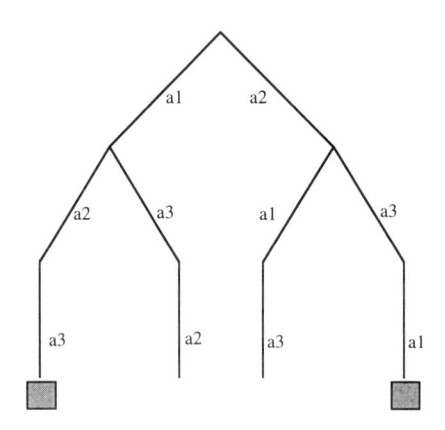

Two solutions can have different tails which lead to solutions with equal objective function values but which cannot produce equivalence relations because they have the same ancestor.

Fig. 4. Schematic depicting ancestral equality.

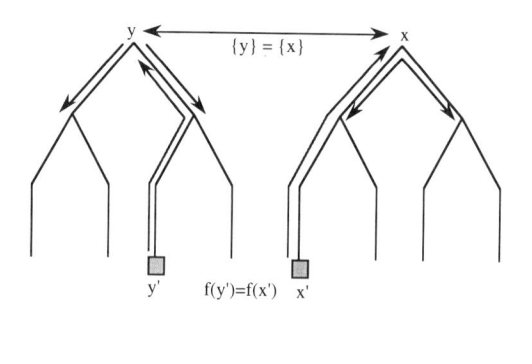

FIG. 5. Pattern of reasoning required to identify examples of equivalence or dominance by the syntactic criterion.

feasible solutions, x' and y', for which $f(x') = f(y')$ and x', y' are the descendants of x, y. We must now reverse the direction of reasoning, and verify that for all the feasible children of x and y that $f(x\omega) = f(y\omega)$, as in Fig. 5.

If we are successful, then we have verified that for the conditions prevailing in the example, the partial solutions, x and y would indeed be equivalent, as far as *Condition-a* is concerned. We now move to *Condition-b*, which ensures that the equivalent node will play an equivalent role in the enumeration as the one that was eliminated. Thus, as we examine the children of x, y, regardless of whether they are members of the feasible set, we would verify that their lower-bound values were equal, and that if we had any existing dominance, or equivalence conditions that the equivalent descendant of x, i.e., $x\upsilon$, participates in the same relationships as does $y\upsilon$.

To make the above discussion more concrete, consider the example branching structure of Fig. 3. In this structure, we have identified an x' and y', which have the same objective function values, and their ancestors, (x, y) which are characterized by the same set of symbols, (i.e., batches). Furthermore, we can see that the children of x, y do indeed satisfy the requirements of *Condition-a* and *Condition-b* and hence, (x, y) would be considered as candidates to develop a new equivalence relationship. If we examine the partial schedules (x, y) as depicted in Fig. 6, our knowledge

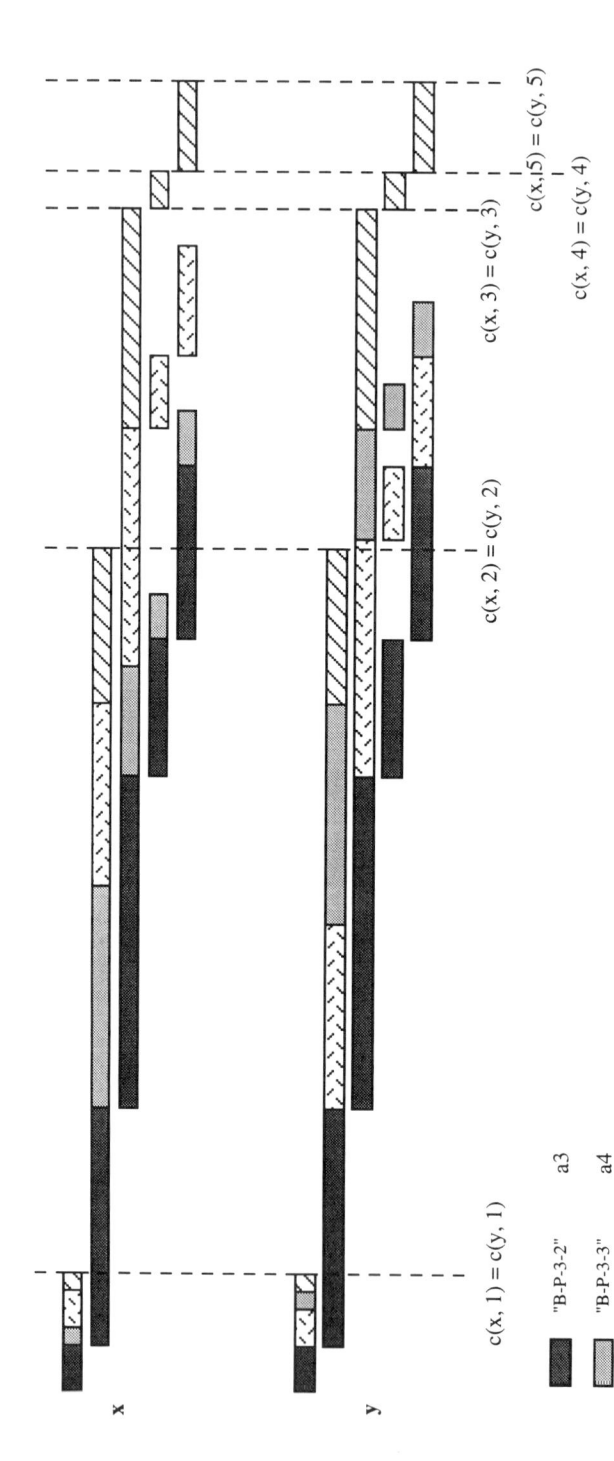

FIG. 6. Potentially equivalent schedules x and y.

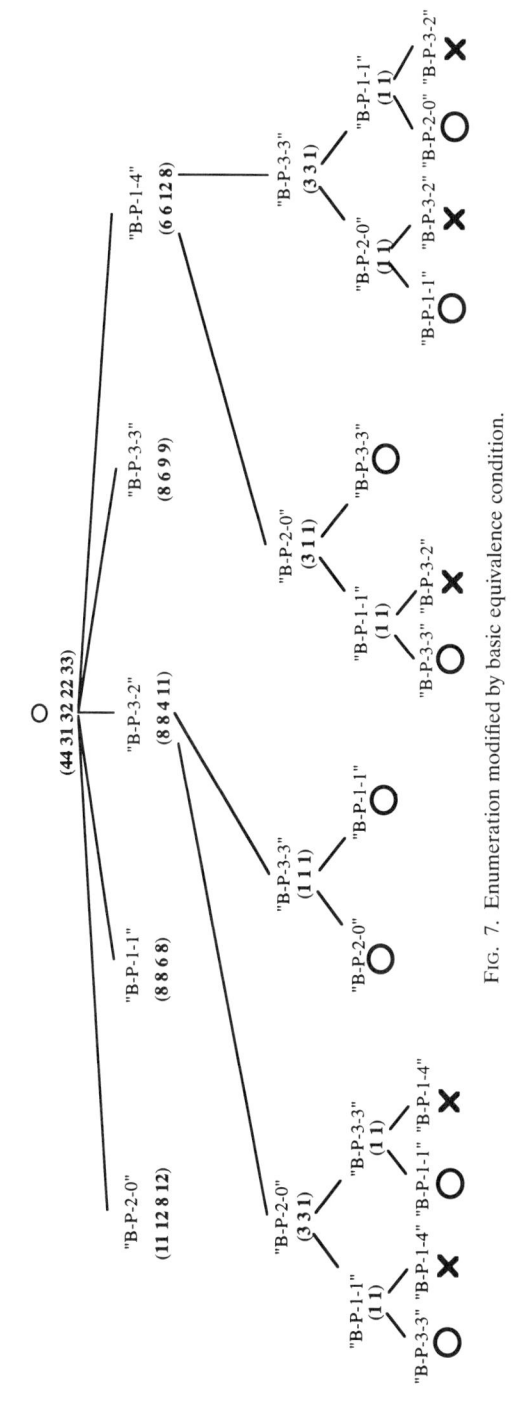

Fig. 7. Enumeration modified by basic equivalence condition.

of flowshop scheduling should enable us to see that indeed (x, y) are equivalent because the completion times on each machine are equal, i.e., $C(x, i) = C(y, i)$ for every machine, i, and hence for each arrangement of the remaining batches, the completion times will be equal, i.e., $C(xu, m) = C(yu, m)$ for all feasible completions, u.

The last step in the preceding argument, the use of our knowledge about flowshop scheduling, turns what had been a mainly *syntactic criterion* over the tree structure of the example, into a criterion based on state variables of (x, y). The state variable values, the completion times of the various flowshop machines, are accessible *before* the subtrees beneath x and y have been generated. Indeed, they determine the relationships between the respective elements of the subtrees (xu, yu). If we can formalize the process of showing that the pair (x, y) identified with our syntactic criterion, satisfies the conditions for equivalence or dominance, we will in the process have generated a "new" equivalence rule.

Using this equivalence rule in the example, we generate the branching structure of Fig. 7, where we can see the number of nodes generated has been reduced from 218 to 211. Our syntactic criterion, combined with ensuring that all the lower bounds are equal, will identify five pairs of potentially equivalent solutions in the example. All five of these pairs can be proved to be equivalent on the basis of a general theory of equivalence. Of these pairs, one would have enabled the elimination of a partial schedule of length three, which would have been expanded to give three more solutions, and the other four pairs each eliminate one solution for a total of seven.

The extension of this scheme to dominance conditions requires us to deal with the problem of *incomplete tree structure*. When certain objective and lower-bound values are unavailable, due to nodes being terminated before full expansion, we cannot be certain that the (x, y) pair satisfies the criterion for dominance. To solve this problem, we have to extend and relax the syntactic criteria. Such extensions are beyond the scope of this chapter, and for more details, the reader is referred to Realff, (1992).

C. Logical Analysis of Problem-Solving Experience

The analysis of problem-solving experience that has taken place so far has been based on finding subproblems within an existing branching structure that, when solved, will produce subtrees that satisfy the definitions of dominance and equivalence. As we have noted, this is insufficient for generating new dominance and equivalence conditions because we want to avoid the subproblem solution, and not use it as the means of proof.

Thus, the next step in the problem-solving analysis is to use information about the *domain of the problem*, in this case flowshop scheduling, and information about dominance and equivalence conditions that is pertinent to the overall *problem formulation*, in this case as a state space, to convert the experience into a form that can be used in the future problem-solving activity.

We can now identify constraints on the "logical analysis" which must be satisfied for it to produce the desired dominance and equivalence conditions.

1. *Validity*. The reasoning involved in this phase must be logically justifiable; that is, the final conclusion should follow deductively from the facts of the example and from the theory of the domain. If we do not ensure this, we may be able to derive conditions on the two solutions that do not respect either the structure of the domain or the example. The use of these conditions could then invalidate the optimum-seeking behavior of the branch-and-bound algorithm.

2. *Sufficiency*. Given the example, the theory we employ to deduce the conditions must guarantee that the conditions on equivalence or dominance will be satisfied, since the example itself is simply an instance of the particular problem structure, and may not capture all the possible variations. The sufficient theory will thus end up being *specialized* by the example, but since we have asserted the need for validity of this process, it will still be guaranteed to satisfy the abstract necessary and sufficient conditions on dominance and equivalence.

D. SUFFICIENT THEORIES FOR STATE-SPACE FORMULATION

Before showing how the logical analysis will be carried out, it is useful to describe the sufficient theory we will be using for the specific flowshop example. This theory is not restricted to flowshop scheduling, but applies to many state-space problems.

We have characterized our equations and variables as inter- and intra-situational. In transitioning from one state to another, we can view the intersituational equations as constraining the values of the next state's variables. Viewed as constraints, we can define the "looseness" or "tightness" of the constraints that values of a state's variables place on the next state. Thus, for example, since the start-times of the units in the next state are constrained to be larger than the end-times of the same unit in the current state, the constraint will be "tighter" if the end-times are greater in one state than in another. The question then arises, what about the intrasituational constraints on the variables, such as the end-times of the previous unit in the same state?

By definition, the intrasituational variables must eventually be expressed in terms of the intersituational variables. It can be shown (Realff, 1992) that as long as we guarantee that the intersituational variables are at least as loosely constrained in x and y, the intrasituational variables will be, also. The final issue concerns intersituational variables that are constrained via an equality constraint.

In these cases there is no well defined notion of a "looser constraint," the choice is then either to force those variables to be equal in x and y, or to find some path from their value to a constraint on another inter- or intrasituational variable and thus be able to show that their values in x, y should obey some ordering based on these other constraints. This topic is the subject of current research, but is not limiting in the flowshop example, since no such constraints exist. Lastly, it is not enough to assert conditions on the state variables in x and y, since we have made no reference to the discrete space of alternatives that the two solutions admit. Our definition of equivalence and dominance constrains us to have the same set of possible completions. For equivalence relationships the previous statement requires that the partial solutions, x and y, contain the same set of alphabet symbols, and for dominance relations the symbols of x have to be equal to, or a subset of those of y. Thus our sufficient theory can be informally stated as follows:

> If every variable that is the output of an intersituational constraint is at least as loosely constrained in a state x and state y, and x has a subset of, or equal set of, the alphabet symbols of y, then x dominates y.

1. General Qualities of Sufficient Theories

In general, we would like our sufficient theory to have the following qualities:

1. *Simplicity.* The complexity of the sufficiency conditions will tend to translate into complex dominance and equivalence conditions. This complexity can take the form of either a large number of predicates to guarantee the conclusion or complex predicates.

2. *Efficacy.* We must balance making the sufficient theories simple versus making them effective. If, for example, the relative ordering of the intrasituational variables or some function of the state variables, such as the difference between the end-times of successive units, is the driving

force behind the problem structure, then the sufficient theory will be unable to explain why the observed relationship between x and y should hold in general.

3. *Generality*. The more abstract the general theory, the wider the class of problems to which it will apply, and hence the higher the likelihood of being able to transfer knowledge acquired in one class of problems to another, related class. However, we have to construct problem domain theories that connect the information from the specific problem-solving experience to the general theories. The more abstract the theory, the more effort required to do this. On the other hand, if the theory is very specific, then it will probably admit few generalizations, and apply in few problem domains.

4. *Modularity*. Since we would like to use the sufficient theory in a variety of contexts and problems, we need a theory that was easy to extend and modify depending on the context. In our state-space formulation the sufficient theory is couched in terms of constraints on variables. This theory gives us the opportunity to modularize its representation, partitioning the information necessary to prove the looseness of one type of constraint from that required to prove the looseness of a different constraint type. The ability to achieve modularity is a function not only of the theory but also of the representation, which should have sufficient granularity to support the natural partitioning of the components of the theory.

The next section will focus on the representation necessary to express this sufficient theory to the computer, so that it can automatically carry out the reasoning associated with analyzing the examples selected by the syntactic criteria presented in this section. Section V will describe the learning methodology, which, using the representation of Section IV, will generate the new dominance and equivalence conditions.

IV. Representation

This section details the different aspects of the representation we have adopted to describe the problem solutions and the new control knowledge generated by the learning mechanism. Throughout the section we will continue to use the flowshop scheduling problem as an illustration. The section starts by discussing the motives for selecting the horn clause form of first-order predicate calculus, and then proceeds to show how the representation supports both the synthesis of problem solutions and their analysis. The section concludes with a description of how the sufficient theory for state-space models can be expressed in a general way using the proposed representation. The "horn clause form" is described in the paragraphs that follow.

It is beyond the scope of this chapter to explain the syntax and semantics of first-order predicate logic, and the reader is referred to Lloyd's text (1987) for a general introduction. However, it is useful to provide some details on the horn clause form.

A logical clause is a disjunction of terms, and a horn clause is one where, at most, one term is positive. In propositional logic, where only literals are allowed horn clauses have the following form:

$$\neg p \vee \neg q \vee \neg r \vee s.$$

This statement is logically equivalent to

$$p \wedge q \wedge r \Rightarrow s.$$

In predicate calculus, where we are allowed function and predicate symbols that take one or more arguments, the form is

$$\neg p(x) \vee \neg q(y) \vee \neg r(x,y) \vee s(x,y),$$

where implicitly, the x, y variables are allowed to range over the entire domain; i.e., they are universally quantified.

We have chosen this representation for a variety of reasons:

1. Theoretically it has been shown (Thayse, 1988) that the DDP formalism is closely related to a simpler form of horn clause logic, i.e., the propositional calculus. This would suggest that we could use the horn clause form to express some of the types of knowledge we are required to manipulate in combinatorial optimization problems. The explicit inclusion of state information into the representation, necessitates the shift from the simpler propositional form, to the first-order form, since we wish to parsimoniously represent properties that can be true, or take different values, in different states. By limiting the form to horn clauses, we are striving to retain the maximum simplicity of representation, whilst admitting the necessary expressive power.

2. First-order predicate calculus admits proof techniques that can be shown to be *sound* and *complete* (Lloyd, 1987). The soundness of the proof technique is important because it ensures that our methodology will not deduce results that are invalid. We are less concerned with completeness, because in most cases, although the proof technique will be complete, the theory of dominance or equivalence we have available will be incomplete for most problems. Restricting the first-order logic to be of horn clause form, enables the employment of SLD resolution, a simpler

proof technique than full resolution (Robinson, 1965). [*Note*: SLD stands for linear resolution with selection function for definite clauses (Lloyd, 1987)].

3. First-order horn clause logic is the representation that has been adopted by workers in the field of *explanation-based learning* (Minton et al., 1990). Hence, using this representation allows us take advantage of the results and algorithms developed in that field to carry out the machine learning task.

A. REPRESENTATION FOR PROBLEM SOLVING

The representation introduced in the previous subsection must now be utilized to express the information derived from the problem-solving experience, and required to derive the new control knowledge. Our first step will be to define the types of predicates we require to manipulate the properties of the branching structure and the theory that is needed to turn those properties into useful dominance and equivalence conditions.

In Section II, we presented the computational model involved in branching from a node, σ, to a node σa_i. In this model, it was necessary to *interpret* the alphabet symbol a_i, and ascribe it to a set of properties. In the same way, we have to interpret σ as a state of the flowshop, and for convenience, we assigned a set of state variables to σ that facilitated the calculation of the lower-bound value and any existing dominance or equivalence conditions. Thus, we must be able to manipulate the variable values associated with state and alphabet symbols. To do this, we can use the distinguishing feature of first-order predicates, i.e., the ability to parameterize over their arguments. We can use two place predicates, or binary predicates, where the first place introduces a variable to hold the value of the property and the second holds the element of the language, or the string of which we require the value. Thus, if we want to extract the lower bound of a state σ, we can use the predicate (*Lower-bound* ?g [σ]) to bind ?g to the value of the lower bound of σ. This idea extends easily to properties, which are indexed by more than just the state itself, for example, unit-completion-times, v, which are functions of both the state and a unit

$$(\text{unit} - \text{completion} - \text{time ?v k } [\sigma])$$

where k can range over the number of units in the flowshop.

We must now connect these various predicates to be able to derive state variables from other state variables. The process of transitioning from one state to another, can be modeled by the deductive form of the horn clauses. This is a common strategy in artificial intelligence, and is formalized by *situational calculus* (Green, 1969).

However, we do not need the full complexity of situational calculus, which allows for multiple operators to be applied to a given state, because we have only one way of going from one state to another, i.e., branching. Branching has the effect of updating the state properties and the partial string, σ. We will "borrow" the notation of Prolog (Clocksin and Mellish, 1984) to indicate a list of items, where $[?a.?\sigma]$ parses the list into the head of the list, $?a$, and the rest of the list, $?\sigma$. Thus, our horn clause *implications* (*rules*) will have the form

$$\left.\begin{array}{l} p(?v_1,\ldots,?\sigma) \\ q(?v_2,\ldots,?\sigma) \\ r(?v_i,\ldots,?a) \\ (?i_1 = h_1(?v_1,\ldots,?v_n)) \\ (?\omega = h_\omega(?v_1,\ldots,?v_n,?i_1,\ldots,?i_m)) \end{array}\right\} \Rightarrow p(?\omega,\ldots,[?a.?\sigma])$$

Notice that the left-hand side of this rule contains two types of clauses. The first type is the variable values of the current state and those necessary to compute the new state, while the second, represented by " = " computes the value of the variable in the new state. This last clause enables the procedural information about how to compute the state variables to be attached to the reasoning. We must, however, be careful about how much of the computation we hide procedurally, and how much we make explicit in the rules. The level to which computation can be hidden will be a function of the theories we employ to try to obtain new dominance and equivalence conditions. If we do not hide the computation, we will be able to explicitly reason about it, and thus may find simplifications or redundancies in the computation that will lead to more computationally efficient procedures.

We can further subdivide the implications of the above form into two categories; *intersituational* and *intrasituational*:

1. The first type of implications involve clauses with $?\sigma$ on the left-hand side and $[?a.?\sigma]$ on the right-hand side, indicating a branching step.
2. The second type of implications involve clauses with the same state on the left- and right-hand sides of the implication, indicating the derivation of a state variable value from other variables in the state.

These implications types correspond closely to the definitions of state variable types we gave in Section II, E. Those variables that appear in the clause of the right-hand side of an implication that is intersituational are intersituational variables; those variables that appear on the right-hand side of an implication that is intrasituational are intrasituational variables.

1. Horn Clauses in the Synthesis of Branching Structures

Let us now illustrate how the horn clause for of predicate logic can be used for the unfolding of the branching structure and the synthesis of new solutions.

In the flowshop example, when we branch, we have to calculate the start-times of the new batch to be scheduled. Here are the rules that can accomplish this:

$$Axiom\text{-}1: (End\text{-}time\ 0\ ?j\ [\ ])$$
$$Axiom\text{-}2: (End\text{-}time\ 0\ 0\ ?\sigma)$$

a. Intersituational Implications (Rules).

$Rule\text{-}1$: $(End\text{-}time\ ?e_1\ 1\ ?\sigma) \Rightarrow (Start\text{-}time\ ?e_1\ 1\ [?a.?\sigma])$

$Rule\text{-}2$:
$$\left. \begin{array}{l} (?j = ?i - 1) \\ (End\text{-}time\ ?e_\sigma\ ?i\ ?\sigma) \\ (End\text{-}time\ ?e_j\ ?j\ [?a.?\sigma]) \\ (\geq ?e_j\ ?e_\sigma) \end{array} \right\} \Rightarrow (Start\text{-}time\ ?e_j\ ?i\ [?a.?\sigma])$$

$Rule\text{-}3$:
$$\left. \begin{array}{l} (?j = ?i - 1) \\ (End\text{-}time\ ?e_\sigma\ ?i\ ?\sigma) \\ (End\text{-}time\ ?e_j\ ?j\ [?a.?\sigma]) \\ (\geq ?e_\sigma\ ?e_j) \end{array} \right\} \Rightarrow (Start\text{-}time\ ?e_\sigma\ ?i\ [?a.?\sigma])$$

There are several important points about these rules.

1. *Explicit conditioning.* We could have written the following rule:

$$\left. \begin{array}{l} (?j = ?i - 1) \\ (End\text{-}time\ ?e_\sigma\ ?i\ ?\sigma) \\ (End\text{-}time\ ?e_j\ ?j\ [?a.?\sigma]) \\ (?s = \max(?e_j\ ?e_\sigma) \end{array} \right\} \Rightarrow (Start\text{-}time\ ?s\ ?i\ [?a.?\sigma])$$

However, by doing so we hide, in the computation of the maximum, a potentially important symbolic constraint between the values of the end-times of the previous unit and the current unit.

2. *Bidirectionality.* During the branching step, we could use these rules to deduce the new value of the start-time, by reasoning from the antecedents to the conclusion, i.e., the facts in the state associated with ?σ are known such as the unit end-times. We can use these facts to establish the new facts in [?a.?σ] by opportunistically applying our rules. However, actually carrying out the updating in this way would be extremely inefficient, and a purely procedural approach is sufficient. During the logical analysis of partial solutions we would start with facts about [?a ?σ] and be able to deduce constraints on the earlier partial solution variable values.

2. Horn Clauses in the Analysis of Branching Structure

The analysis of the branching structure turns the preceding deduction process around. We have all the facts available to us at the end of the solution synthesis, i.e., at the end of solving a particular problem. Our task is to select and connect subsets of those facts to prove new results that are useful for deriving new control information. In essence, we have to *turn facts about solutions and partial solutions at lower levels in the tree, into constraints on the properties of states and alphabet interpretations higher in the tree.*

For example, if our goal were to show that the end-time of the schedule, x, was equal to a particular value, but we wanted to verify this not for x directly, but for some ancestor of x, then we could use the following implication, in addition to our conditions for start-time deduction (see Section IV, A, 1, a) to carry out the reasoning.

a. Intrasituational Implication.

$$\text{Rule-4} \left. \begin{array}{l} (\textit{Start-time } ?s \ ?j \ [?a.?\sigma]) \\ (\textit{Proc-time } ?p \ ?j \ ?a) \\ (= ?e (+ ?s \ ?p)) \end{array} \right\} \Rightarrow (\textit{End-time} ?e \ ?j \ [?a.?\sigma]))$$

Figure 8 shows the trace of the application of *Axiom*-1, *Axiom*-2, *Rule*-1, *Rule*-2, *Rule*-3, and *Rule*-4 in our specific scheduling problem. The trace consists of a repeated pattern of rule applications *Rule*-4 followed by either *Rule*-2 or *Rule*-3. At each step the intrasituational rule converts an end-time to a start-time, and then the start-time is matched to the

INTELLIGENCE IN NUMERICAL COMPUTING 587

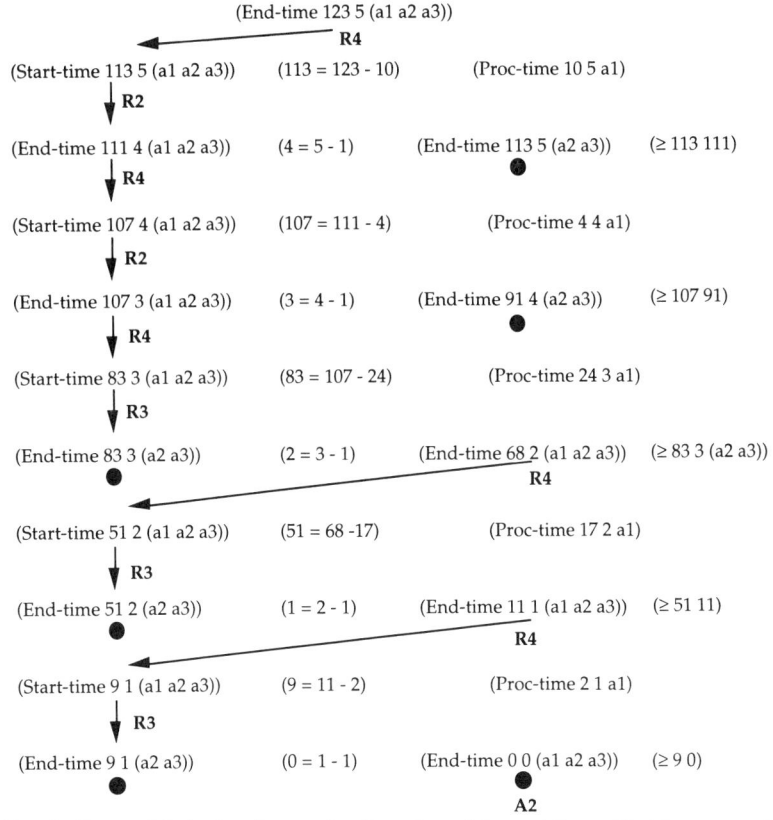

```
                        (End-time 123 5 (a1 a2 a3))
                    ◄─────────────────
                              R4
(Start-time 113 5 (a1 a2 a3))   (113 = 123 - 10)   (Proc-time 10 5 a1)
         │ R2
         ▼
(End-time 111 4 (a1 a2 a3))     (4 = 5 - 1)        (End-time 113 5 (a2 a3))    (≥ 113 111)
         │ R4                                              ●
         ▼
(Start-time 107 4 (a1 a2 a3))   (107 = 111 - 4)    (Proc-time 4 4 a1)
         │ R2
         ▼
(End-time 107 3 (a1 a2 a3))     (3 = 4 - 1)        (End-time 91 4 (a2 a3))     (≥ 107 91)
         │ R4                                              ●
         ▼
(Start-time 83 3 (a1 a2 a3))    (83 = 107 - 24)    (Proc-time 24 3 a1)
         │ R3
         ▼
(End-time 83 3 (a2 a3))         (2 = 3 - 1)        (End-time 68 2 (a1 a2 a3))  (≥ 83 3 (a2 a3))
         ●                                   ◄─────────────
                                                      R4
(Start-time 51 2 (a1 a2 a3))    (51 = 68 -17)      (Proc-time 17 2 a1)
         │ R3
         ▼
(End-time 51 2 (a2 a3))         (1 = 2 - 1)        (End-time 11 1 (a1 a2 a3))  (≥ 51 11)
         ●                                   ◄─────────────
                                                      R4
(Start-time 9 1 (a1 a2 a3))     (9 = 11 - 2)       (Proc-time 2 1 a1)
         │ R3
         ▼
(End-time 9 1 (a2 a3))          (0 = 1 - 1)        (End-time 0 0 (a1 a2 a3))   (≥ 9 0)
         ●                                                 ●
                                                          A2
```

FIG. 8. Trace of deduction for proving the value of the end-time of batch a_1 on unit 5.

appropriate intersituational rule, to generate another end-time. This process is the reversal of the calculation of the state variables performed during the branching, where end-times were derived from start-times. The other rule antecedents fix the relationships between the relative sizes of end-times, and enable calculations to be performed. These conditions must be true of the example for the proof to be successful. We stopped the proof at the predicates

$$(\textit{End-time } 113\ 5[a_2 a_3]), (\textit{End-time } 83\ 2[a_2 a_3]),$$
$$(\textit{End-time } 51\ 2[a_2 a_3]), (\textit{End-time } 9\ 1[a_2 a_3])$$

Why did we not continue the deduction process beyond these particular predicates? Also, we continued to expand the end-time predicate,

$$(\text{end-time } 111\ 4[a_1 a_2 a_3])$$

which did not constrain the start-times. This lead to the derivation of a number of constraints between end-times that would appear to be unnecessary.

The first of these issues is known as *operationality* (Minton et al., 1990) and will be discussed in Section V. The second problem requires us to equip the deduction process with a means to know when deductions are irrelevant, or to change the way the conditions are expressed to avoid the problem altogether. In general, neither of these solutions can be accomplished in a domain independent way. However, the latter solution confines the domain dependency to the conditions themselves, rather than the deduction mechanism, and thus is preferred.

B. Representation for Problem Analysis

The focus of the representation so far, has been on giving the form of rules, which enable us to reason about the values of state variables. This, however, is only one part of the overall reasoning task. We must also represent the theoreies we are going to use to derive the new control knowledge.

The goal of the reasoning is to prove that two partial solutions are equivalent to one another, or that one dominates the other. To do this, we will start with conditions that contain

$$(equivalent\ ?x?y), (dominates\ ?x?y)$$

and terminate with predicates that are easy to evaluate within the branch-and-bound procedure.

The type of theories we will be using to prove dominance and/or equivalence of solutions will not be specific to the particular problem domain, but will rely on more general features of the problem formulation. Thus, for our flowshop example, we will not rely on the fact that we are dealing with processing times, end-times, or start-times, to formulate the general theory. The general theory will be in terms of sufficient statements about the underlying mathematical relationships, as described in Section III.

For example, consider trying to prove that one solution, x, will dominate another solution, y, if they both have scheduled the same batches, but the end-times of each machine are earlier in the partial schedule (partial solution), x, than in y. The general theory will not be couched in these terms, but more abstractly, in terms of the properties of binary operators, such as max, min, $+$, $-$. Specifically, the implications

$$A \leq B \wedge C \leq D \Rightarrow A + C \leq B + D$$
$$A \leq B \wedge C \leq D \Rightarrow \max(A, C) \leq \max(B, D)$$

would enable us to deduce the requisite sufficient conditions for dominance–equivalence relationships. The tree of algebraic manipulations and implications, shown in Fig. 9, is such a case in point.

We can thus think of our predicates as falling into two classes: (1) the *formulation-specific*, and (2) the *problem-specific*. The implications fall into three classes: (1) those that interconnect the general concepts of the formulation, (2) those that connect the general concepts of the formulation to the specific details of the problem, and finally (3) those that enable reasoning about the specific details of the problem. We have already described the predicates necessary for reasoning within the flowshop problem; thus the rest of this section will focus on the general predicates and their interconnection with the specific problem details.

1. Predicates for Problem Analysis

To support the analysis of the solutions we will introduce several new predicate types. We have already described the basic state model in Section II, E. Explicitly manipulating parts of this model is an important component of the reasoning required to derive new control conditions. We will introduce the predicate, "form?," which will enable us to analyze the form of the state update rules, or constraints between variables. Thus, for our flowshop example we have two basic types of constraints:

$$(form? \; ?c \; (?d \geq ?a))$$
$$(form? \; ?c \; (?d = + \; ?a?b))$$

These predicate structures can then be bound to the various instances of the constraints in the problem. For example, we would include among our facts declarations of the forms of the constraints, such as

$$(form? \; c_1 \; (s_{i+11} \geq e_{i1}))$$
$$(form? \; c_2 \; (s_{i+12} \geq e_{i+11}))$$

In addition to being able to identify the types of constraints, we must also be able to identify the types of variables. We introduce two predicates, "Intersituational?" and "Intrasituational?" to identify the variable types.

The definition of the abstract form of the constraints between states must now be linked to the actual values of the variables in the constraints

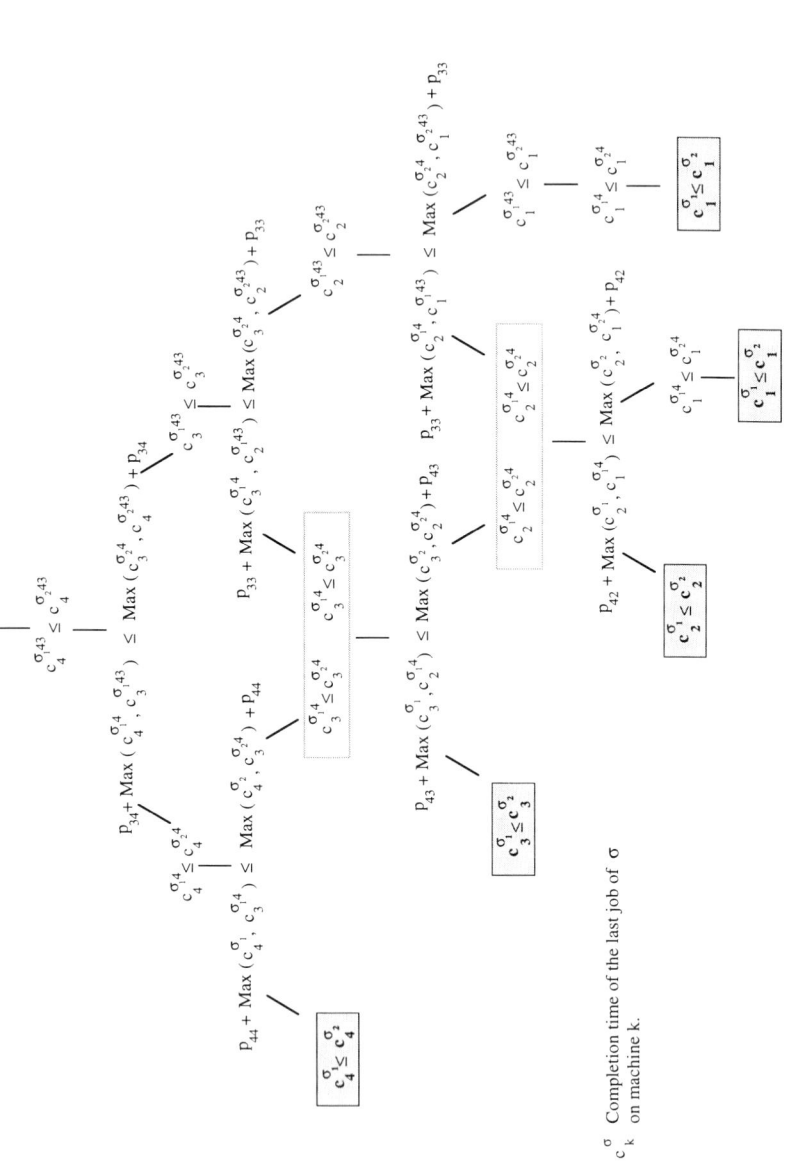

FIG. 9. Tree showing algebraic manipulations necessary to derive sufficient conditions for dominance or equivalence.

themselves, for different states. For this we use the predicates

$$(variable - in - con?\ ?c\ ?var\ ?partial - solution)$$
$$(State - variable - value\ ?val\ ?var).$$

These predicates could be created for the various states that are explored during the branch-and-bound algorithm, or an appropriate subset that has been identified during the example identification.

We then use a predicate for each variable type to be able to pattern match from the variable to its value:

$$(State - variable - value\ ?val\ (end\text{-}time\ ?val\ ?unit\ ?state))$$
$$(State - variable - value\ ?val\ (start\text{-}time\ ?val\ ?unit\ ?state)).$$

These predicates are clearly dependent on the specific problem. We will assume that they are available as basic facts, but we could continue to analyze them using the specific theory of the problem, to turn the start- and end-times of one state into constraints on processing times, and start- and end-times of other states.

We need to introduce two more types of predicates to support the sufficient theory used in the analysis. As we have stated in Section III, D, the sufficient theory rests on being able to prove that the constraints on one state are looser than those on another. The predicate that is used to express this is "looser − constraint − on − variable?," which takes the form:

$$(looser - constraint - on - variable?\ ?var\ ?x\ ?y)$$

where $?x$ and $?y$ will be bound to particular partial solutions and the variable will be one that occurs on the left-hand side of the constraint. To prove that variables are more loosely constrained, we need to compare variable values in different states. Thus, we introduce a new predicate, "*less-equal*" which implements the predicate, "*looser-constraint-on-variable.*" This predicate has the form:

$$(less\text{-}equal?\ ?a\ ?b\ ?x\ ?y)$$

where $?a\ ?b$ will be variables in the constraints and $?x\ ?y$ will be the partial solutions.

2. Expression of Sufficient Theory

With the basic predicate types in place we can now define the various implications that will allow us to express the sufficient theory. There are two key steps that have to be made. The first is to take an intersituational variable and figure out what the constraint on the variable is, which of the current state variables are involved in the constraint, and how these variables should be ordered for the constraint to be looser in x than in y. In our example there is a single constraint type, the greater than constraint, leading to the following rule:

$$\left.\begin{array}{l} (form?\ ?c\ (?d \geq ?a)) \\ (Intersituational?\ ?a) \\ (less\text{-}equal?\ ?a\ ?a\ ?x\ ?y) \end{array}\right\} \Rightarrow (looser\text{-}constraint\text{-}on\text{-}variable\ ?d\ ?x\ ?y)$$

For the second step we define predicates that analyze the relationships between variables, recognizing that we are interested only in comparing the actual numerical values of variables associated with solutions, $?x$ and $?y$, which are explicitly declared to be "operational." Thus, we create the following rule:

$$\left.\begin{array}{l} (Variable\text{-}in\text{-}con?\ ?a\ ?v_1\ ?x) \\ (Variable\text{-}in\text{-}con?\ ?b\ ?v_1\ ?y) \\ (Operational\ ?v_1) \\ (Operational\ ?v_2) \\ (State\text{-}variable\text{-}value\ ?val_1\ ?v_1) \\ (State\text{-}variable\text{-}value\ ?val_2\ ?v_2) \\ (>=\ ?val_1\ ?val_2). \end{array}\right\} \Rightarrow (less\text{-}equal?\ ?a\ ?b\ ?x\ ?y)$$

We can now use the preceding implications (rules) to help build the general sufficient theory. As we stated in Section III, D, we have to ensure that all the intersituational variables of the next state are more loosely constrained in x than in y. Thus, our top-level implication (rule) is simply

$$\left.\begin{array}{l} (Output\text{-}intersituational\text{-}variables?\ ?v) \\ (looser\text{-}constraints\text{-}on\text{-}variables?\ ?v\ ?x\ ?y) \end{array}\right\} \Rightarrow (Dominates?\ ?x\ ?y)$$

Then we create a recursive definition of the second predicate to test each variable individually:

$$\left.\begin{array}{l} (looser\text{-}constraint\text{-}on\text{-}variable?\ ?var\ ?x\ ?y) \\ (looser\text{-}constraints\text{-}on\text{-}variables?\ ?rest\ ?x\ ?y) \end{array}\right\} \Rightarrow (looser\text{-}constraints\text{-}on\text{-}variables?\ (?var\ ?rest)\ ?x\ ?y)$$

The specific value of the first predicate would have been entered into the system from the results of the analysis of the structure of the con-

TABLE II

PREDICATES REQUIRED FOR PROOF OF DOMINANCE CONDITION

(*Dominates?* ?x ?y)
(*Output-intersituational-variables?* ?v)
(*looser-constraints-on-variables?* ?vars ?x ?y)
(*looser-constraint-on-variable* ?d ?x ?y)
(*Intersituational* ?a)
(*less-equal?* ?a ?b ?x ?y)
(*Variable-in-con?* ?b ?v_1 ?y)
(*Operational* ?v_1)
(*State-variable-value* ?val_1 ?v_1)
(*form?* ?c (?d?rel?a))

straints between two states in the branching structure. This is a predicate that connects the general theory to the specific details of the problem. It can be done manually or by the computer deducing the structure based on the definitions of variable types we gave in Section II, E.

Then if there are no variables, the following predicate is satisfied:

(*looser-constraints-on-variables?* ( ) ?x ?y)

This completes the representation of the sufficient theory required for the flowshop example. It consists of about 10 different predicates listed in Table II and configured in four different implications (rules). These predicates have an intuitive appeal, and are not complex to evaluate, thus the sufficient theory could be thought of as being "simple." The theory is capable of deriving the equivalence–dominance condition in flowshop problem. It is, however, expressed in terms that could be applied to any problem with that type of constraint. Thus it has *generality*, and since we can add new implications to deal with new constraint types, it has *modularity*.

Expressing this theory shifts the emphasis of creating specific knowledge for each problem formulation, to developing "pieces" of theory for more general problems. The method allows the computer to put these pieces together, based on the specific details of the problem, and the opportunities that the problem solving reveals.

V. Learning

In the previous sections, we presented three of the four components necessary to carry out the improvement of problem-solving performance for flowshop scheduling, using branch-and-bound algorithms. The first of these dealt with the declarative representation of the branch-and-bound algorithms, including the representation of both the solutions and control mechanisms. It also introduced a metric to quantify the efficiency of a branch-and-bound algorithm. The second component dealt with the methods that are needed in order to analyze the experience gained from the solution of specific problems and convert it into new knowledge of generic value. The third component covered the representation needed to analyze and generalize the selected portions of the acquired problem-solving experience.

In this section, we will briefly outline the method of *explanation-based learning* that we use to carry out the synthesis of new control conditions. This method has become well established within the AI community, and for more details, see Minton et al. (1990), Dejong (1990), and Mitchell et al. (1986).

The ultimate goal of the learning process is to transform a problem-solving experience into useful control knowledge that can be applied in future problem-solving episodes. We have already stated that the methodology should be sound, which translates to no dominance or equivalence rules being introduced that would violate the optimum-seeking behavior of the branch-and-bound algorithm. In addition, there are several other requirements that the learning methodology should try to include

1. *Generalizations derived from a few problem-solving instances.* Solving branch-and-bound problems is computationally expensive. Thus we would like to be able to achieve improvements in problem solving as rapidly as possible.
2. *Branching structure generalization.* It is unlikely that we will be solving problems with exactly the same length of alphabet all the time. Thus, the underlying branching structure on which the control conditions act is subject to change. It is important to be able to allow for these changes without having to re-learn the control rules.

A. Explanation-Based Learning

The following definition of explanation-based learning is taken from Minton et al. (1990).
Given:

1. Target concept definition. A concept definition describing the concept to be learned.
2. Training example. An example of the target concept.

3. Domain theory. A set of rules and facts to be used in explaining how the training example is an example of the target concept.
4. Operationality criterion. A predicate over concept definitions, specifying the form in which the learned concept must be expressed.

Determine:

A generalization of the training example that is a sufficient concept description for the target concept and that satisfies the operationally criterion.

The purpose of this section is to try and link these abstract components to the learning task which we have defined.

1. Target Concept Definition

The targets of our learning are new dominance and equivalence conditions. During problem solving each existing dominance or equivalence will pick out a subset of all the pairs of DDPs that appear in the branching structure. In the flowshop example, the pairs of DDPs are representative of partial schedules that share certain features, and whose properties obey certain relationships, such as having equal end-times and having the same set of batches scheduled. Thus, the target concept is defined by the set of all instances that belong to it; in other words, *our target concept is a dominance or equivalence condition, where the instances are all the pairs of DDPs for which the condition is true.*

Discrete decision processes (DDPs) are defined by the three parameters Σ, S, f. It is important to understand the *coverage* of the dominance and equivalence rules in terms of how much we can change the structure of the DDP before we invalidate the condition. It should be clear that any change to S or f will change the target concept definition, since we may no longer be able to guarantee that the necessary conditions on dominance and equivalence are satisfied. For example, if we were to redefine the objective of the flowshop problem to include some weighted value of the mean flowtime, or introduce a new property such as a lateness or earliness penalty, we could not be sure that $f(x) < f(y)$. Similarly, if we were to include extra conditions on the feasibility of a schedule, such as forbidding certain batches from following each other, we would again potentially disrupt the previous ordering on $f(x), f(y)$.

Having demonstrated that S, f must remain constant, we must now define how Σ is allowed to change. If we examine the necessary conditions on dominance and equivalence, Σ is not explicitly mentioned, but we are required to ensure that for all subsequent DDPs, certain properties hold. This suggests that, provided the *interpretation* of the alphabet remains the same, we can allow its length and thus the symbols it contains to vary. In our scheduling example, this corresponds to allowing additional batches or batch types to be added to the alphabet, but we would not be allowed to change the interpretation of the alphabet symbol from being associated with a single batch to being a batch type.[2]

In summary, the branch-and-bound algorithm as defined in this chapter assumes that the semantics of "objective function," "feasibility," and "branching operation" are fixed with respect to the problem class. As long as their interpretations are not changed, the derived equivalence and dominance rules would remain valid.

Our last concern is for any change in the control information itself. Such change can be caused by a change in the lower-bound function, or a change in the dominance and/or equivalence conditions. In the former case, we must again abandon our existing dominances and equivalences, unless the altered lower-bound function, g', satisfies the condition

$$g'(x) < g'(y) \Rightarrow g(x) < g(y),$$

or, unless we abandon lower-bound consistency.

Changes in dominance conditions can, in very special circumstances, lead to possible expansion of more nodes, but *only* in situations where we have relaxed our conditions on dominance (for details, see Realff (1992)).

2. Training Example

A training example is an instance of the target concept, which in our case is a pair of DDPs, and their enumeration via the branch-and-bound algorithm. The information about the training example that the learning algorithm manipulates is dependent on the sufficient theory. For each DDP the problem-specific predicates will be constructed using variables, their values, and the constraints associated with the DDP. These problem-specific predicates given the values from DDPs constitute the *facts* about the example.

3. Domain Theory

The domain theory consists of the horn clauses that represent

1. The general implications of the sufficient theory (i.e., the state-space sufficient theory).
2. The implications that link the general theory to the specific problem structure
3. The facts associated with the training example.

---

[2] Note, however, in this case, this change would make a change in S also.

4. Operationality Criterion

In an earlier section, we had alluded to the need to stop the reasoning process at some point. The operationality criterion is the formal statement of that need. In most problems we have some understanding of what properties are easy to determine. For example, a property such as the processing time of a batch is normally given to us and hence is determined by a simple database lookup. The optimal solution to a nonlinear program, on the other hand, is not a simple property, and hence we might look for a simpler explanation of why two solutions have equal objective function values. In the case of our branch-and-bound problem, the operationality criterion imposes two requirements:

1. The predicates themselves must be easy to evaluate. Thus, we will restrict ourselves to simple comparison predicates, ($<$, $\leq$, $+$, $\geq$, $>$), and predicates that extract properties from alphabet symbols and DDPs.
2. The predicates have to express fact about the DDPs for which we are trying to prove dominance or equivalence. This restriction is clearly necessary to avoid declaring the explanation complete before we have expressed the conditions at the appropriate level in the branching structure.

In choosing the definition of operationality, we are deciding the trade-off between the number of predicates required to evaluate the condition versus the complexity of each predicate. This tradeoff further reduces to the cost of matching the facts to the predicates, versus the cost of evaluating a given match. If there are many different ways to match the facts to the predicates, but only one is successful, we will tend to have high match cost, unless we can find some ordering of the predicates, such that the unsuccessful ones are eliminated early. This topic has been the subject of much discussion; see Minton et al. (1990) for details.

To express operationality within our representation framework, we will use two methods:

1. Implicitly, we can express the fact that a predicate is operational by not including it on the right-hand side of an implication. In this case, it can never be further expanded or explained, and hence if it is not operational, we will fail to generate a satisfactory explanation. This definition makes it difficult to distinguish between the operationality of facts at one level of the branching structure versus another.
2. Explicitly, we can express the operationality by including a predicate (*Operational*), which does not appear as a consequent. We can then assert that certain facts are operational by including (*Operational fact* $- i$) in the database.

This explicit inclusion of operationality allows us to declare explicitly some facts about the pair (x, y), such as their state variable values, as operational. This will not permit the explanation to stop at other partial solutions, whose states have not been declared operational; thus we will use this approach.

B. EXPLANATION

So far we have described all the inputs to the explanation-based learning procedure. We must now describe a few of the basic feature of the algorithm itself.

The first step of the explanation-based learning paradigm is the construction of an explanation of why the training example is an example of the target concept. This explanation will be constructed by backward chaining from the instantiated target concept definition, through the horn clause implications, to reach a set of facts that are true of the example, and that satisfy the operationality criterion. In the backward chain process, we *unify* (Lloyd, 1987) the successively generated subgoals with the consequences of the horn clause implication.

This is high-level description of the explanation process can be illustrated in the flowshop example. To instantiate the target concept, we use the partial solution strings that represent x and y; call them σ_x, σ_y. Thus, our target concept, for example, becomes

$$(Dominates? \; \sigma_x \; \sigma_y).$$

We now search our horn clause implications for one whose consequent unifies with (Dominates? $\sigma_x \; \sigma_y$). The unification algorithm returns both whether a unification exists and the bindings generated by the algorithm. The bindings of a unification are the substitutions from the specific example for the variables in the consequent. Thus, to unify (Dominates? $\sigma_x \; \sigma_y$) with (Dominates? ?x ?y), we can do so by binding ?x to σ_x and ?y to σ_y. Having found an implication whose consequent matches the goal, we backward-chain to the antecedents of the implication, and set these up as subgoals. In these antecedents, we carry the bindings of the consequent through to the corresponding variables of the antecedent. Thus, if the implication were

$$\left. \begin{array}{l} (\textit{Intersituational-variables?} \; ?v) \\ (\textit{looser-constraints-on-variables?} \; ?v \; ?x \; ?y) \end{array} \right\} \Rightarrow (\textit{Dominates} \; ?x \; ?y)$$

then the subgoals would simply be

(*Intersituational-variables*? ?v)
(*looser-constraints-on-variables*? ?v σ_x σ_y)

The antecedents are ordered in such a way that bindings, which are necessary to solve certain subgoals, are found before the subgoals are expanded. Thus, in the preceding case, binding for ?v will be sought before trying to solve the subgoal (*Looser-constraints-on-variables*? ?v ?σ_x ?σ_y). Some of the antecedents may not appear as the consequences of other implications, and to satisfy them, we need to search for matching facts in the database. In the preceding example, we would have entered the intersituational variables of the problem formulation as a fact, and hence the subgoal (*Intersituational-variables*? ?v) will be solved by binding ?v to a list of those variables. Figure 10 details the resulting explanation, including the unifications or bindings.

The explanation thus consists of a set of implications that connect the original goal, and subgoals, to the facts of the example. This set of implications is interconnected via the bindings that hold between the subgoal and the consequent of the implication used to solve it. The preceding qualitative description of the form of an explanation can be formalized in graph-theoretical terms (Mooney, 1990). For our purposes, the important feature of the explanation is its *structure*. The structure is the pattern of the inference and the bindings that enable us to connect the objects of the original goal to the facts at the leaves of the explanation.

1. Specific Explanation

The specific explanation structure for the flowshop problem is given in Fig. 10. In the example we have assumed that the sufficient condition is satisfied by having all the end-times of x less than or equal to those of y. Thus the proof begins by selecting the appropriate variable set, and proceeds to prove that each variable is more loosely constrained in x than in y. The intersituational variables in the flowshop problem are the start-times of the next state.

If we assume that the variables are ordered from the first unit to the last, we will begin by examining the start-times on the first unit. The start-times are not operational, but the end-times on the units for partial solution x, y are. The start-times of the first unit in the next state, which are equal to the end-times of the first unit in x and y are analyzed by using one of the less-equal implications. The end-times, which are operational, are then compared.

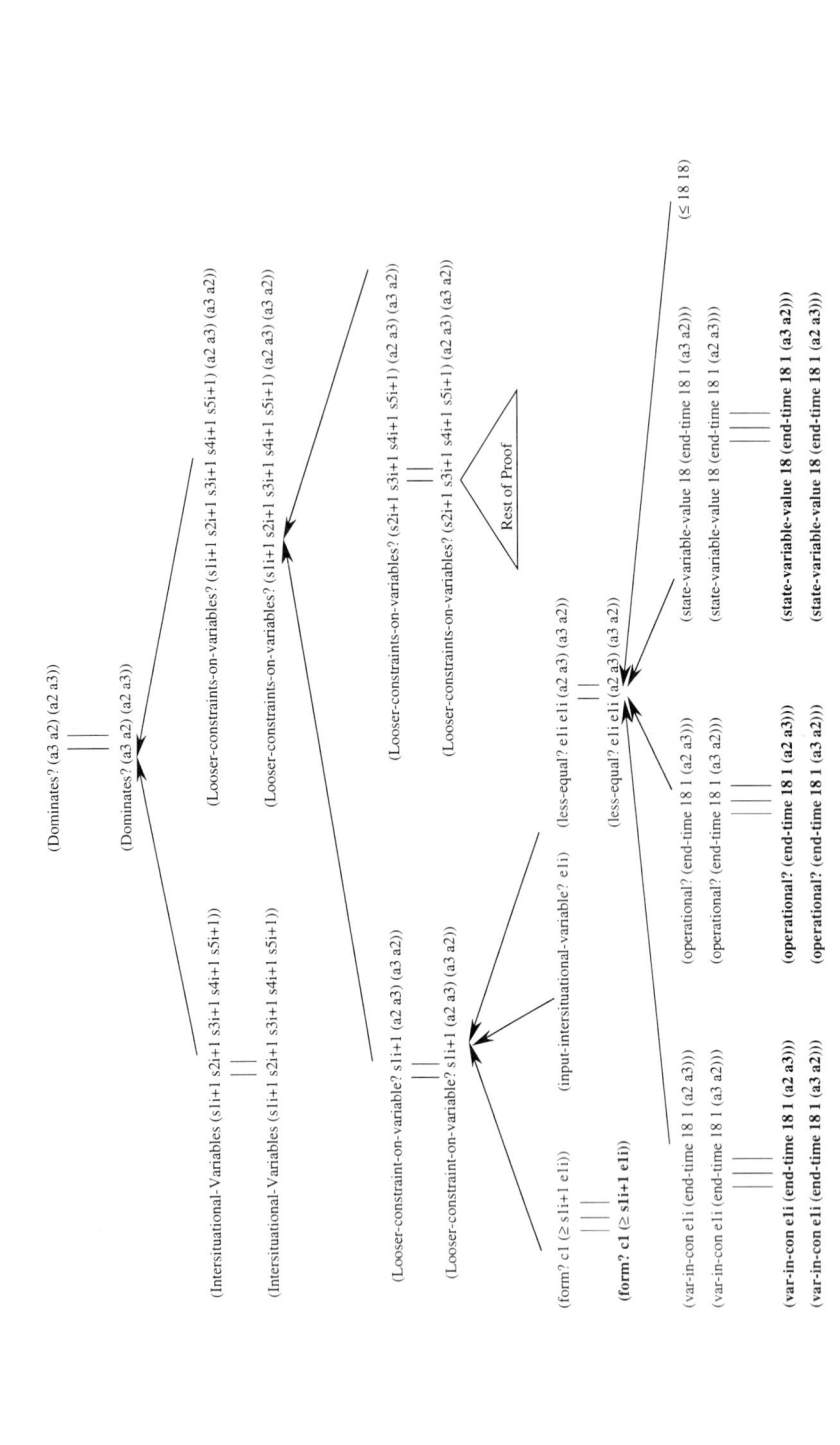

FIG. 10. Proof of simple dominance condition.

Facts are in bold.

A similar analysis is carried out for the start-times of the second unit. Note that during the solution of the subgoal, it is possible that the constraint involving the end-time of the previous unit will be tried. This will fail since the end-time is not an input-intersituational variable. This process is repeated for each subsequent start-time, and in each case the subgoal is resolved the same way.

C. Generalization of Explanations

According to Mooney (1990), the process of generalization has an input an instance of an explanation; and as output, the explanation, divorced from the specific facts of the example, but retaining the structure implied by it.

We will divide the process of generalization into two parts. In the first, we will consider only the generalization of the explanation while leaving its structure unchanged. In the second, we will consider the generalization of the structure itself.

1. Variable Generalization

During the solution of a specific example we take the implications (rules) and instantiate them with the specific facts of the example. Thus part of the explanation might look like

$$\left.\begin{array}{l} (\textit{Variable-in-con } e_{i1} \ (\textit{end-time } e_{i1} \ 18 \ (a_2 \ a_3))) \\ (\textit{Variable-in-con } e_{i1} \ (\textit{end-time } e_{i1} \ 18 \ (a_3 \ a_2))) \\ (\textit{Operational? } (\textit{end-time } e_{i1} \ 18 \ (a_2 \ a_3))) \\ (\textit{Operational? } (\textit{end-time } e_{i1} \ 18 \ (a_3 \ a_2))) \\ (\textit{State-variable-value } 18 \ (\textit{end-time } e_{i1} \ 18 \ (a_2 \ a_3))) \\ (\textit{State-variable-value } 18 \ (\textit{end-time } e_{i1} \ 18 \ (a_3 \ a_2))) \\ (< = 18 \ 18) \end{array}\right\}$$

$\Rightarrow (\textit{less-equal? } e_{i1} \ e_{i1} \ (a_2 \ a_3)(a_3 \ a_2))$

and this would connect to the facts of the example. The explanation for why the variable $S_{i+1,1}$ is no more constrained in the solution $(a_2 \ a_3)$ than in $(a_3 \ a_2)$ is given in Fig. 11, which includes the specific facts and the relevant bindings of variables. From this specific instance of explanation, we generate the explanation structure, which removes the dependency on

FIG. 11. Explanation with original bindings.

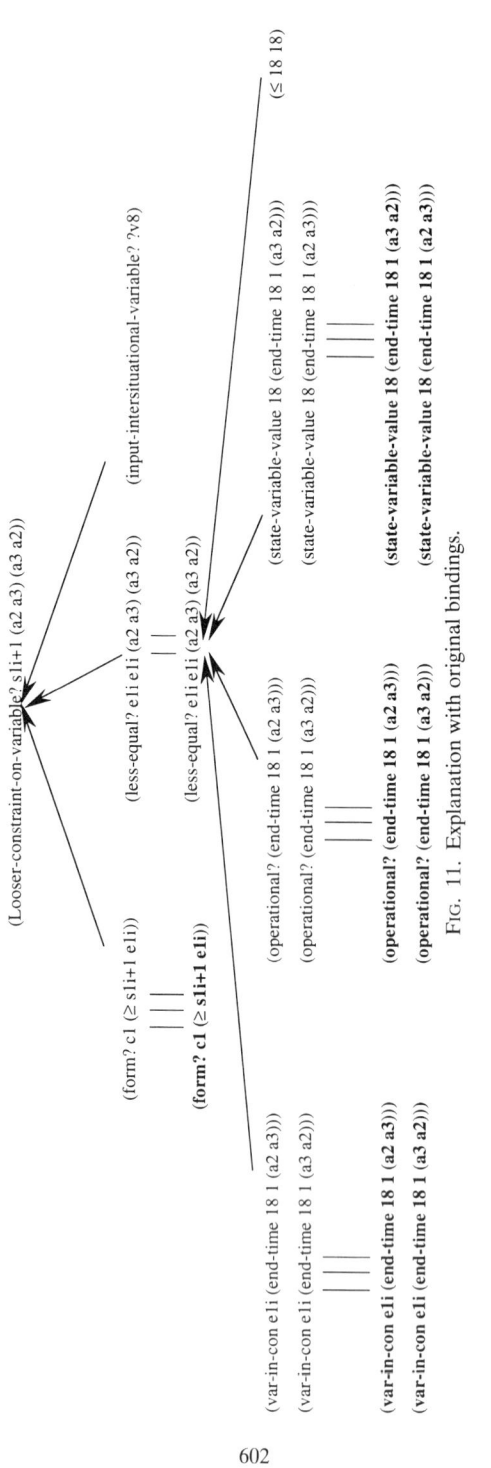

(Looser-constraint-on-variable? ?v1 ?v2 ?v3)

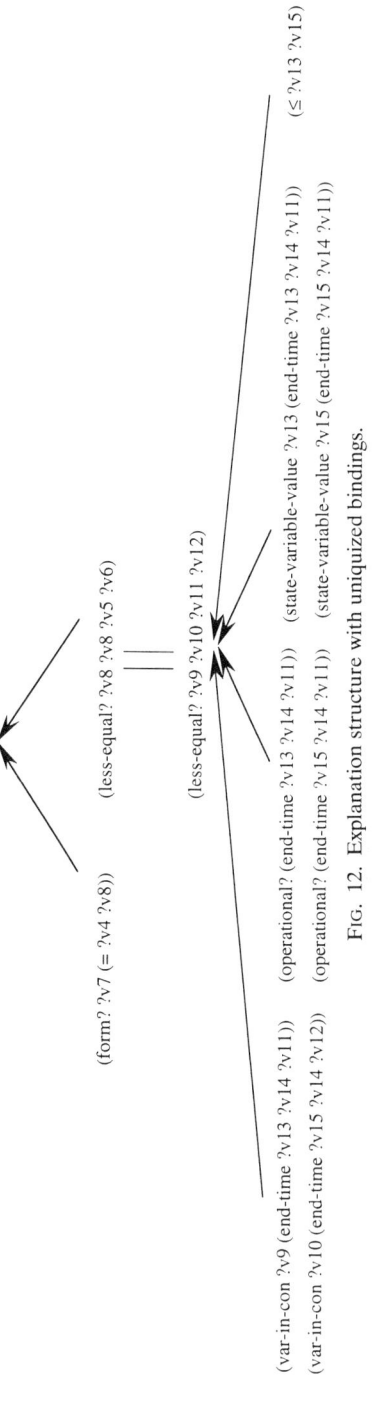

(Looser-constraint-on-variable? ?v4 ?v5 ?v6)

(form? ?v7 (= ?v4 ?v8))

(less-equal? ?v8 ?v5 ?v6)

(less-equal? ?v9 ?v10 ?v11 ?v12)

(operational? (end-time ?v13 ?v14 ?v11))

(operational? (end-time ?v15 ?v14 ?v11))

(state-variable-value ?v13 (end-time ?v13 ?v14 ?v11))

(state-variable-value ?v15 (end-time ?v15 ?v14 ?v11))

($\leq$?v13 ?v15)

(var-in-con ?v9 (end-time ?v13 ?v14 ?v11))

(var-in-con ?v10 (end-time ?v15 ?v14 ?v12))

FIG. 12. Explanation structure with uniquized bindings.

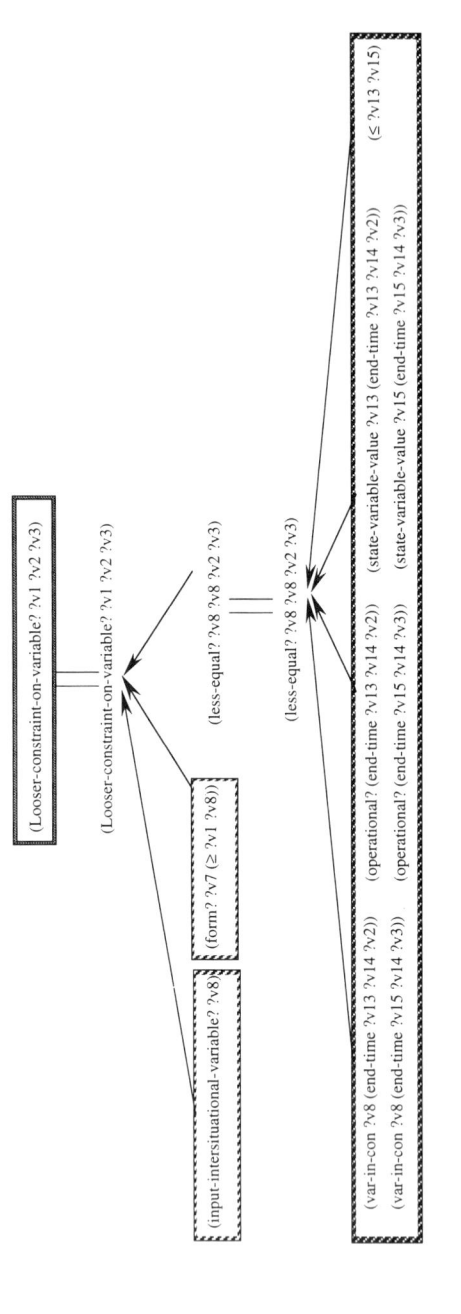

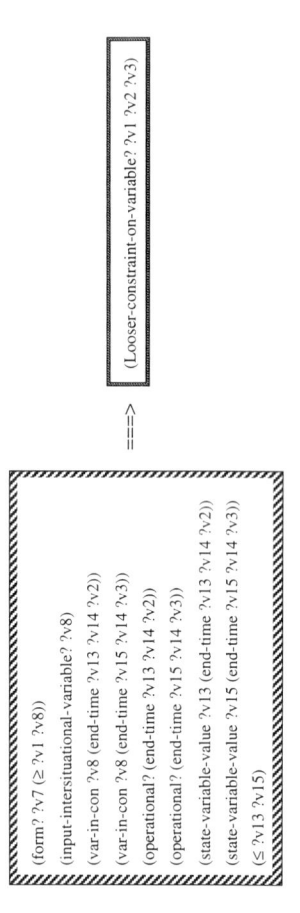

Final Condition - The leaves of the explanation and the top level goal.
FIG. 13. Explanation structure with unified bindings and the final rule.

TABLE III

GENERALIZATION ALGORITHM

Let γ be the null substitution { } for each equality between p_i and p_j in the explanation structure do
  Let θ be the *most-general-unifier* of p_i and p_j
  Let γ be $\gamma\theta$
  for each pattern p_{kN} in the explanation structure do
    replace p_k with $p_k\gamma$

---

the specific facts, and makes the arguments of the predicates unique (Fig. 12). Finally, we compute the most *general unifier*, so that at each antecedent consequent match the predicates are identical. The resulting generalized explanation is given in Fig. 13. Generalization of variables thus requires the computation of the most general unifier, which is the same as finding the unification, along with an algorithm for achieving the generalization, given in Table III, taken from Mooney (1990).

The purpose of creating and generalizing the explanation was to generate a way of identifying whether an instance—in this case, a pair of partial solutions—was a member of the target concept, i.e., dominance. The generalized explanations forms a "trace" from the difficult-to-evaluate predicate, at the root, *Dominates?* to a set of easier-to-evaluate predicates at the leaves. To form our condition, we gather up the leaves of the generalized explanation, and use them as antecedents to a new implication:

$Leaf_1 \wedge Leaf_2 \wedge \ldots \Rightarrow goal\text{-}predicate.$

In the preceding example:

$(form?\ ?con_1\ (?a \geq ?b))$
$(Input\text{-}intersituational\text{-}variable?\ ?b)$
$(Variable\text{-}in\text{-}con?\ ?b\ (end\text{-}time\ ?val_1\ ?j\ ?x))$
$(Variable\text{-}in\text{-}con?\ ?b\ (end\text{-}time\ ?val_2\ ?j\ ?y))$
$(Operational\ (end\text{-}time\ ?val_1\ ?j\ ?x))$
$(Operational\ (end\text{-}time\ ?val_2\ ?j\ ?y))$
$(State\text{-}variable\text{-}value\ ?val_1\ (end\text{-}time\ ?val_1\ ?j\ ?x))$
$(State\text{-}variable\text{-}value\ ?val_2\ (end\text{-}time\ ?val_2\ ?j\ ?y))$
$(<=\ ?val_1\ ?val_2)$

$\Rightarrow (Looser\text{-}constraint\text{-}on\text{-}variable\ ?a\ ?x\ ?y).$

In this case, we have omitted one intermediate step of reasoning, but in general, there could be many steps between the original goal and the operational predicates. Any future application of the condition will not be required to perform those intermediate steps, or any of the associated search for implications with matching consequences.

2. Generalizations of the Proof Structure

The generalization procedure described above carefully preserves the structure of the proof; it does not attempt to take into account any repetitive structure, which might itself be capable of being generalized. In solving branch-and-bound problems, repetitive structure can occur very easily, since we are performing roughly similar functions as we branch from a parent node to a child, whatever level we are at in the tree.

Furthermore, in the context of flowshop scheduling we have repetitive calculations being performed not only at each level of the branching tree but also within a given node, since each unit essentially has its start-time and end-time calculated by identical procedures. The difficulty we face is understanding when generalization of the proof structure is justified, and when it is simply coincidental that certain elements of the proof have been repeated.

We will adopt the procedure due to Shavlik (1990), in which the justification for generalizing the proof structure is the existence of a recursive pattern of the application of the implications. A recursive pattern implies that at some point in proving a goal a subgoal of the same type, i.e., the same predicate, is generated which contains different arguments. This type of proof structure generalization will enable us to generalize certain facets of the problem structure. Since we have defined the parsing of the list of output intersituational variables recursively, we will be able to generalize the number of items in the list, or in this case the number of start-times. Generalizing the number of start-times implicitly enables us to have an arbitrary number of units in the flowshop. Figure 14 presents the overall structure of the proof.

The way we have stated the domain theory for the state-space representation has enabled us to avoid making explicit reference to the alphabet symbol properties. However, if in other formulations we need to refer to these properties, we would again use a recursive parsing of the list of symbols to enable generalization over the size of the alphabet.

The method that Shavlik developed to carry out the structure generalization is complex, and requires the introduction of an extension to the horn clause deduction scheme to allow the repetitive application of a

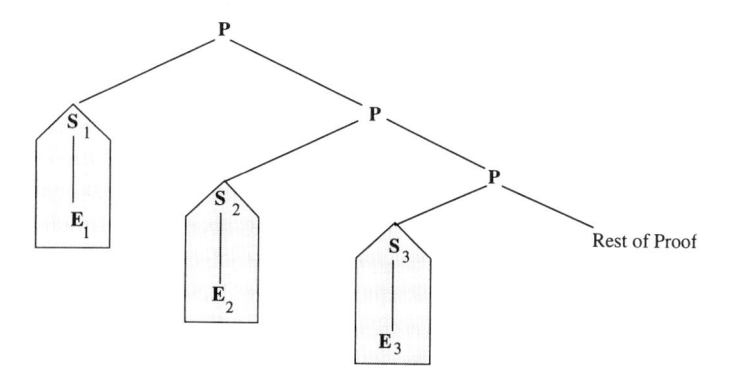

FIG. 14. Abstract structure of the proof with repeated subgoals.

S   Start times of the next state - Intersituational Variables

E   End times of the current state, which are operational

Proof Structure Repeated but not in a sub-goal relationship

group of implications without going through the general backward chaining procedure. The details can be found in Shavlik (1990).

VI. Conclusions

This chapter has presented a methodology for integrating machine learning into the branch-and-bound algorithm. The methodology has been shown to be capable of acquiring a simple dominance condition used in flowshop scheduling algorithms. In Realff (1992) the application of the methodology to a more general scheduling problem (Kondili et al. 1993) is presented. The general problem is cast as an MILP and a sufficient theory of dominance developed for MILP is used to find a dominance rule. Thus we have developed and represented sufficient theories for dominance in both a state-space problem-solving formulation and the traditional optimization formulation of an MILP, both of which cover a wide range of problems of interest to chemical engineers.

We believe this work can be extended in several directions. First, we have shown how to acquire the dominance condition expressed in first-order logic; the next step is to use automatic programming concepts (Barstow, 1986) to convert this expression back into the underlying programming language of the branch-and-bound algorithm, to speed up its application within problem solving. Second, we have concentrated on the theoretical aspects of improving the problem-solving performance. In addition to the theoretical improvements in efficiency, we must demonstrate empirical improvements in the execution time of the algorithm, or in the size of the problems tackled. The computer would *experiment* with the dominance and equivalence rules that it acquired, noting the changes in execution time solving the problem with or without the rules. It would try to detect features of the numerical data that make a given problem harder or easier to solve, using an appropriate empirical learning method such as that presented by Saravia (second chapter in this volume). This experience could then be converted into rules for selecting the appropriate configuration of the branch-and-bound algorithm.

This approach has opened up the possibility of a new type of optimization system, one that can learn from its own experience, test its conclusions, and configure itself to best solve the particular problem, based on its structure and on its data. By reducing some of the algorithm designer's synthetic burden of constructing new algorithms, or new extensions to existing general-purpose methods for each problem, allows the designer to concentrate on developing more widely applicable knowledge embodied in the sufficient theories of the system. It takes advantage of both the solution technology and problem-solving architectures of operations research used to *calculate* optimal answers, and the *reasoning* methods of artificial intelligence: a synthesis of a traditional numerical procedure and symbolic deductive techniques.

References

Abelson, H., Eisenberg, M., Halfant, M., Katzenelson, J., Sacks, E., Sussman, G.J., and Yip, K., Intelligence in scientific computing. *Commun. ACM* **32** (1989).
Baker, K.R., "Introduction to Sequencing and Scheduling." Wiley, New York 1974.
Baker, K.R., A comparative study of flow-shop algorithms. *Oper. Res.* **23**(1) (1975).
Barstow, D., A perspective on automatic programming, *In* "Readings in Artificial Intelligence and Software Engineering" (C. Rich and R.C. Waters, eds.), pp. 537–539. Morgan Kaufmann, San Mateo, CA, 1986.
Clocksin, W.F., and Mellish, C.S., "Programming in Prolog." Springer-Verlag, Berlin (1984).
Dantzig, G.B., "Linear Programming and Extensions." Princeton University Press, Princeton, NJ, 1963.

Dejong, G., and Mooney, R., Explanation-based learning: An alternate view. *Mach. Learn.* **1**, 145–176 (1986).
Garey, M.R., and Johnson, D.S., "Computers and Intractability: A Guide to the Theory of NP-Completeness." Freeman, New York, 1979.
Green, C.C., Application of Theorem proving to problem solving. *Proc. Int. Joint Conf. Art. Intell. 1st*, Washington, DC, 1969, pp. 219–239 (1969).
Ibaraki, T., The power of dominance relations in branch bound algorithms. *JACM* **24**(2), 264–279 (1977).
Ibaraki, T., Branch and bound procedure and state-space representation of combinatorial optimization problems. *Inf. Control* **36**, 1–27 (1978).
Jackson, J.R., Scheduling a Production Line to Minimize Maximum Lateness," Res. Rep. No. 43, Management Science Research Project. University of California, Los Angeles, 1955.
Karp, R.M., and Held, M. Finite-state processes and dynamic programming. *SIAM J. Appl. Math.* **15**, 698–718 (1967).
Kondili, E., Pantalides, C.C., and Sargent, R.H.W., A general algorithm for short-term scheduling of batch operations. I. MILP formulation. *Comput. Chem. Eng.* **17** (2) 211–228 (1993).
Kumar, V., and Kanal, L.N., A general branch and bound formulation for understanding and synthesizing and/or tree search procedures. *Art. Intell.* **21** 179–198 (1983).
Lagweg, B.J., Lenstra, J.K., and Rinnooy Kan, A.H.G., A general bounding scheme for the permutation flow-shop problem. *Oper. Res.* **26**(1) (1978).
Lloyd, J.W., "Foundations of Logic Programming," 2nd ed., Springer-Verlag, Berlin, 1987.
Minton, S., Carbonell, J., Knoblock, C., Kuokka, D., Etzioni, O., and Gil, Y., Explanation-based learning: A problem solving perspective in machine learning. *In* "Machine Learning: Paradigms and Methods" (J.G. Carbonell, ed). Massachusetts Institute of Technology/Elsevier, Cambridge and London, 1990.
Mitchell, T., Keller, R., and Cedar-Cabelli, S., Explanation-based generalization: A unifying view. *Mach. Learn.* **1**, 47–80 (1986).
Mooney, R.J., "A General Explanation-Based Learning Mechanism and its Application to Narrative Understanding." Morgan Kaufmann, San Mateo, CA (1990).
Nemhauser, G.L., and Wolsey, L.A., "Integer and Combinatorial Optimization." Wiley, New York (1988).
Nilsson, N.J., "Principles of Artificial Intelligence." Palo Alto, CA, 1980.
Rajagopalan, D., and Karimi, I.A., Completion times in serial mixed-storage multiproduct processes with transfer and set-up times. *Comput. Chem. Eng.* **13**(1/2), 175–186 (1989).
Realff, M.J., "Machine Learning for the Improvement of Combinatorial Optimization Algorithms: A Case Study in Batch Scheduling." Ph.D Thesis, MIT., Cambridge, MA, 1992.
Robinson, J.A., A machine-oriented logic based on the resolution principle, *JACM* **12**(1), 23–41 (1965).
Shah, N., Pantelides, C.C., and Sargent, R.W.H., A general algorithm for short-term scheduling of batch operations. II Computational issues. *Comput. Chem. Eng.* **17**(2), 229–244 (1993).
Shavlik, J.W., "Extending Explanation-Based Learning by Generalizing the Structure of Explanations." Morgan Kaufmann, San Mateo, CA, 1990.
Simon, H.A., Search and reasoning in problem solving. *Art. Intell.* **21**, 7–29 (1983).
Szwarc, W., Elimination methods in the $m \times m$ sequencing problem. *Nav. Res. Logist. Q.* **18**, 295–305 (1971).
Thayse, A., "From Standard Logic to Logic Programming. Wiley, New York, 1988.

Wiede, W., and Reklaitis, G.V., Determination of completion times for serial multiproduct processes. I-III. *Comput. Chem. Eng.*, **11**(4), 337–368 (1987).

Yip, K., "KAM: A System for Intelligently Guiding Numerical Experimentation by Computer." MIT Press, Cambridge, MA, 1992.

INDEX

A

Acceptable separation valve set, 342, 369
　generating, flowsheet modifications, 369–371
Advanced System for Computations in Engineering Design, 7–8
Alanine, synthesis, algorithm application, 181–182
Ammonia, synthesis, application of algorithm for construction of direct mechanisms, 161–166
Aniline
　production, reaction-based hazard identification, 217–221
　reaction path to, 220
Artificial intelligence, 437–438
Aspiration levels, identification and refinement, 411
Axiomatic theory of design, 95

B

Branch-and-bound algorithm
　relative efficiency, 564–566
　specification, 563–564
Branch-and-bound strategy, 557–563
　dominance test, 561–562
　equivalence test, 562–563
　flowshop problem, 559
　formal statement of branching, 558–559
　lower-bound function, 559–561
Branching, as state updating, 566–568

C

Chebyshev approximation problem, 467–468
Chemical plants, purging from offending chemicals, 342–343
　sources and routes, flowsheet modifications, 371

Chemical structures, LCR modeling elements, 14–17
Chlorine, removal from refrigerants, 298–299
Chloropropane, interactive design, 292
Classification decision trees, inductive construction, 392–394
Computational model, design principles, 117–122
　constraint propagation, 118–119
　context-based design, 122
　hierarchical planning, 118
　specification refinement into implementations, 118
　unified transformational design, 120–122
ConceptDesigner, 139–144
　architecture, 139–143
　design plan module, 140–141
　implementation, 143–144
　modeling objects module, 142
　user interface module, 142–143
Conceptual process designs, 93–145; *see also* HDL
　automation issues, 98–103
　　benefits of mechanized models, 100
　　CAD drawbacks, 99
　　designer, 101
　　high-level, design-oriented languages, 101–102
　　human–computer interface, 103
　　object-oriented representation, 102–103
　　planner, 100–101
　　scheduler, 101
　ConceptDesigner, 139–144
　generic design process, 104
　hierarchical approach, 103–122
　　computational model, 117–122
　　goal structures and transformational model, 107–117
　　intermediate milestones, 105–106
　　planning of process design evolution, 104–107
　previous approaches and limitations, 97–98
　underdefined problem, 96–97
Constraint transformation algorithm, 339

D

Decision making, supervisory control layer, 381–382
Decision trees, inductive learning through, 541–543
Decision unit, 419–420
Deductive reasoning, identification of reaction-based hazards, 221–253
　constraints, 223–224
　fault-tree construction, 238–241
　input variables, 225–226
　level-1-gate, 238–239
　methodological framework, 222–224
　preventive mechanism assessment, 235–238
　reaction quench, 241–253
　recursive tracing of variable-influence links, 223–224
　technology type, 233–235
　top-level events, 235–237
　unforeseen disturbances, 235–236
　variable-influence diagram construction, 227–232
　variable-influence pathway characterization, 232–235
　variables as causes or effects, 225–227
Design errors, 189
Dilation parameter, discretized, 512
Discrete decision process, 555–557, 595
Domain theory, 596
Dominance test, 561–562
Drug design, interactive, 301–304
Dyadic wavelet transform, 512

E

Electrical resistivity, polymer packaging materials, 287
Engineering design, 94–95
Engineering science of knowledge-based design, 95–96
Equivalence test, 562–563
Ethane, pyrolysis, LCR modeling, 64–72
Excitation, persistency, 446

F

Failure handling, object-oriented, 138
Fault-tree, construction, 238–241
Feature extraction, data compression, 530
First intractability theorem, 337
Flowshop problem, 552–535
　branch-and-bound strategy, 559
　example, 570–573
　lower-bound function, 568–570
　solution methodology, 553–555
　specific explanation structure, 599–601
Functional estimation problem, 441–451
　approximation properties, 449–450
　decisions involved in, 445–448
　error bounds, 450–451
　expected risk functional, 444–445
　generalization
　　error, sources, 448
　　versus generality tradeoff, 447
　ill-posed, 442
　inductiveness, 442
　learning algorithm, 465–471
　　derivation of model, 467–468
　　error bound derivation, 409–471
　　structural adaptation algorithm variations, 468–469
　learning problem, formulation, 451–465
　　algorithm derivation, 453–454
　　error threshold selection, 457–459
　　expected risk functional convergence, 459–461
　　functional space representation, 462–465
　　function space selection, 454–457
　　localization in space and frequency, 455
　　multiresolution analysis, 462–464
　　multiresolution decomposition of input space, 455–457
　　problem statement, 452–453
　　wavelet properties, 464–465
　mathematical description, 444–448
　neural network solution, 449–451
　overfitting, 447
　radial basis function networks, 451–452
　regression function, 480–481
　smoothness of approximating curve, 442–443

G

Generalization algorithm, 605
Generalization error, sources, 448
Generate-and-test paradigm, 267–270
　combinatorial explosion, 269–270
　design constraints, 268–270

Glass transition temperature, polymer
  packaging materials, 287
Group contributions, additivity, 291–292
Group vectors, 292

H

Hazard analysis
  incompleteness of methodologies, 192–193
  predictive, 190–192
  proposed methodology, 194–195
  traditional approaches, premises, 193–194
Hazards
  appearance when transferring process from
    laboratory to commercial scale,
    188–189
  identification, 189–190
    algorithm, inductive, 211–214
    equipment-based methodologies, 214–215
    methods, 190–191
    modeling language role, 198–205
    reaction-based, 195–209; *see also*
      Deductive reasoning; Inductive
      reasoning
      mapping of equipment into
        thermodynamic state space,
        204–205
      MODEL.LA, 200–204
      system foundations, 196–198
  preventive mechanisms, assessment,
    235–238
HDL, 122–139; *see also ConceptDesigner*
  design alternative, management, 138–139
  design tasks, semantic relationships,
    132–134
  human–machine interaction, 134–138
  modeling design tasks, 129–134
  modeling elements, 123–127
    Compound, 126
    Flowsheet, 125–126
    Generic Unit, 124
    Generic Variable, 126
    Port, 125
    Project, 126
    Reaction, 126–127
    semantic relationships, 127–129
    Stream, 125
  multifaceted modeling of process design
    state, 123–129
  object-oriented failure handling, 138
  task elements, 129–131

Hierarchical Design Language; *see* HDL
Horn clauses, in branching structure
  analysis, 586–588
  synthesis, 585–586
Human–computer interface, 103
Human–machine interaction, 134–138
Hybrid phenomena theory, 9–10

I

Inductive learning, through decision trees,
  541–543
Inductive reasoning, identification of reaction-
  based hazards, 209–221
  algorithm, 211–214
  aniline production, 217–221
  identification of root causes, 214–217
  methodology, 210–212
  properties, 214–217
Inflexion points, detection, 518–519
Influence graphs, construction, 340–341
Interval arithmetic, metacontributions and,
  273–275

K

Kraft pulp plant
  bottom-up approach, 428–430
  decision variables, 427
  final decision policy comparison, 430–431
  operational analysis, 426–431
  overview, 418
  top-down approach, 428

L

Laminar mechanisms, 154
Language for Chemical Reasoning; *see* LCR
LCR, 13–35, 199
  assumptions, 58
  case study, ethane pyrolysis, 64–72
    initialization, 65–67
    pathway generation, 65, 67–72
  contextual reaction models, creation, 58–64
    communication between, 60–62
    structural compatibility, 62–64
  explanation of notation, 34
  generation of reactions creating
    thermodynamic states, 205–209

614 INDEX

LCR (*Continued*)
  infeasible species, generation, 206–207
  model class decomposition digraph, 50–53
  modeling elements, 13–26
    ab-initio operator, 19, 24–25
    atom, 14–16
    bond, 15–16
    chemical behavior, 18–19, 22–24
    chemical structures, defining, 14–17
    composite-operator, 19–20, 25–26
    context, 21
    quantitative relationships, description, 21
    reaction-environment, 19
    reactions and pathways, 20–21
    reactive behavior of chemicals, defining, 17–20
    subclasses, 21–26
  modeling object
    atom-bond-configuration, 15–17, 22, 207, 209
    pathway, 207, 209
  modeling object extension, 36–50
    ab-initio-operator hierarchical tree, 42–46
    atom-bond-configuration hierarchical tree, 36–39
    chemical behavior hierarchical tree, 39–42
    composite-operator hierarchical tree, 47–50
  reaction pathways, generation and representation, 53–58
  semantic relations among modeling elements, 26–33
    aggregation/disaggregation, 30–31
    commutativity, 32
    entity-attribute, 27
    merging, 32–33
    properties, 31–33
    specialization, 27–28
    specification, 28–30
    transitivity, 32
  syntax, 33–35
Learning, 377–432, 438; *see also* Functional estimation problem
  with categorical performance metrics, 389–396
    classification decision trees, 392–394
    classification techniques, 390
    methodology, 391
    operating strategies for octane number, 394–396
    problem statement, 389–391
    search procedure, 391–396

  complex systems with internal structure, 417–431
    bottom-up approach, 424–425
    conventional and alternative problem definitions, 420–423
    final problem statement, 423
    as networks of interconnected subsystems, 419–420
    problem statement, 417–423
    pulp plant operational analysis, 426–431
    search procedures, 424–426
    top-down approach, 425–426
  continuous performance metrics, 396–408
    alternative problem statements and solutions, 398–401
    methodology and search procedures, 403–405
    problem statement, 396–398
    pulp digester, 405–408
    Taguchi loss functions as continuous quality cost models, 401–403
  data available on routine basis, 382–383
  empirical; *see* Neural networks
  explanation-based, 594–598
    domain theory, 596
    explanation
      construction, 598–601
      generalization, 601–607
        with original bindings, 601–602
        with unified bindings, 604–605
        with uniquized bindings, 603, 605
    operationality criterion, 597–598
    proof structure generalizations, 606–607
    specific explanation, 599–601
    target concept definition, 595–596
    training example, 596
    variable generalization, 601–606
  framework to describe procedures, 384–388
    departures from previous approaches, 385–388
    generic formalism, 385
    hyperrectangles as solution format, 386
    interval analysis nomenclature, 386–387
  inductive through decision trees, 541–543
  plants lacking credible first-principles models, 382
  primary goal of approaches, 380
  problem definition flexibility, 383
  problem statement, 381
  prototypical application examples, 383–384

supervisory control layer of decisionmaking, 381–382
systems with multiple operational objectives, 408–416
  aspiration level identification and refinement, 411
  categorical performance variables, 409–413
  continuous performance variables, 409
  operational analysis of plasma etching unit, 413–416
  problem definition, 411
  Learning algorithm, 465–471
  applications, 471–479
    two-dimensional function with distinguished localized features, 477–479
    estimation, 471–474
    reaction and flowrate of heat transfer fluid in continuous-stirred-tank reactor, 474–477
  derivation of
    error bound, 409–471
    model, 467–468
  structural adaptation algorithm variations, 468–469
Learning scheme, 452
Lysine, synthesis, algorithm application, 182–183

M

Management of operational quality, 379
MCDD, 50–53
Metacontribution, 272–274
  interval arithmetic and, 273–275
Metagroups, 271–272
Methanol, synthesis, application of algorithm for construction of direct mechanisms, 163–173
Modal truth criterion, for quantitative constraints, 344–345
MODASS, 8–9
Model class decomposition digraph, 50–53
Modeling, 438
  computer-aided, 2–3
MODeling ASSistant, 8–9
Modeling languages, 1–89; *see also* HDL; LCR; MODEL.LA
  domain-specific, 11
  MODEL.LA, 73–78

requirements, 13
role in hazards identification, 198–205
Modeling systems
  ASCEND, 7–8
  in chemistry, 10–13
  critique of, 10
  hybrid phenomena theory, 9–10
  MODASS, 8–9
  OMOLA, 8
  premises, 3–6
    automatic generation and modification of models, 5–6
    declarative knowledge articulation, 3–4
    hierarchical and multiview representation of entities, 4–5
    separation of declarative and procedural knowledge, 4
    simulation and reasoning, 6
  process simulation, 7–10
MODEL.LA, 73–78, 199
  computer-aided implementation, 87–89
  Generic-Variable, 331–332
  hazards identification, reaction-based, 200–204
  hierarchies of modeling subclasses, 76–78
  modeling elements, 73–75, 329
  Modeling-Scope, 332
  modeling object attributes, 202–204
  phenomena-based modeling of processing systems, 78–89
    chemical engineering science hierarchies of modeling elements, 79–83
    formal construction, 82, 845
    multifaceted modeling, 82, 84, 86–87
  semantic relationships, 75–76
  syntax, 78
Model truth criterion, 336–337
Molecular abstractions
  formation strategies, 278–279
  successive, searching through, 275–278
Molecular stoichiometry, 174
Molecule, design, 257–308
  automatic synthesis, 267–290
    generate-and-test paradigm, 267–270
    polymers as packaging materials, 284, 286–290
    refrigerant design, 283–285
    search algorithm, 271–283
  desired chemical, 262–264
  final evaluation, 266–267

616 INDEX

Molecule, design (*Continued*)
  general framework, 264–267
  interactive synthesis, 290–304
    extraction solvents, 299–301
    illustration, 291–296
    pharmaceuticals, 301–304
    refrigerants, 296–299
    search space dimensionality reduction, 293–294
    utilization, 295–296
  Molecule-Designer, 304–307
  optimization formulations, 259
  physical properties, estimation, 260–262
  previous work, 260–264
  problem formulation, 264
  procedure, 265
  representation and enumeration of molecules, 265–266
  screening of molecules, 266
  selection of desired chemicals, 262
  target transformation, 264–265
Molecule-Designer, 304–307
  description, 304–305
  group-contribution section, 305
  interactive design section, 306
  molecule evaluation, 307
  problem formulation section, 305
  target transformation section, 305–307

N

Neural networks, 437–482; *see also* Functional estimation problem
  applications in process control, 439
  as nonlinear regression techniques, 439
  universal approximation property, 439
Nodes, 200
  terminal, 211–212
Nonmonotonic planning, 323, 334–351
  algorithm for downward propagation of temporal constraints, 338–339
  constraint generation on temporal ordering of primitive operations, 341–342
  Constraint Transformation Algorithm, 339
  constraint violation identification, 341
  construction of hierarchical models and definition of operating states, 351–358
    goal state definition, 358
    initial state
      completeness, 356–357
      consistency, 356–357
      object-oriented description, 352, 354
      operating state definition, 353, 356–357
    features, 335–336
    handling constraints
      on mixing of chemicals, 339–343
      quantitative, 343–348
      on temporal ordering of operational goals, 337–339
    identification of clobberers of quantitative constraints, 345–347
    influence graphs, construction, 340–341
    operator models and complexity, 336–337
    plan modification operation selection, 347–348
    procedure, 334–335
    purging plant from offending chemicals, 342–343
    switchover procedure, 359–368
      abstraction of operators, 362–363
      constraints, 360
      constraints-2 and -3 transformation, 366
      constraint-4 transformation, 366–367
      initial and goal states, 361–362
      mixing constraint transformation, 363–365
      operating situation requiring, 359–360
      purge/evacuation procedure generation, 364–366
      synthesis, 367–368
    synthesis of operating procedures, 348–351
      plan synthesis, 350–351
      problem formulation, 349–350
    truth criterion, quantitative constraints, 344–345
Numerical computing, intelligent, 549–608
  ancestral equality, 574–575
  branch-and-bound framework, 555–570
    branch as state updating, 566–568
    brand-and-bound algorithm
      relative efficiency, 564–566
      specification, 563–564
    brand-and-bound strategy, 557–563
    discrete decision process, 555–557
    flowshop lower-bounding scheme, 568–570
  definition, 551
  discrete decision processes, 595
  explanation-based learning; *see* Learning, explanation-based
  flowshop problem, 552–535
    example, 570–573
    solution methodology, 553–555

identification of control information, 551–552
problem-solving experience
  definition and analysis, 573–578
  logical analysis, 578–579
  representation, 581–593
    horn clauses, in branching structure
      analysis, 586–588
      synthesis, 585–586
    predicates for problem analysis, 589, 591
    for problem analysis, 588–593
    for problem solving, 583–588
    sufficient theory expression, 591–593
  successive filtering of branching structure, 573
  sufficient theories for state-space formulation, 579–581

O

Object-oriented MOdeling LAnguage, 8
Object-oriented programming, conceptual process designs, 102–103
Occam's razor, 442
Octane number, operating strategies for, 394–396
OMOLA, 8
Operating procedures, 313–375; *see also* Nonmonotonic planning
  branch-and-bound paradigm, 323
  domain-dependent planning theory, 316
  domain-independent planning theory, 316
  hierarchical modeling, 324–334
    conditional operators, 327
    functional operators, 327
    maintaining consistency among descriptions, 332, 334
    operating state description, 330–333
    operations, 325–329
    process behavior, 329–334
    process topology description, 329–330
    STRIPS-operator, 325–326
  nonmonotonic planning, 323
  planning methodology components, 318–324
    constraint character, 321–322
    problem statement, 319–321
    search procedures, 322–324
  previous approaches to synthesis, 316–318
  revamping process designs to ensure feasibility, 368–374

ASVS generation, flowsheet modifications, 369–371
case studies, 372–374
purge source and route flowsheet modifications, 371–372
set of constraints, 315–316
state transition networks, 317
Operationality criterion, 597–598
Organic synthesis, computer-aided, 11–12
Oxygen, permeability, polymer packaging materials, 287, 290

P

Performance metrics
  categorical, 389–396
    systems with multiple operational objectives, 409–410
  continuous, 396–408
    systems with multiple operational objectives, 409
Pharmaceuticals, interactive design, 301–304
Physical properties
  estimation, 260–262
  impact on economics of processes, 258–259
Plasma etching unit, operational analysis, 413–416
  categorical performance variables, 415
  continuous performance variables, 415–416
  system characterization, 413–415
Polymers, as packaging materials, automatic design, 284, 286–290
  constraints, 286
  electrical resistivity, 287
  glass transition temperature, 287
  oxygen permeability, 287, 290
  physical properties, 286
  thermal conductivity, 287
Process data, compression, 493–494
Process trends, 485–546; *see also* Wavelet decomposition
  ad hoc treatment, 490–492
  compression of process data, 493–494
  content, 488–490
  data compression, 527–535
    example, 532–535
    inaccuracies due to end effects, 531
    in real time, 530–531
    selecting mother wavelet and compression criteria, 531–532
    through feature extraction, 530

Process trends (*Continued*)
  through orthonormal wavelets, 527–529
  formal representation, 495–507
    episodes, 498–499
    from quantitative to qualitative, 496–498
    scale space filtering, 500–507
    second-order zero crossings, 505–506
    structure of scale, 503–505
    triangular episode, 499–500
  generalization, 525–526
  generating multiscale descriptions, 523–525
  hierarchy of process operational tasks, 490
  multiscale representation, 500–502
  representation, 491–492
  temporal pattern recognition, 492–493, 535–545
    generating generalized descriptions, 538–540
    inductive learning through decision trees, 541–543
    learning input–output mappings, 537
    pattern matching of multiscale descriptions, 540
    pattern recognition with single input variable, 543–545
    qualitatively equivalent patterns, 538
Pulp digester, continuous performance metrics applications, 405–408

Q

Quality cost models, Taguchi loss functions, 401–403
Quality loss coefficient, 402
Quantitative reasoning; *see* Symbolic and quantitative reasoning

R

Radial basis function networks, 451–452
Reaction pathways; *see also* Symbolic and quantitative reasoning
  hierarchical structure, 148–149
  synthesis, 149–150
Reaction quench, reaction-based hazard identification, 241–253
  breaking cause-and-effect relationship loop, 245
  elements, 241
  fault tree implementation, 252–253
  root cause diagram construction, 247–248
  structural incidence matrix, 242–243
  system boundary, 243–244
  technology types, 247–249
  topological fault tree, 249–251
  variable-influence diagram, 245–247
Reactive behavior, chemicals, LCR modeling elements, 17–20
Reasoning; *see also* Deductive reasoning; Inductive reasoning
  models, 6
  in time; *see* Process trends
Refrigerants
  automatic design, 283–285
  interactive design, 296–299
Regression, 438
  function, 480–481

S

Scale space filtering
  process trends, 500–507
  properties, 506–507
Scale-up, 189
Search algorithm, 271–283
  evaluation, 279–283
  interval arithmetic and metacontributions, 273–275
  metagroups, 271–272
  metamolecule evaluation, 272–273
  searching through successive molecular abstractions, 275–278
    molecular abstraction formulation strategies, 278–279
Second intractability theorem, 337
Sodium chromoglycate, 302
Solvents, extraction, interactive design, 299–301
Streams, 200
STRIPS-Operator, 325–329
Structural adaptation algorithm, variations, 468–469
Structural incidence matrix, 227–230
  reactor-quench example, 242–243
Structural minimization, 454
Sufficient theory
  expression, 591–593
  state-space formulation, 579–581
Symbolic and quantitative reasoning, 147–185
  application domains, 150
  biochemical pathways, 169, 173–183
    active set update, 178

alanine synthesis, 181–182
algorithm, 176–179
 computation efficiency, 179–180
 constraint formulation, 175–176
 initialization and reaction processing, 177
 intermediate metabolite elimination, 178
 lysine synthesis, 182–183
 number of combinations for intermediates, 177
 pathway processing, 179
 synthesis problem features, 173–175
catalytic reaction systems, 151–169
 active set update, 159
 algorithm features, 159–160
 algorithm structure, 155–159
 ammonia synthesis, 161–166
 direct mechanisms, 153–154
 initialization, 156
 intermediate and terminal species, 151–152
 intermediate elimination and selection, 158–159
 methanol synthesis, 163–173
 number of combinations for intermediates, 157–158
 overall mechanisms, 152–153
 previous work on mechanism construction, 154–155
 reaction directionality importance, 166–167, 174
 interpretation of pathway design, 150–151
 synthesis algorithm extensions, 183–185

T

Taguchi loss functions, as continuous quality cost models, 401–403
Temporal patterns, recognition in process trends, 492–493, 535–545
 generating generalized descriptions, 538–540
 inductive learning through decision trees, 541–543
 learning input–output mappings, 537
 pattern matching of multiscale descriptions, 540
 pattern recognition with single input variable, 543–545
 qualitatively equivalent patterns, 538
Thermal conductivity, polymer packaging materials, 287

Thermodynamical entropy, 529
Thermodynamic states, reaction generation, 205–209
Time
 dyadic discretization, 513–514
 uniform discretization, 514–515
Top–down induction of decision trees, 393
Transformational model, 107–117
 general separation structure, 113–116
 initialize project definition, 108–109
 input–output structure design, 110–111
 liquid separation subsystem, 116–117
 plant complex structure specification, 109–110
 recycle structure design, 112–113
Truth criterion
 of complete plans, 336
 quantitative constraints, 344–345
TWEAK, 368

U

Unification algorithm, 598

V

Variable-influence diagrams, construction, 227–232
Variable-influence pathways, characterization, 232–235

W

Watson relation, 274
Wavelet
 dyadic transform, 512
 first-order, 519
 mother, selection, 531–532
 orthonormal data compression through, 527–529
 properties, 464–465
Wavelet decomposition, 507–526
 extraction of multiscale temporal trends, 516–526
  algorithm for extraction of trends, 521–523
  generating multiscale descriptions, 523–525
  inflexion point detection, 518–519

Wavelet decomposition (*Continued*)
   process trend generalization, 525–526
   translationally invariant representation of variables, 517–518
   wavelet interval–tree of scale, 519–521
  of pressure signal, 532–533
  reconstruction of compressed signal from, 532, 534
  theory, 508–516
   decomposition and reconstruction of functions, 511
   discretization of scale, 511–513
   dyadic discretization of time, 513–514
   practical considerations, 515–516
   resolution in time and frequency, 508, 510–511
   uniform descretization of time, 514–515
Wavelet interval–tree of scale, 519–521
Wave-Net solution; *see* Neural networks

Z

Zero crossings, second-order, 505–506

CONTENTS OF VOLUMES IN THIS SERIAL

Volume 1

J. W. Westwater, *Boiling of Liquids*
A. B. Metzner, *Non-Newtonian Technology: Fluid Mechanics, Mixing, and Heat Transfer*
R. Byron Bird, *Theory of Diffusion*
J. B. Opfell and B. H. Sage, *Turbulence in Thermal and Material Transport*
Robert E. Treybal, *Mechanically Aided Liquid Extraction*
Robert W. Schrage, *The Automatic Computer in the Control and Planning of Manufacturing Operations*
Ernest J. Henley and Nathaniel F. Barr, *Ionizing Radiation Applied to Chemical Processes and to Food and Drug Processing*

Volume 2

J. W. Westwater, *Boiling of Liquids*
Ernest F. Johnson, *Automatic Process Control*
Bernard Manowitz, *Treatment and Disposal of Wastes in Nuclear Chemical Technology*
George A. Sofer and Harold C. Weingartner, *High Vacuum Technology*
Theodore Vermeulen, *Separation by Adsorption Methods*
Sherman S. Weidenbaum, *Mixing of Solids*

Volume 3

C. S. Grove, Jr., Robert V. Jelinek, and Herbert M. Schoen, *Crystallization from Solution*
F. Alan Ferguson and Russell C. Phillips, *High Temperature Technology*
Daniel Hyman, *Mixing and Agitation*
John Beek, *Design of Packed Catalytic Reactors*
Douglass J. Wilde, *Optimization Methods*

Volume 4

J. T. Davies, *Mass-Transfer and Interfacial Phenomena*
R. C. Kintner, *Drop Phenomena Affecting Liquid Extraction*
Octave Levenspiel and Kenneth B. Bischoff, *Patterns of Flow in Chemical Process Vessels*
Donald S. Scott, *Properties of Concurrent Gas–Liquid Flow*
D. N. Hanson and G. F. Somerville, *A General Program for Computing Multistage Vapor–Liquid Processes*

Volume 5

J. F. Wehner, *Flame Processes–Theoretical and Experimental*
J. H. Sinfelt, *Bifunctional Catalysts*
S. G. Bankoff, *Heat Conduction or Diffusion with Change of Phase*
George D. Fulford, *The Flow of Liquids in Thin Films*
K. Rietema, *Segregation in Liquid–Liquid Dispersions and Its Effect on Chemical Reactions*

Volume 6

S. G. Bankoff, *Diffusion-Controlled Bubble Growth*

John C. Berg, Andreas Acrivos, and Michel Boudart, *Evaporation Convection*
H. M. Tsuchiya, A. G. Fredrickson, and R. Aris, *Dynamics of Microbial Cell Populations*
Samuel Sideman, *Direct Contact Heat Transfer between Immiscible Liquids*
Howard Brenner, *Hydrodynamic Resistance of Particles at Small Reynolds Numbers*

Volume 7

Robert S. Brown, Ralph Anderson, and Larry J. Shannon, *Ignition and Combustion of Solid Rocket Propellants*
Knud Østergaard, *Gas–Liquid–Particle Operations in Chemical Reaction Engineering*
J. M. Prausnitz, *Thermodynamics of Fluid–Phase Equilibria at High Pressures*
Robert V. Macbeth, *The Burn-Out Phenomenon in Forced-Convection Boiling*
William Resnick and Benjamin Gal-Or, *Gas–Liquid Dispersions*

Volume 8

C. E. Lapple, *Electrostatic Phenomena with Particulates*
J. R. Kittrell, *Mathematical Modeling of Chemical Reactions*
W. P. Ledet and D. M. Himmelblau, *Decomposition Procedures for the Solving of Large Scale Systems*
R. Kumar and N. R. Kuloor, *The Formation of Bubbles and Drops*

Volume 9

Renato G. Bautista, *Hydrometallurgy*
Kishan B. Mathur and Norman Epstein, *Dynamics of Spouted Beds*
W. C. Reynolds, *Recent Advances in the Computation of Turbulent Flows*
R. E. Peck and D. T. Wasan, *Drying of Solid Particles and Sheets*

Volume 10

G. E. O'Connor and T. W. F. Russell, *Heat Transfer in Tubular Fluid–Fluid Systems*
P. C. Kapur, *Balling and Granulation*
Richard S. H. Mah and Mordechai Shacham, *Pipeline Network Design and Synthesis*
J. Robert Selman and Charles W. Tobias, *Mass-Transfer Measurements by the Limiting-Current Technique*

Volume 11

Jean-Claude Charpentier, *Mass-Transfer Rates in Gas–Liquid Absorbers and Reactors*
Dee H. Barker and C. R. Mitra, *The Indian Chemical Industry–Its Development and Needs*
Lawrence L. Tavlarides and Michael Stamatoudis, *The Analysis of Interphase Reactions and Mass Transfer in Liquid–Liquid Dispersions*
Terukatsu Miyauchi, Shintaro Furusaki, Shigeharu Morooka, and Yoneichi Ikeda, *Transport Phenomena and Reaction in Fluidized Catalyst Beds*

Volume 12

C. D. Prater, J. Wei, V. W. Weekman, Jr., and B. Gross, *A Reaction Engineering Case History: Coke Burning in Thermofor Catalytic Cracking Regenerators*
Costel D. Denson, *Stripping Operations in Polymer Processing*
Robert C. Reid, *Rapid Phase Transitions from Liquid to Vapor*
John H. Seinfeld, *Atmospheric Diffusion Theory*

CONTENTS OF VOLUMES IN THIS SERIAL 623

Volume 13

Edward G. Jefferson, *Future Opportunities in Chemical Engineering*
Eli Ruckenstein, *Analysis of Transport Phenomena Using Scaling and Physical Models*
Rohit Khanna and John H. Seinfeld, *Mathematical Modeling of Packed Bed Reactors: Numerical Solutions and Control Model Development*
Michael P. Ramage, Kenneth R. Graziano, Paul H. Schipper, Frederick J. Krambeck, and Byung C. Choi, *KINPTR (Mobil's Kinetic Reforming Model): A Review of Mobil's Industrial Process Modeling Philosophy*

Volume 14

Richard D. Colberg and Manfred Morari, *Analysis and Synthesis of Resilient Heat Exchanger Networks*
Richard J. Quann, Robert A. Ware, Chi-Wen Hung, and James Wei, *Catalytic Hydrometallation of Petroleum*
Kent David, *The Safety Matrix: People Applying Technology to Yield Safe Chemical Plants and Products*

Volume 15

Pierre M. Adler, Ali Nadim, and Howard Brenner, *Rheological Models of Suspensions*
Stanley M. Englund, *Opportunities in the Design of Inherently Safer Chemical Plants*
H. J. Ploehn and W. B. Russel, *Interactions between Colloidal Particles and Soluble Polymers*

Volume 16

Perspectives in Chemical Engineering: Research and Education

Clark K. Colton, *Editor*

Historical Perspective and Overview

L. E. Scriven, *On the Emergence and Evolution of Chemical Engineering*
Ralph Landau, *Academic–Industrial Interaction in the Early Development of Chemical Engineering*
James Wei, *Future Directions of Chemical Engineering*

Fluid Mechanics and Transport

L. G. Leal, *Challenges and Opportunities in Fluid Mechanics and Transport Phenomena*
William B. Russel, *Fluid Mechanics and Transport Research in Chemical Engineering*
J. R. A. Pearson, *Fluid Mechanics and Transport Phenomena*

Thermodynamics

Keith E. Gubbins, *Thermodynamics*
J. M. Prausnitz, *Chemical Engineering Thermodynamics: Continuity and Expanding Frontiers*
H. Ted Davis, *Future Opportunities in Thermodynamics*

Kinetics, Catalysis, and Reactor Engineering

Alexis T. Bell, *Reflections on the Current Status and Future Directions of Chemical Reaction Engineering*
James R. Katzer and S. S. Wong, *Frontiers in Chemical Reaction Engineering*
L. Louis Hegedus, *Catalyst Design*

Environmental Protection and Energy

John H. Seinfeld, *Environmental Chemical Engineering*
T. W. F. Russell, *Energy and Environmental Concerns*
Janos M. Beer, Jack B. Howard, John P. Longwell, and Adel F. Sarofim, *The Role of Chemical Engineering in Fuel Manufacture and Use of Fuels*

Polymers

Matthew Tirrell, *Polymer Science in Chemical Engineering*
Richard A. Register and Stuart L. Cooper, *Chemical Engineers in Polymer Science: The Need for an Interdisciplinary Approach*

Microelectronic and Optical Materials

Larry F. Thompson, *Chemical Engineering Research Opportunities in Electronic and Optical Materials Research*
Klavs F. Jensen, *Chemical Engineering in the Processing of Electronic and Optical Materials: A Discussion*

Bioengineering

James E. Bailey, *Bioprocess Engineering*
Arthur E. Humphrey, *Some Unsolved Problems of Biotechnology*
Channing Robertson, *Chemical Engineering: Its Role in the Medical and Health Sciences*

Process Engineering

Arthur W. Westerberg, *Process Engineering*
Manfred Morari, *Process Control Theory: Reflections on the Past Decade and Goals for the Next*
James M. Douglas, *The Paradigm After Next*
George Stephanopoulos, *Symbolic Computing and Artificial Intelligence in Chemical Engineering: A New Challenge*

The Identity of Our Profession

Morton M. Denn, *The Identity of Our Profession*

Volume 17

Y. T. Shah, *Design Parameters for Mechanically Agitated Reactors*
Mooson Kwauk, *Particulate Fluidization: An Overview*

Volume 18

E. James Davis, *Microchemical Engineering: The Physics and Chemistry of the Microparticle*
Selim M. Senkan, *Detailed Chemical Kinetic Modeling: Chemical Reaction Engineering of the Future*
Lorenz T. Biegler, *Optimization Strategies for Complex Process Models*

Volume 19

Robert Langer, *Polymer Systems for Controlled Release of Macromolecules, Immobilized Enzyme Medical Bioreactors, and Tissue Engineering*
J. J. Linderman, P. A. Mahama, K. E. Forsten, and D. A. Lauffenburger, *Diffusion and Probability in Receptor Binding and Signaling*
Rakesh K. Jain, *Transport Phenomena in Tumors*

R. Krishna, *A Systems Approach to Multiphase Reactor Selection*
David T. Allen, *Pollution Prevention: Engineering Design at Macro-, Meso-, and Microscales*
John H. Seinfeld, Jean M. Andino, Frank M. Bowman, Hali J. L. Forstner, and Spyros Pandis, *Tropospheric Chemistry*

Volume 20

Arthur M. Squires, *Origins of the Fast Fluid Bed*
Yu Zhiqing, *Application Collocation*
Youchu Li, *Hydrodynamics*
Li Jinghai, *Modeling*
Yu Zhiqing and Jin Yong, *Heat and Mass Transfer*
Mooson Kwauk, *Powder Assessment*
Li Hongzhong, *Hardware Development*
Youchu Li and Xuyi Zhang, *Circulating Fluidized Bed Combustion*
Chen Junwu, Cao Hanchang, and Liu Taiji, *Catalyst Regeneration in Fluid Catalytic Cracking*

Volume 21

Christopher J. Nagel, Chonghun Han, and George Stephanopoulos, *Modeling Languages: Declarative and Imperative Descriptions of Chemical Reactions and Processing Systems*
Chonghun Han, George Stephanopoulos, and James M. Douglas, *Automation in Design: The Conceptual Synthesis of Chemical Processing Schemes*
Michael L. Mavrovouniotis, *Symbolic and Quantitative Reasoning: Design of Reaction Pathways through Recursive Satisfaction of Constraints*
Christopher Nagel and George Stephanopoulos, *Inductive and Deductive Reasoning: The Case of Identifying Potential Hazards in Chemical Processes*
Kevin G. Joback and George Stephanopoulos, *Searching Spaces of Discrete Solutions: The Design of Molecules Possessing Desired Physical Properties*

Volume 22

Chonghun Han, Ramachandran Lakshmanan, Bhavik Bakshi, and George Stephanopoulos, *Nonmonotonic Reasoning: The Synthesis of Operating Procedures in Chemical Plants*
Pedro M. Saraiva, *Inductive and Analogical Learning: Data-Driven Improvement of Process Operations*
Alexandros Koulouris, Bhavik R. Bakshi, and George Stephanopoulos, *Empirical Learning through Neural Networks: The Wave-Net Solution*
Bhavik R. Bakshi and George Stephanopoulos, *Reasoning in Time: Modeling, Analysis, and Pattern Recognition of Temporal Process Trends*
Matthew J. Realff, *Intelligence in Numerical Computing: Improving Batch Scheduling Algorithms through Explanation-Based Learning*

ISBN 0-12-008522-4